Topics in Elementary Mathematics

6

Topics in Elementary Mathematics

Abe Mizrahi
Indiana University Northwest
Michael Sullivan
Chicago State University

HOLT, RINEHART AND WINSTON, INC.
New York Chicago San Francisco Atlanta
Dallas Montreal Toronto London Sydney

Library of Congress Catalog Card Number: 77-145747

SBN: 03-084688-9

Printed in the United States of America
4 3 2 1 038 1 2 3 4 5 6 7 8 9

To our wives, Caryl and Mary

Preface

The purpose of *Topics in Elementary Mathematics* is to provide a sound mathematical training for students intending to teach at the elementary school level. Although intended primarily for an Elementary Education curriculum, the contents of *Topics in Elementary Mathematics* are also suitable for satisfying General Education requirements. Throughout, the authors' aim has been to maintain a clear, easily readable, accurate presentation. That is to say, definitions and theorems are motivated, precisely stated, and fully explained in a simple language the student can understand. The authors feel that an accurate knowledge of subject is a must for competent teaching at the elementary level. Five years of experimentation have shown that students who have mastered the contents of this book are very well prepared for appropriate methodology courses in the teaching of mathematics. The level of presentation is appropriate for students with no previous college training in mathematics.

The content of *Topics in Elementary Mathematics* adheres to the Level I recommendations of the Committee of Undergraduate Programs in Mathematics (CUPM), and, in some areas, exceeds those recommendations. The material contained in the text is adequate for a two semester sequence of courses each of three credits (or the equivalent). Chapters 1 through 7 form a foundation for subsequent chapters. Chapters 8, 9, 10, and 11 are independent of each other and may be covered in any order. Chapter 8 may be covered immediately following Chapter 5, if so desired.

Chapter 7, in particular, contains many exercises which complement and complete a study of the real numbers. Furthermore, if desired, a more formal construction of the integers and rational numbers than that found in Chapters 4–6 is given in Chapter 9. In addition, many chapters contain starred sections which may be omitted at the discretion of the instructor without destroying the continuity of the course. Starred problems refer to the material given in starred sections. Although the proofs of most results are given in the text, they can easily be postponed at the discretion of the instructor.

The authors wish to express their sincere appreciation to the many mathematicians, mathematics educators, and students who have helped make this project a reality. In particular, we are thankful for the constructive criticisms and helpful suggestions of Dr. Irwin Feinstein, Professor of Mathematics at the University of Illinois—Chicago Circle and Dr. Ralph Seifert, Professor of Mathematics at Indiana University in Bloomington. In addition, the authors wish to express their gratitude to Lydia Beidel for her patience in typing the final draft.

The present book is truly a collaborative effort. The order of authorship signifies only alphabetical precedence. We assume equal responsibility for any strengths and weaknesses and will welcome any comments or suggestions on improving the contents.

Abe Mizrahi
Michael Sullivan

Gary, Indiana
Chicago, Illinois
May 1971

Contents

Topics in Elementary Mathematics

1 Logic

1.1 INTRODUCTION

The purpose in studying logic is twofold: first, for its intrinsic value in the study of inductive reasoning; and second, as a tool in mathematics for abstractly establishing facts from a given premise or hypothesis.

It is not our goal to systematize and codify the principles of reasoning, but rather to introduce the beginning student to the basic principles of logic—its language and symbols, and their applications. With this approach we hope to increase the reasoning proficiency of the reader.

1.2 PROPOSITIONS

Logic has its roots in the study of language. Any developed language, for example English, possesses a number of words and phrases with a distinct function. These special words and phrases have a bearing in the truth or falsity of the assertion in which they occur.

English sentences have been classified into four categories: declarative, interrogative, exclamatory, and imperative. It is assumed that the student can easily recognize the classification in which a particular sentence belongs. This ability on the part of the student is necessary, since the study of mathematics deals mainly with *declarative statements*.

DEFINITION 1.2.1

A *proposition* is a declarative statement that can be meaningfully classified as true or false.

EXAMPLE 1.2.1

The statement *the lines* l_1 *and* l_2 *are parallel* is a proposition since the lines are either parallel or not parallel.

EXAMPLE 1.2.2

The price of lettuce in Chicago rose on July 3, 1965

Note that this proposition is a perfectly good one, even though few of us know if it is true.

EXAMPLE 1.2.3

The earth is round

This statement records a possible fact about reality. However, some people will classify the above proposition as true and some as false, depending on their interpretation of the word *round*. If it is taken to mean *spherical*, then the above proposition is false, for the earth has mountains, valleys, and so on. If, however, it is taken to mean *roughly spherical*, then the proposition is true. Thus we see that the classification of propositions as true or false is an idealization. *Actually, the truth values of propositions will fluctuate as we keep sharpening our interpretation of the words they contain.*

EXAMPLE 1.2.4

Does Helen have green eyes?

This is not a proposition because it is a question, not a statement.

EXAMPLE 1.2.5

Where is George? What a beauty!

Just as in Example 1.2.4, these are not statements.

EXAMPLE 1.2.6

It is raining

This is a proposition. However, its truth or falsity depends upon two conditions: a specific location and a specific time. Thus, we must be sure that when we utter *it is raining*, we either state or assume that a definite location and a definite time are known.

EXAMPLE 1.2.7

$$2 + 7 = 6$$

This is a false proposition, but it is a proposition.

We note that certain propositions have different connotations to different people. We also see that the classification of propositions into two categories, false and true, *rests on the tacit and ideal assumption that the meaning of the words assumed in the propositions are perfectly clear.* This problem, however, is avoided in mathematics through the use of the principles of *logic.*

Consider the proposition *it is cold and the sun is shining.* This sentence is obtained by joining the two propositions *it is cold* and *the sun is shining* with the word *and.*

DEFINITION 1.2.2

A proposition that is a combination of two or more propositions is called a *compound proposition.* The words used to combine given propositions into compound propositions are called *connectives.*

Examples of connective words are *not, or, and, if· · ·then, if and only if.*

DEFINITION 1.2.3

If we let p and q denote any two propositions, then p *and* q is a compound proposition called the *conjunction of* p *and* q. We denote this symbolically by

$$p \wedge q \tag{1.2.1}$$

The statement $p \wedge q$ is true when *both* p *and* q are true, and is false otherwise.

EXAMPLE 1.2.8

Consider the two statements

p: London is the capital of Great Britain
q: $2 + 2 = 4$

The conjunction of p and q is

$p \wedge q$: London is the capital of Great Britain and $2 + 2 = 4$

Since both p and q are true statements, the compound statement $p \wedge q$ is also true.

EXAMPLE 1.2.9

Consider the two statements

p: London is the capital of France
q: $2 + 2 = 4$

The conjunction of p and q is

$p \wedge q$: London is the capital of France and $2 + 2 = 4$

Since the statement p is false, the compound statement $p \wedge q$ is false, even though q is true.

DEFINITION 1.2.4

If p and q are any propositions, then the compound proposition p *or* q, denoted by the symbol

$$p \vee q \tag{1.2.2}$$

is called *the inclusive disjunction of p and q*.

The statement $p \vee q$ is true if both p and q are true, or if q is true and p false, or if p is true and q false. It is false if both p and q are false.

EXAMPLE 1.2.10

Consider the two statements:

p: the triangle is a right triangle
q: the equation has no solution

The inclusive disjunction of p and q is

$p \vee q$: the triangle is a right triangle or the equation has no solution

DEFINITION 1.2.5

For any two propositions p and q, the *exclusive disjunction of p and q*, read as p *or* q, is denoted by

$$p \veebar q \tag{1.2.3}$$

The compound proposition $p \veebar q$ is true whenever p is true and q is false or whenever p is false and q is true. It is false whenever p is false and q is false or whenever p is true and q is true.

Thus *or* is a connective used to denote two different ideas. This word is unusually sensitive in the sense that its meaning changes with the context. It is possible to interpret the compound proposition p *or* q in two ways:

(1) p or q, or both ($p \vee q$)
(2) p or else q, that is, p or q, but *not* both ($p \veebar q$)

The operations of forming $p \vee q$ and $p \veebar q$ are called the operations of *disjunction.*

EXAMPLE 1.2.11

Consider the two statements:

p: he is working
q: he is watching television

The inclusive disjunction of p and q is

$p \vee q$: he is working or he is watching television, or both

The exclusive disjunction of p and q is

$p \veebar q$: he is working or he is watching television, but not both

EXAMPLE 1.2.12

Consider the compound proposition:

p: this summer I will date Carol or Barbara

The use of the connective *or* is not clear here. If the *or* means inclusive disjunction, then both girls may be dated this summer. If the *or* means exclusive disjunction, then only one of the girls may be dated.

Thus, in mathematics, we will be very precise in the use of the connective *or* so that a misunderstanding cannot occur.

Another operation performed quite frequently on propositions is that of *negation.*

DEFINITION 1.2.6

If p is any proposition, then the *negation of p*, denoted by

$$\sim p \tag{1.2.4}$$

and read as *not p*, is a proposition such that

(1) if p is true, $\sim p$ is false
(2) if p is false, $\sim p$ is true

EXAMPLE 1.2.13

Let p stand for the proposition

p: the board is green

Then $\sim p$ is the proposition

$\sim p$: the board is not green

Note that the proposition *the board is white* is *not* a correct negation of *the board is green.*

Mathematical statements often concern sets of objects rather than a single object. A *quantifier* is a word or phrase telling how many (quant-) of the objects of the set are involved in the statement.

The quantifiers *each, any, every,* and *all* may be used interchangeably.

EXAMPLE 1.2.14

The following statements have the same meaning:

p: all men are greedy
r: every man is greedy
s: each man is greedy
t: any man is greedy

The quantifiers *some* and *there exists* may be used interchangeably.

EXAMPLE 1.2.15

Consider the statement:

p: all men are greedy

The negation of this statement is

$\sim p$: there exists a man who is not greedy

or

$\sim p$: some men are not greedy

The negation of the statements

p: there exists a man who is not greedy

or

p: some men are not greedy

is

$\sim p$: all men are greedy

Exercise 1.2

Determine whether each of the following is a proposition

1. The price of milk rose on September 1, 1968.
2. The gross national product will be less than one trillion dollars in 1990.
3. What a day!
4. Boys are tall.
5. Jack is smart.

6. Look at the ground.

7. Why is the grass green?

8. I am passing this course.

9. Miss Brooks has blue eyes.

10. George Washington was a good president.

Negate the following.

11. Jack is not married.

12. John is tall.

13. Jack is smart and ugly.

14. John is tall and fat.

15. Jack is neither smart nor ugly.

16. No men are mean.

17. All soldiers are bloodthirsty.

18. Some soldiers are bloodthirsty.

19. No soldiers are bloodthirsty.

20. Some soldiers are not bloodthirsty.

21. Let p be "Jack studies mathematics" and let q be "Jack studies physics." Give a simple verbal sentence that describes each of the following:

(a)	$p \vee q$	(e)	$\sim p \vee \sim q$
(b)	$p \wedge \sim q$	(f)	$\sim(\sim p \wedge \sim q)$
(c)	$\sim(\sim p)$	(g)	$p \veebar q$
(d)	$p \wedge q$		

1.3 TRUTH TABLES

For a given proposition p we are interested in determining whether the proposition is true or false; i.e., we seek for the *truth value* of p. To do this let us denote *true* by "T" and *false* by "F." We can determine the truth value by a truth table which spells out the truth value of a compound proposition in each of the possible truth value cases.

For example, consider the conjunction of any two propositions p and q. Recall that $p \wedge q$ is false if either p is false or q is false, or if both p and q are false. There are four truth value possibilities for the compound statement:

Case 1: p is true and q is true
Case 2: p is true and q is false
Case 3: p is false and q is true
Case 4: p is false and q is false

The four truth value cases for $p \wedge q$ are gathered together in Fig. 1.1. For convenience, the possibilities for p and for q will always be listed in this order:

	p	q	$p \wedge q$
Case 1	T	T	T
Case 2	T	F	F
Case 3	F	T	F
Case 4	F	F	F

FIG. 1.1

The truth values of $\sim p$ are illustrated in Fig. 1.2.

p	$\sim p$
T	F
F	T

FIG. 1.2

Using the previous definitions of inclusive disjunction and exclusive disjunction, the truth tables for $p \vee q$ and $p \veebar q$ are shown in Fig. 1.3.

p	q	$p \vee q$	$p \veebar q$
T	T	T	F
T	F	T	T
F	T	T	T
F	F	F	F

FIG. 1.3

In addition to using the connectives $\vee$, $\wedge$, $\underline{\vee}$, and $\sim$ separately to form compound propositions, we can use them together to form more complex assertions. Let us denote by $P(p, q, \cdots)$ a compound proposition, where $p, q, \cdots$ are the components making the compound proposition. The truth value of a compound proposition depends exclusively upon the truth values of its components; that is, the truth value of a compound proposition is known once the truth values of its components are known. Truth tables will be used to demonstrate this relationship.

For example, the truth table for $(p \vee q) \vee (\sim p)$, which involves two components p and q, is given in Fig. 1.4.

p	q	$p \vee q$	$\sim p$	$((p \vee q) \vee (\sim p)$
T	T	T	F	T
T	F	T	F	T
F	T	T	T	T
F	F	F	T	T

FIG. 1.4

Observe that the first columns of the table are for the component propositions $p, q, \cdots$, and that there are enough rows in the table to allow for all possible combinations of T and F (for two components p and q, four rows are necessary, for three, eight are necessary, and for n, 2^n rows will be necessary). We notice that there are five columns. Each column corresponds to a "stage" in constructing the compound proposition. Note that each stage depends on its predecessors. In essence, the first two columns and the last one are of interest to us; the other two are just intermediate ones leading to the final stage. Hence, let us analyze the compound proposition to observe these stages carefully.

First construct the following table:

p	q	$p \vee q$	$\sim p$	$(p \vee q) \vee \sim p$
T	T			
T	F			
F	T			
F	F			

Observe that the compound proposition is written on the top row to the right of its components and that there is a column under each component or connective. Truth values are then entered in the truth table in various steps as follows:

p	q	$p \vee q$	$\sim p$	$(p \vee q) \vee \sim p$
T	T	T		
T	F	T		
F	T	T		
F	F	F		

Stage 1

p	q	$p \vee q$	$\sim p$	$(p \vee q) \vee \sim p$
T	T	T	F	
T	F	T	F	
F	T	T	T	
F	F	F	T	

Stage 2

p	q	$p \vee q$	$\sim p$	$(p \vee q) \vee \sim p$
T	T	T	F	T
T	F	T	F	T
F	T	T	T	T
F	F	F	T	T

Stage 3

The truth table of the compound proposition then consists of the original columns under the variables and the fifth column entered into the table, that is, the last stage.

EXAMPLE 1.3.1

Determine the truth table $(p \vee \sim q) \wedge p$. Notice that the component parts of this proposition are p, $\sim q$, and $p \vee \sim q$, and the proposition is the conjunction of $p \vee \sim q$ and p. The truth table is

p	q	$\sim q$	$p \vee \sim q$	$(p \vee \sim q) \wedge p$
T	T	F	T	T
T	F	T	T	T
F	T	F	F	F
F	F	T	T	F

FIG. 1.5

EXAMPLE 1.3.2

Determine the truth table for the compound proposition $(p \vee \sim q) \wedge \sim p$. The component parts of this proposition are p, $\sim q$, $p \vee \sim q$, and $\sim p$.

p	q	$\sim q$	$p \vee \sim q$	$\sim p$	$(p \vee \sim q) \wedge \sim p$
T	T	F	T	F	F
T	F	T	T	F	F
F	T	F	F	T	F
F	F	T	T	T	T

FIG. 1.6

EXAMPLE 1.3.3

Determine the truth table for $\sim(p \vee q) \vee (\sim p \wedge \sim q)$.

p	q	$\sim p$	$\sim q$	$p \vee q$	$\sim(p \vee q)$	$(\sim p \wedge \sim q)$	$\sim(p \vee q) \vee (\sim p \wedge \sim q)$
T	T	F	F	T	F	F	F
T	F	F	T	T	F	F	F
F	T	T	F	T	F	F	F
F	F	T	T	F	T	T	T

FIG. 1.7

The following is an example of a truth table involving three components, p, q, and r.

EXAMPLE 1.3.4

Determine the truth table of $p \wedge (q \vee r)$.

p	q	r	$q \vee r$	$p \wedge (q \vee r)$
T	T	T	T	T
T	T	F	T	T
T	F	T	T	T
T	F	F	F	F
F	T	T	T	F
F	T	F	T	F
F	F	T	T	F
F	F	F	F	F

FIG. 1.8

Exercise 1.3.1

Construct a truth table for the following compound propositions:

1. $(p \vee \sim q)$
2. $\sim p \vee \sim q$
3. $\sim p \wedge \sim q$
4. $\sim p \wedge q$
5. $\sim(\sim p \wedge q)$
6. $(p \vee \sim q) \wedge \sim p$
7. $\sim(\sim p \vee \sim q)$
8. $(p \vee \sim q) \wedge (q \wedge \sim p)$
9. $(p \vee \sim q) \wedge p$
10. $p \wedge (q \vee \sim q)$
11. $(p \vee q) \wedge (p \wedge \sim q)$
12. $(p \wedge \sim q) \vee (q \wedge \sim p)$
13. $\sim(p \vee q) \vee (\sim p \wedge \sim q)$
14. $(p \wedge q) \vee (p \wedge r)$
15. $(p \wedge \sim q) \vee r$
16. $(\sim p \vee q) \wedge \sim r$

DEFINITION 1.3.1

If two propositions a and b have the same truth values in every possible case, the propositions are called *logically equivalent*. This situation is denoted by $a \equiv b$.

EXAMPLE 1.3.5

$\sim(p \wedge q)$ is logically equivalent to $\sim p \vee \sim q$.

p	q	$p \wedge q$	$\sim(p \wedge q)$	$\sim p$	$\sim q$	$(\sim p \vee \sim q)$
T	T	T	F	F	F	F
T	F	F	T	F	T	T
F	T	F	T	T	F	T
F	F	F	T	T	T	T

FIG. 1.9

are equivalent. However, it is already known that

if q, then p
if not p, then not q

are equivalent.

Thus, the three compound statements

only if p, then q
if q, then p
if not p, then not q

are equivalent to each other.

Thus we have another rule for forming the converse of the implication *if p then q*. That is, to form the converse of *if p, then q*, replace *if* by *only if*.

DEFINITION 1.5.1

The connective *if and only if* is called the *biconditional connective* and is denoted by $\leftrightarrow$. If p and q are two propositions, and if

$$p \rightarrow q \qquad \text{and} \qquad q \rightarrow p \tag{1.5.1}$$

then we say p *if and only if* q and we write

$$p \leftrightarrow q \tag{1.5.2}$$

The statement $\leftrightarrow$ may be read as p *if and only if* q, or as p *is equivalent to* q, or as p *implies* q *and* q *implies* p, or as p *is necessary and sufficient for* q. An abbreviation for $p \leftrightarrow q$ is p *iff* q, read as p *if and only if* q.

We can restate Definition 1.5.1 by using the truth table shown in Fig. 1.17.

p	q	$p \leftrightarrow q$
T	T	T
T	F	F
F	T	F
F	F	T

FIG. 1.17

EXAMPLE 1.5.1

Consider problems 6 and 7 in Exercise 1.4. Let p and q denote the statements

p: it is cloudy
q: it is raining

Then problem 6 states that

if it is raining, then it is cloudy

and problem 7 is

if it is cloudy, then it is raining

Note that the statements if it is cloudy, it is raining; if it is not raining, then it is not cloudy; and only if it is cloudy, then it is raining are equivalent.

Also, assuming the two implications $p \rightarrow q$ and $q \rightarrow p$ are both correct from a physical point of view, then we conclude that a necessary and sufficient condition for rain is that it be cloudy. In this case, we say

it is raining if and only if it is cloudy.

The truth table for the biconditional connective $\leftrightarrow$ for two statements p and q is illustrated by the truth table given previously in Fig. 1.17.

The converse of the implication $p \rightarrow q$ is the implication $q \rightarrow p$. It is a common error to confuse the two, or to assert that one follows from the other. The student should be *very* careful that he does not conclude the truth of the converse of an implication from the implication itself.

Exercise 1.5

1. Construct a truth table for each of the following:

(a) $\sim p \vee (p \wedge q)$
(b) $\sim p \wedge (p \vee q)$
(c) $p \vee (\sim p \wedge q)$
(d) $(p \vee q) \wedge \sim q$
(e) $\sim p \rightarrow q$
(f) $(p \vee q) \rightarrow p$
(g) $\sim p \vee p$
(h) $p \wedge \sim p$
(i) $p \wedge (p \rightarrow q)$

2. Prove that the following only have truth value T:

(a) $p \wedge (q \wedge r) \leftrightarrow (p \wedge q) \wedge r$
(b) $p \wedge (p \vee q) \leftrightarrow p$
(c) $p \vee (p \wedge q) \leftrightarrow p$

3. Let p be *the examination is hard* and q be *the grades are low*. Write symbolically each of the following sentences:

(a) If the examination is hard, the grades are low.
(b) The examination is not hard nor are the grades low.
(c) The grades are not low, and the examination is not hard.
(d) The examination is not hard, and the grades are low.
(e) The grades are low only if the examination is hard.

1.6 TAUTOLOGIES AND ARGUMENTS

In Section 1.3 we saw how it was possible to obtain compound propositions from simple propositions by using connectives. By using symbols of grouping, such as parentheses and brackets, we can form more complicated propositions. We denote by $P(p, q, \cdots)$ a compound proposition, where $p, q, \cdots$ are components making the compound proposition. Some examples of such propositions are

$$\sim(p \wedge q), \quad \sim p \wedge \sim q, \quad p \wedge \sim q,$$
$$(p \wedge \sim q) \vee (\sim p \wedge q) \qquad (1.6.1)$$

Ordinarily when we write a compound proposition, we cannot be certain of the truth of that proposition unless we know the truth or falsity of the component propositions. There are certain compound propositions which are true regardless of the truth or falsity of the component propositions. Compound propositions that have this characteristic are called *tautologies.*

DEFINITION 1.6.1

A *tautology* is a compound proposition $P(p, q, \cdots)$ which is always true.

Examples of tautologies are

$$p \vee \sim p, \qquad p \rightarrow (p \vee q) \qquad (1.6.2)$$

Figures 1.18 and 1.19 show the truth tables of these tautologies.

Let $P(p, q, \cdots)$ represent a compound composition. Now if we substitute for p a proposition h which is logically equivalent to p, we obtain a compound proposition $P(h, q, \cdots)$ which is logically equivalent to

p	$\sim p$	$p \vee \sim p$
T	F	T
F	T	T

FIG. 1.18

p	q	$p \vee q$	$p \rightarrow (p \vee q)$
T	T	T	T
T	F	T	T
F	T	T	T
F	F	F	T

FIG. 1.19

$P(p, q, \cdots)$. This is a reasonable assumption since the truth of $P(p, q, \cdots)$ depends on the truth of the component propositions $p, q, \cdots$, and *h is true whenever p is true,* and *h is false whenever p is false.* This leads us to assert the following law.

Law of Substitution

If $h \equiv p$ and we substitute for p in the compound proposition $P(p, q, \cdots)$, then

$$P(p, q, \cdots) \equiv P(h, q, \cdots) \tag{1.6.3}$$

The usefulness of this principle in deriving tautologies is apparent. If we have a tautology and substitute for some of the component propositions other propositions which are logically equivalent to them, we obtain a *new* tautology.

DEFINITION 1.6.2

It is our desire to use some of the concepts introduced thus far in checking the validity of *arguments.*

By *argument* we mean the assertion that a certain statement called the *conclusion* follows from the other statements called *premises.* An argument is valid if and only if the conjunction of the premises implies (yields, has as a consequence) the conclusion; if an argument is false it is called a *fallacy.*

It is important to realize that the truth of the conclusion is irrelevant as a test for the validity of the argument. A true conclusion does not necessarily assume a valid argument, nor does a valid argument necessarily assume a true conclusion.

We are now in a position to discuss how to determine whether a given proposition is true or false. Up until this time, we were concerned with the various truth values that could be assigned to the proposition

$$\text{if } p, \text{ then } q \tag{1.6.4}$$

for all possible truth values of p and of q.

Now we shall discuss how to demonstrate the conclusion q of a proposition $p \rightarrow q$ is true whenever p is true.

We shall limit our discussion to two types: direct proof and indirect proof. In a *direct proof* a chain of propositions is reasoned to, beginning with the given proposition and leading to the desired conclusion.

EXAMPLE 1.6.1

Suppose it is true that

either John rakes the lawn or John is punished

and

John was not punished

We would like to prove that

John raked the lawn

A valid argument in the form of a direct proof can be given as follows:

Let p and q denote the propositions

p: John rakes the lawn
q: John is punished

We can write the premise as

$$p \vee q \qquad \text{and} \qquad \sim q \tag{1.6.5}$$

Since $\sim q$ is true and since $p \vee q$ is true, then q is false and either p is true or q is true. Thus, p must be true.

More examples of direct proof are given later, but before we look at them, let us discuss two laws of logic which are quite useful in a direct proof.

Law of Detachment

If the implication

$$p \to q \tag{1.6.6}$$

is true and if p is true, then q must be true.

See Fig. 1.13 for an illustration of this law.

Law of Syllogisms

Let p, q, r be three propositions. If

$$p \to q \qquad \text{and} \qquad q \to r \tag{1.6.7}$$

then

$$p \to r \tag{1.6.8}$$

The truth table shown in Fig. 1.20 illustrates this law.

p	q	r	$p \to q$	$q \to r$	$p \to r$	$(p \to q \wedge q \to r) \to (p \to r)$
T	T	T	T	T	T	T
T	T	F	T	F	F	T
T	F	T	F	T	T	T
T	F	F	F	T	F	T
F	T	T	T	T	T	T
F	T	F	T	F	T	T
F	F	T	T	T	T	T
F	F	F	T	T	T	T

FIG. 1.20

EXAMPLE 1.6.2

Suppose it is true that

it is snowing

and

if it is warm, then it is not snowing

and

if it is not warm, then I cannot go swimming

We want to prove that

I cannot go swimming

is a true statement.

A valid argument for this is given as a direct proof below.

Let p, q, r represent the statements

p: it is snowing
q: it is warm
r: I can go swimming

Our premise is that the propositions

$$p, \qquad q \to \sim p, \qquad \sim q \to \sim r \tag{1.6.9}$$

are true. We want to prove that

$$\sim r \tag{1.6.10}$$

is true.

Since $q \to \sim p$ is true, its contrapositive

$$p \to \sim q \tag{1.6.11}$$

is also true. Using the law of syllogisms, we see that

$$p \to \sim r \tag{1.6.12}$$

since $p \to \sim q$ and $q \to \sim r$. But we know p is true. By the law of detachment, $\sim r$ is true.

EXAMPLE 1.6.3

Suppose it is true that

if Jan comes, so will Jill

and

if Sandy does not come, then Jill will not come

We want to show that

if Jan comes, then Sandy will come

Let p, q, r denote the propositions

p: Jan is coming
q: Jill will come
r: Sandy is coming

Then we know that

$$p \rightarrow q, \qquad \sim r \rightarrow \sim q \tag{1.6.13}$$

Since $\sim r \rightarrow \sim q$ is true, then $q \rightarrow r$ is also true. Thus,

$$p \rightarrow q, \qquad q \rightarrow r \tag{1.6.14}$$

By the law of syllogisms, it is true that

$$p \rightarrow r \tag{1.6.15}$$

That is, if Jan comes, then Sandy will come.

EXAMPLE 1.6.4

Suppose it is true that

if I enjoy studying, then I will study

and

I will do my homework or I will not study

and

I will not do my homework

We want to show that

I do not enjoy studying.

Let p, q, r denote the three propositions:

p: I enjoy studying
q: I will study
r: I will do my homework

Then we know that $\sim r$ and $r \vee \sim q$ are true and that

$$p \rightarrow q \tag{1.6.16}$$

We want to prove that $\sim p$ is true.

Since $\sim r$ is true, then r is false. Also, either r or $\sim q$ is true. Thus, $\sim q$ is true. From (1.6.16),

$$\sim q \rightarrow \sim p \tag{1.6.17}$$

is true. Thus, $\sim p$ must be true.

The above are all examples of proving statements by means of a direct proof. Note that this technique is essentially one of proceeding by means of the laws of logic from the premise to the desired conclusion.

Suppose it is desired to show that a proposition q is true. To prove that q is true by the method of an *indirect proof*, it is necessary to show that $\sim q$ is true and showing that this assumption, when combined with other statements of the premise, leads to a logical contradiction.

The following two examples illustrate indirect proof.

EXAMPLE 1.6.5

Prove the result of Example 1.6.1 using an indirect proof.

Since we wish to prove that p is true, we shall make the assumption that p is false or that $\sim p$ is true, hoping that this leads to a contradiction—for, if it does, then p must be true.

We know that $\sim p$ and $\sim q$ are true and that

$$p \vee q \tag{1.6.18}$$

is true. Thus, either p is true or q is true. But both p and q are false. This is impossible. Thus, we have reached a contradiction, which means that p must be true.

EXAMPLE 1.6.6

Suppose that it is true that

if I am lazy, then I do not study

and

I study or I enjoy myself

and

I do not enjoy myself

Prove that

I am not lazy

Let p, q, r be the statements,

p: I am lazy
q: I study
r: I enjoy myself

Then, we know that $\sim r$ is true and that

$$p \to \sim q, \qquad q \vee r \tag{1.6.19}$$

are true.

We want to show that $\sim p$ is true.

If we use an indirect proof, then we assume that p is true. Since p is true and $p \to \sim q$ is true, then $\sim q$ is true. Thus, q is false. Also, r is false. But either q or r is true. This is impossible. This contradiction means that the assumption that p is true is a false one. Thus, $\sim p$ is true.

Exercise 1.6

Prove the following, first by using a direct proof, and then by an indirect proof.

1. When it rains, John does not go to school. John is going to school. Show that it is not raining.
2. If I do not go to work, I will go fishing. I will not go fishing. Show that I will go to work.
3. If Smith is elected president, Kuntz will be elected secretary. If Kuntz is elected secretary, then Brown will not be elected treasurer. Smith is elected president. Show that Brown is not elected treasurer.
4. Joe is a D student and Mary is an A student. If Joe is not a D student, then John is a D student. If Mary is an A student, then John is not a D student. Show that John is not a D student.

1.7 MATHEMATICAL STRUCTURES

In the study of a geometry, one quickly finds out that some objects are impossible to define. For example, consider the concept of a *point*. If we try to define point as an object without dimensionality, we must ask ourselves what is meant by the dimension of an object. In trying to define dimensionality, we will soon be led back to the concept of a point. Thus, in reality, all we have done is define a point as a point. To avoid such *circular* definitions in geometry, points are treated as *undefined objects*. That is, we do not attempt to define them (since it is not possible).

Suppose we agree that in our geometry we have two undefined objects, *point* and *line*. We must now talk about relationships between these two objects. Again, we are led to speak of an *undefined relation* between them, the relation of a point being *on* a line.

Next, before we can begin to study this geometry, we must ask what properties are obeyed by points, lines, and this relation *on*. Thus, we shall list two assumptions obeyed by points and lines in our geometry.

(I) On two distinct points, there is one and only one line.
(II) If l is a line, there is a point not on this line.

These two statements, which we assume are true, are called *axioms* of our geometry.

In our geometry, we shall want *definitions*. These are names for concepts involving points and lines. For example, we could say two lines are *parallel* if there is no point on both of them.

We are now prepared to prove results concerning points and lines in our geometry, using the laws of logic discussed earlier.

For example, if our geometry has two distinct points, it must have at least three distinct points and at least three lines.

Let A and B represent two distinct points. By assumption (I), there is a line L_{AB} on A and B. By assumption (II), there is a point C not on L_{AB}. Now, assumption (I) guarantees that the points B and C and A and C will determine lines L_{BC} and L_{AC}. Thus, we have shown there are three points A, B, C and three lines L_{AB}, L_{BC}, L_{AC} in our geometry.

The above result is a *theorem* of our geometry, derived by logic from our axioms. We sometimes refer to a theorem as a *property* or a *law*.

The above description of a geometry was an example of a *mathematical structure.*

A *mathematical structure* consists of

(1) undefined objects
(2) undefined relations between these objects
(3) axioms involving these objects and relations
(4) definitions
(5) theorems

In Chapter Nine, other examples of a mathematical structure are discussed in detail.

★1.8[1] AN APPLICATION TO THE THEORY OF SWITCHING NETWORKS

Let us discuss two simple but important applications of the concepts introduced thus far to the theory of switching networks. The purpose of this example is not to illustrate switching networks, but to show the use of symbolic logic to solve problems that arise in a simple situation.

By a switching network we will refer to a collection of wires capable of carrying an electric current, and numbers of switches each capable of creating a gap in the wire.

An example of such a network is a single wire capable of transmitting a current from one end to the other.

A less simple example of a switching network is the addition of a switch in the wire. Refer to Fig. 1.21. When switch P is closed current flows. An open switch creates a gap in the wire so that no current flows. This simple network suggests two statements, *switch P is closed* and *current flows.* Denote *P is closed* by p and *current flows* by q. Notice that the truth value of q depends on the truth value of p (see Fig. 1.22). Examining Fig. 1.22,

[1] Optional problems or sections are designated by a ★.

we see that $p \equiv q$; that is, in this example the proposition *current flows* is logically the same as the statement *P is closed.*

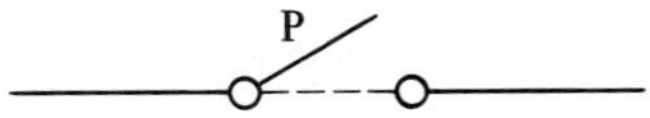

FIG. 1.21

p	q
T	T
F	F

FIG. 1.22

Exercise 1.8

★1. Write the following symbolically.

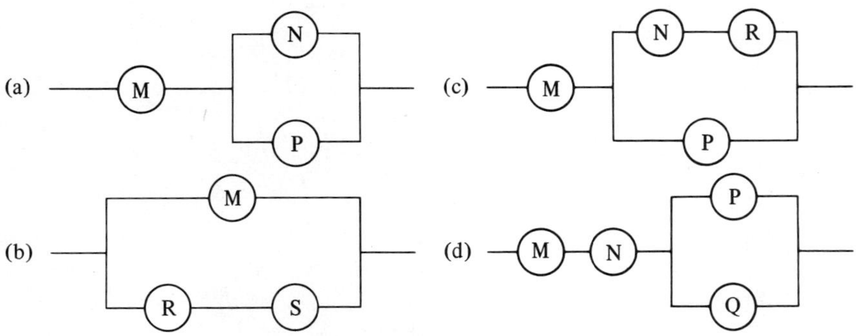

★2. Make diagrams of the switching networks of the following:

(a) $p \wedge q \wedge r$

(b) $(p \vee q) \wedge r$

(c) $(p \wedge q) \vee (p \wedge r)$

(d) $p \vee q \vee r$

(e) $p \wedge [q \vee (s \wedge t)]$

(f) $(p \vee q) \wedge (p \vee s)$

2 Introduction to the Theory of Sets

2.1 INTRODUCTORY DEFINITIONS

There are many examples of sets of which the student is already aware. For example, a student may speak of being a member of a certain class, say Mathematics 101. In this case, Mathematics 101 is a set and a student enrolled in this course is an element of the set. A student *not* enrolled in Mathematics 101 is *not* a member of the set. If no students register for Mathematics 101, then the set is called an *empty set*, that is, a set having *no* elements.

The above concepts of *set* and *element of a set* are *described*[1] more precisely in the following paragraphs.

A *set* S is a collection of objects, considered as a whole. The objects of a set S are called *elements of* S or *members of* S. A set that has no elements is called the *empty set*, or *null set*, or *vacuous set*, and is denoted by the symbol $\varnothing$.

Suppose S is a set and a is an element of the set S. Then we may write

$$a \in S \tag{2.1.1}$$

read as *a is an element of S* or *a is in S*. If a is *not* an element of the set S,

[1] The concept of a set cannot be defined. It is used here as an undefined term as discussed in Section 1.7 of Chapter 1.

then we write

$$a \notin S \tag{2.1.2}$$

read as *a is not an element of S* or *a is not in S*.

Ordinarily, a set S can be written in one of two ways. These two ways of writing sets are illustrated by the following examples.

EXAMPLE 2.1.1

Consider the set D of digits. Here the elements of the set D are

$$0, 1, 2, 3, 4, 5, 6, 7, 8, 9 \tag{2.1.3}$$

In this case, we write

$$D = \{0, 1, 2, 3, 4, 5, 6, 7, 8, 9\} \tag{2.1.4}$$

read as *D is the set consisting of the elements* 0, 1, 2, 3, 4, 5, 6, 7, 8, 9. Here, we are actually *listing* or *displaying* the elements of the set D.

Another way of writing this same set D of digits in the decimal system is to write

$$D = \{x \mid x \text{ is a digit}\} \tag{2.1.5}$$

read as *D is the set of all x such that x is a digit.* Here we define the set D by describing a property that every element of D has and that no element not in D can have, namely the property of being a digit. This way of writing a set is called *set-builder* notation.

Of course, when it is possible, either of the above two methods of describing a set and its elements may be used. Sometimes, however, it is impossible to list all elements of a set, because of the nature of the set. Examples of such occurrences appear quite often in the study of elementary mathematics.

EXAMPLE 2.1.2

Consider the set W, whose elements are the days in a week. Here we may actually list the elements of W by writing

$$W = \{\text{Monday, Tuesday, Wednesday, Thursday, Friday, Saturday, Sunday}\} \tag{2.1.6}$$

In set-builder notation, W is written as

$$W = \{x \mid x \text{ is a day of the week}\} \tag{2.1.7}$$

It is important for the student to notice that elements of a set are *not* repeated. A given element is entered as a member of a set only once; there is never a duplication of an element in a set. Thus, we would never write

$$\{3, 2, 2\} \tag{2.1.8}$$

but rather

$$\{3, 2\} \tag{2.1.9}$$

Also, because a set is a collection of objects considered as a whole, when considering a set, the order in which the elements of a set are listed does not make any difference. Thus, the three sets

$$\{2, 3, 4\}, \qquad \{2, 4, 3\}, \qquad \{3, 4, 2\} \tag{2.1.10}$$

are all the *same* set. The elements of a set, not the order in which the elements are written, distinguish sets.

Let us agree, henceforth, to adopt the convention that upper case letters such as $A, B, \cdots$ will represent sets and lower case letters, $a, b, \cdots$ will represent elements of sets.

Exercise 2.1

Write each of the following sets in two ways.

1. The set A of digits from 6 to 9 inclusive.
2. The set B of vowels in the English alphabet.
3. The set C of digits in your age (in years).

In the following problems, replace the $*$ by either $\in$ or $\notin$, whichever is correct, or neither.

4. $3 * \{2, 3, 7\}$
5. $\varnothing * \{1, 3\}$
6. $1/2 * \{x \mid x \text{ is a digit}\}$
7. $10 * \{x \mid x \text{ is a digit}\}$
8. $4 * \{2, 1, 6\}$
9. Find a set whose elements cannot be listed, but which can be described using the set-builder notation.
10. Find a set whose elements can be listed, but not described by using set-builder notation.

2.2 RELATIONS BETWEEN SETS

When we see two people, we can usually relate them in many ways. We say that they have the same color eyes, or one is taller than the other, or that one is heavier than the other.

Relative to sets, we will discuss *four* ways of relating two sets. If we look at the sets

$$A = \{1, 2\}, \qquad B = \{1, 2, 3\} \tag{2.2.1}$$

we notice that the elements of the set A are also elements of the set B. This is one way of comparing two sets.

Suppose we consider the following two sets.

$$A = \{1, 2\}, \qquad B = \{4, 5\} \tag{2.2.2}$$

If we look at these two sets, we notice that each of these sets possesses the same number of elements. This is easily seen by *matching* or *corresponding* elements of these sets. If we use the double arrow $\leftrightarrow$ to denote this match, then

$$1 \leftrightarrow 4, \qquad 2 \leftrightarrow 5 \tag{2.2.3}$$

Here we would say that the sets A and B are *equivalent*.

These *relations between sets* are defined more carefully below.

Definition 2.2.1

Let A and B be two nonempty sets. We say that A *is equal to* B, written as

$$A = B \tag{2.2.4}$$

if and only if A and B have the *same* elements.

If two sets A and B are *not* equal, we write

$$A \neq B \tag{2.2.5}$$

Definition 2.2.2

Let A and B be two nonempty sets. We say that A *is equivalent to* B, written as

$$A \sim B \tag{2.2.6}$$

if and only if there can be found a one-to-one correspondence between the elements of A and B. By one-to-one correspondence is meant that to each element in A, there can be found one element in B; and each element in B can be matched to exactly one element in A.

If two sets A and B are *not* equivalent, we write

$$A \nsim B \tag{2.2.7}$$

There is the following relationship between the concept of equality and that of equivalence.

If two sets A and B are equal, so that they have the *same* elements, it is easy to see that A and B must also be equivalent.

The necessary one-to-one correspondence is found merely by associating or matching that element in A with the identical element of B. Since the elements of A and B are the same, the elements of the two sets A and B are in one-to-one correspondence, and hence are equivalent.

Suppose we consider the converse. That is, suppose two sets A and B are equivalent. Are they then equal? Or can two sets be equivalent without being equal? The following example gives us the answer.

EXAMPLE 2.2.1

Let two sets A and B be given by

$$A = \{1, 2, 3\}, \qquad B = \{7, 8, 9\} \tag{2.2.8}$$

Clearly, A is not equal to B. However, since we can match the elements of A and B by

$$1 \leftrightarrow 7, \qquad 2 \leftrightarrow 8, \qquad 3 \leftrightarrow 9 \tag{2.2.9}$$

it is seen that A is equivalent to B.

The following result has thus been established.

If two sets A and B are equal, then A and B are equivalent. The converse is *not always* true. Thus, two sets A and B may be equivalent, and yet be unequal.

DEFINITION 2.2.3

Let A and B be two nonempty sets. We shall say that A *is a subset of* B or that A *is contained in* B or that B *contains* A, written as

$$A \subseteq B \quad \text{or} \quad B \supseteq A \tag{2.2.10}$$

if and only if every element of A is also an element of B.

If a set A is not a subset of a set B, we write

$$A \nsubseteq B \tag{2.2.11}$$

The definition that A is a subset of B could have been given equivalently as *there are no elements in set A that are not also elements in set B*. Of course, $A \subseteq B$ if and only if whenever $x \in A$, then $x \in B$ for all x. This latter way of interpreting the meaning of $A \subseteq B$ is useful for obtaining various laws that sets obey.

DEFINITION 2.2.4

Let A and B be two nonempty sets. We shall say that A *is a proper subset of* B or that A *is properly contained in* B or that B *properly contains* A, written as

$$A \subset B \quad \text{or} \quad B \supset A \tag{2.2.12}$$

if and only if every element of the set A is also an element of set B, but there is at least one element in set B that is *not* in set A.

The student should note that A *is a proper subset of* B means that there are *no* elements of A that are not also elements of B, but there is at least one element of B that is not in A.

If a set A is *not* a proper subset of a set B, we write

$$A \not\subset B \tag{2.2.13}$$

The following example illustrates the uses of the four relations of $=$, $\sim$, $\subset$, and $\subseteq$ just defined.

EXAMPLE 2.2.2

Consider three sets A, B, and C given by

$$A = \{1, 2, 3\}, \quad B = \{1, 2, 3, 4, 5\}, \quad C = \{1, 2, 3\} \tag{2.2.14}$$

The following are some of the relations between pairs of these sets:

$$\begin{aligned} &A = C \\ &A \sim C \\ &A \subseteq B \\ &A \subset B \\ &B \supset C \\ &A \subseteq C \\ &C \subseteq A \end{aligned} \tag{2.2.15}$$

In comparing the two definitions given above of *subset* and *proper subset,* the student should notice that if a set A is a subset of a set B, then either A is a proper subset of B or else A equals B. That is,

$$A \subseteq B \text{ if and only if either } A \subset B \text{ or } A = B \tag{2.2.16}$$

Also, if A is a proper subset of B, we can infer that A is a subset of B, but A does not equal B. That is,

$$A \subset B \text{ if and only if } A \subseteq B \text{ and } A \neq B \tag{2.2.17}$$

The distinction that is made between *subset* and *proper subset* is rather subtle and is quite important. Let us try to describe this distinction further by using some familiar everyday examples.

We often hear "my weight *does not exceed* 200 pounds." What this means is the weight in question equals 200 pounds or is less than 200 pounds. If A is a subset of B, then A equals B or A is a proper subset of B. If one's weight is less than 200 pounds, then his weight does not exceed 200 pounds and does not equal 200 pounds. Thus, if A is a proper subset of B, we know that A is a subset of B, but $A \neq B$.

We can think of the relation $\subset$ as a refinement of the relation $\subseteq$. On the other hand, the relation $\subseteq$ is an extension of $\subset$ in the sense that $\subseteq$ may include equality, whereas with $\subset$, equality cannot be included.

Because of the way the relation $\subseteq$ (is a subset of) has been defined, it is easy to see that for any set A,

$$\varnothing \subseteq A \tag{2.2.18}$$

The reason for this is that there is no element of the set $\varnothing$ which is not also in A since the empty set $\varnothing$ has no elements.

Also, if A is any nonempty set, that is, any set having at least one element, then

$$\varnothing \subset A \tag{2.2.19}$$

The following proposition gives some properties which the relation $\subseteq$ (is a subset of) obeys.

Theorem 2.2.1

The relation $\subseteq$ between two sets satisfies the three laws listed below.

(I) *Reflexive Law:* Every set is a subset of itself. That is, for any set A

$$A \subseteq A \tag{2.2.20}$$

(II) *Anti-Symmetric Law:* If a set A is a subset of a set B and if the set B is a subset of set A, then the two sets A and B are equal. That is,

$$\text{if } A \subseteq B \text{ and if } B \subseteq A, \text{ then } A = B \tag{2.2.21}$$

(III) *Transitive Law:* If A is a subset of B and if B is a subset of C, then A must be a subset of C. That is,

$$\text{if } A \subseteq B \text{ and if } B \subseteq C, \text{ then } A \subseteq C \tag{2.2.22}$$

Later on, we shall study other relations. So that we can describe such relations better, we shall use the term *partial ordering* to describe any relation having the above *three properties.*

PROOF: (I) Since every element in a set A must be in set A, we can write

$$A \subseteq A \tag{2.2.23}$$

(II) If $A \subseteq B$, then every element in set A must be in set B. Since $B \subseteq A$, then every element in set B must be in set A. This means that A and B have the *same* elements. That is to say,

$$A = B \tag{2.2.24}$$

(III) If $A \subseteq B$, then every element in set A is in set B. If $B \subseteq C$, then every element in set B must be in set C. Hence, every element in set A must be in set C. That is,

$$A \subseteq C \tag{2.2.25}$$

As will be seen later in the chapter, the anti-symmetric law is *very* important for showing the equality of two sets. To actually prove that two sets A and B are equal, first show that any element in set A is in set B (so that $A \subseteq B$), then show that any element in set B must be in set A (so that $B \subseteq A$). Once this has been done, the equality of A and B is a consequence of the anti-symmetric law.

Thus far, we have talked about comparing pairs of sets by means of the relations $\subset$ and $\subseteq$. We have said A is a subset of B, or A equals B, or B is a subset of A. For some pairs of sets, none of these relations can be used for comparison. This situation is illustrated by the following example.

EXAMPLE 2.2.3

Consider two sets A and B whose elements are

$$A = \{1, 2, 3\}, \qquad B = \{3, 4\} \tag{2.2.26}$$

Notice that not every element in A is in B. Thus,

$$A \not\subset B \tag{2.2.27}$$

Also, notice that not every element in B is in A. Then

$$B \not\subset A \tag{2.2.28}$$

Finally, A and B are *not* equal. They have different elements. Thus

$$A \neq B \tag{2.2.29}$$

Thus, the sets A and B given by equation (2.2.26) are *not comparable*. That is, we cannot say about A and B that either (1) A is a subset of B, or (2) A is equal to B, or (3) B is a subset of A. When we can say one of the above three about two sets A and B, then A and B are said to be *comparable*.

DEFINITION 2.2.5

The *universal set* U is defined as that set consisting of all elements under consideration.

Thus, if A is any set A and if U is the universal set U, then every element in A must be in U (since U consists of all elements under consideration). Hence, we may write

$$A \subseteq U \tag{2.2.30}$$

for *any* set A.

Here, we introduce the notation

$$\varnothing \subseteq A \subseteq U \tag{2.2.31}$$

to mean *the empty set* $\varnothing$ *is a subset of* A *and* A *is a subset of the universal set* U.

The above concept can be extended to any three sets. If it is true that

$$A \subseteq B \quad \text{and} \quad B \subseteq C \tag{2.2.32}$$

then this fact may be written in a shortened way by writing

$$A \subseteq B \subseteq C \tag{2.2.33}$$

read as A is a subset of B and B is a subset of C.

Exercise 2.2

Replace the asterisk $(*)$ by those relations $=$, $\sim$, $\subset$, $\subseteq$, $\supset$, $\supseteq$, that give a true statement.

1. $\{1, 3, 7\} * \{1, 3\}$
2. $\{4, 9\} * \{9, 10, 4\}$
3. $\{5, 7\} * \{5, 8\}$
4. $\{0, 1, 4\} * \{0, 1, 5, 8\}$
5. $\varnothing * \{1, 3\}$
6. $\{0\} * \{1, 3\}$
7. $\{5, 8, 9, 15\} * \{9\}$
8. $\{2, 3\} * \{2, 3, 6, 8, 0\}$
9. $\{2, 3\} * \{2, 3\}$
10. If $A \subset B$ and $B \subset C$, what can you conclude? Why?

11. Find two sets that are not comparable.

12. Show that the relation = (is equal to) between sets has the following properties:

    (I) *Reflexive Law:* $A = A$ for all sets A.
    (II) *Symmetric Law:* If $A = B$, then $B = A$.
    (III) *Transitive Law:* If $A = B$ and if $B = C$, then $A = C$.

    Any relation having the above three properties is called an *equivalence relation.*

13. Give an example of a partial ordering relation and of an equivalence relation.

14. Which of the following (if any) are partial ordering relations? Which are equivalence relations?

    (a) "is the father of"
    (b) "is the son of"
    (c) "is an uncle of"
    (d) "is a brother-in-law of"
    (e) "is a friend of"

15. If the universal set is the set of people, let A denote the subset of fat people, let B denote the subset of bald people, and let C denote the subset of bald and fat people. Write down several correct relations involving A, B, and C.

2.3 OPERATIONS WITH SETS

In the last section, we discussed relations between sets. In this section, we shall introduce operations which are performed on sets. The student is already familiar with many operations that are performed on numbers. For example, we speak of "taking the square root of a number" or "squaring a number." These are examples of *unary operations,* since they are operations performed on only *one* number. Also, we speak of "adding two numbers" or "multiplying two numbers." These are examples of *binary operations,* since they involve *two* numbers.

DEFINITION 2.3.1

Let A and B be any two sets. The *union of A with B,* written as

$$A \cup B \tag{2.3.1}$$

is defined to be that set consisting of those elements *either* in A *or* in B *or* in both A and B. That is, using set-builder notation,

$$A \cup B = \{x \mid x \in A \text{ or } x \in B\} \tag{2.3.2}$$

The student should notice the similarity between the concept of the union of two sets and the inclusive disjunction of two statements from Chapter One. Thus, we can write

$$A \cup B = \{x \mid x \in A \vee x \in B\} \tag{2.3.3}$$

Definition 2.3.2

Let A and B be any two sets. The *intersection of A with B*, written as

$$A \cap B \tag{2.3.4}$$

is defined as the set consisting of those elements that are *both* in A *and* in B. Thus,

$$A \cap B = \{x \mid x \in A \text{ and } x \in B\} \tag{2.3.5}$$

Just as with the union of two sets, the intersection of two sets is much like the conjunction of two statements. That is,

$$A \cap B = \{x \mid x \in A \wedge x \in B\} \tag{2.3.6}$$

The student should recognize that to find the intersection of two sets, A and B, means to find the elements *common* to A and B.

Definition 2.3.3

If two sets A and B have *no* elements in common, that is, if

$$A \cap B = \varnothing \tag{2.3.7}$$

then A and B are called *disjoint sets.*

example 2.3.1

Consider the three sets

$$A = \{1, 3, 5\}, \qquad B = \{3, 4, 5, 6\}, \qquad C = \{6, 7\} \tag{2.3.8}$$

Then

$$\begin{aligned} A \cup B &= \{1, 3, 5\} \cup \{3, 4, 5, 6\} = \{1, 3, 4, 5, 6\} \\ A \cap B &= \{1, 3, 5\} \cap \{3, 4, 5, 6\} = \{3, 5\} \\ A \cap C &= \{1, 3, 5\} \cap \{6, 7\} = \varnothing \end{aligned} \tag{2.3.9}$$

Here A and C are disjoint sets.

The next result gives us a relationship between the operations of union and intersection. This theorem says that the intersection of two sets is always a subset of the union of the two sets. That is, every element of the intersection is also an element of the union.

Theorem 2.3.1

For any two sets A and B, we have

$$A \cap B \subseteq A \cup B \tag{2.3.10}$$

PROOF: If $x \in A \cap B$, then $x \in A$ and $x \in B$. But if $x \in A$ *and* $x \in B$, then $x \in A$ *or* $x \in B$. This says that $x \in A \cup B$. Hence, any element in $A \cap B$ is in $A \cup B$. That is, $A \cap B \subseteq A \cup B$.

Theorem 2.3.2

If A and B are two sets, and if

$$A \cup B = \varnothing \tag{2.3.11}$$

then

$$A = \varnothing \quad \text{and} \quad B = \varnothing \tag{2.3.12}$$

PROOF: If $A \cup B = \varnothing$, then there are *no* elements either in A or in B. This says that A has no elements and B has no elements. Hence, $A = \varnothing$ and $B = \varnothing$.

The following theorem states that the intersection of a set A with any other set is a subset of the set A. It further says that a set A is always a subset of the union of A with any set. Thus, every element both in a set A and in some other set is also in A. Finally, every element in a set A is also in either A or any other set.

Although these statements are intuitively obvious, we shall give a formal proof of their validity.

Theorem 2.3.3

Let A and B be two sets. Then

$$A \cap B \subseteq A, \qquad A \subseteq A \cup B \tag{2.3.13}$$

PROOF: Let $x \in A \cap B$. Then $x \in A$ and $x \in B$ so that $x \in A$. Hence, any element in $A \cap B$ must be in A. That is,

$$A \cap B \subseteq A \tag{2.3.14}$$

In Theorem 2.3.4, we state some further facts that are true about sets. This one says that if a set A is a subset of a set B, the intersection of A and B (those elements both in A and in B) is A and the union of A and B (those elements either in A or in B) is B.

Theorem 2.3.4

Let A and B be any two sets. If

$$A \subseteq B \tag{2.3.15}$$

then

$$A \cap B = A, \qquad A \cup B = B \tag{2.3.16}$$

PROOF: By Theorem 2.3.3, we know that

$$A \cap B \subseteq A \tag{2.3.17}$$

Now, let $x \in A$. Then $x \in A$ and $x \in B$ since every element in A must be in B. This says that $x \in A \cap B$. Thus, every element in A is in $A \cap B$. Hence,

$$A \subseteq A \cap B \tag{2.3.18}$$

If we apply the anti-symmetric law to equations (2.3.17) and (2.3.18), we see that

$$A = A \cap B \tag{2.3.19}$$

The proof of the other relation in (2.3.16) is similar.

Next, we shall define a third binary operation involving sets. This one is quite similar to the concept of subtraction of numbers.

DEFINITION 2.3.4

Let A and B be any two sets. The *relative complement of A with respect to B*, written as

$$B - A \tag{2.3.20}$$

is defined as that set consisting of elements that are in B, but not in A. Thus, in set-builder notation,

$$B - A = \{x \mid x \in B \text{ and } x \notin A\} \tag{2.3.21}$$

Sometimes the relative complement of A with respect to B is called the *difference* $B - A$.

Thus, to find the relative complement of a set A with respect to a set B, merely write down the elements of set B and delete from it those elements that are in A.

EXAMPLE 2.3.2

If

$$A = \{1, 3, 5\}, \qquad B = \{1, 2\} \tag{2.3.22}$$

then

$$B - A = \{2\} \tag{2.3.23}$$

and

$$A - B = \{3, 5\} \tag{2.3.24}$$

The above shows that the relative complement of B with respect to A is not always the same as the relative complement of A with respect to B. In other words,

$$B - A \neq A - B \tag{2.3.25}$$

Thus far, three operations involving sets have been introduced. These three operations of union $\cup$, intersection $\cap$, and difference $-$ are examples of binary operations.

We are now ready to introduce an operation on sets involving only one set—that is, a *unary* operation.

DEFINITION 2.3.5

Let A be any set. The *absolute complement of* A, written as

$$\bar{A} \quad (\text{or } A', \text{ or } c(A), \text{ or } -A) \tag{2.3.26}$$

is defined as the set consisting of elements in the universal set U which are not in A. Thus, $\bar{A} = \{x \mid x \notin A\}$.

It is noted that taking the absolute complement of A is the same as finding the relative complement of A with respect to the universal set U. That is,

$$\bar{A} = U - A \tag{2.3.27}$$

Also, from (2.3.21) and Definition 2.3.5, we can conclude

$$B - A = B \cap \bar{A} \tag{2.3.28}$$

EXAMPLE 2.3.3

If U = universal set = $\{0, 1, 2, 3, 4\}$ and if $A = \{0, 1, 2\}$, then

$$\bar{A} = \{3, 4\} \tag{2.3.29}$$

Also, we see that

$$U - A = \{0, 1, 2, 3, 4\} - \{0, 1, 2\} = \{3, 4\} \tag{2.3.30}$$

EXAMPLE 2.3.4

Find the most general conditions on two sets A and B so that

$$A \cup B = B \tag{2.3.31}$$

Here it is known that the elements $A \cup B$ and those of B are the same. But the elements of $A \cup B$ consist of all elements either in A or in B. Now, if we demand that elements either in A or B are the same as those in B, we must have every element in A also in B. For, if A had elements not found in B, then $A \cup B$ would contain elements not found in B, and hence $A \cup B$ and B would not be equal. Thus, if $A \cup B = B$, then $A \subseteq B$.

EXAMPLE 2.3.5

For a general nonempty set A, find the most general conditions on a set X such that

$$X \cup A = U \tag{2.3.32}$$

Here, we want to find a set X such that the union of X with A is the universal set. We can rephrase the question as follows: What elements must set X have if those elements either in X or in A are all elements under consideration? It is clear now that X must at least have those elements *not* in A. For, if X did not have an element found in $\bar{A}$, then $X \cup A$ could not be the universal set. However, it does no harm for X to contain elements besides those found in $\bar{A}$. We can state these conditions on X by writing

$$\bar{A} \subseteq X \tag{2.3.33}$$

EXAMPLE 2.3.6

If U = universal set = $\{1, 2, 3, 4\}$ and if A and B are two nonempty sets, find all possibilities for A if

$$A \cap B = \{3\} \quad \text{and} \quad B = \{2, 3\} \tag{2.3.34}$$

Here, because $A \cap B \subseteq A$, we know that the element 3 must be in set A. Since $B = \{2, 3\}$, then if 2 were an element of A, we would have to have 2 in their intersection. Since $2 \notin A \cap B$, then $2 \notin A$. Thus, the only possible elements for A to have are

$$A = \{3\}, \qquad A = \{1, 3\},$$
$$A = \{3, 4\}, \qquad A = \{1, 3, 4\} \tag{2.3.35}$$

Exercise 2.3

1. If A is any set, evaluate the following:

   (a) $A \cup \varnothing$ (b) $A \cap \varnothing$ (c) $A \cup U$ (d) $A \cap U$

2. If U = universal set = $\{x \mid x \text{ is a digit}\}$ and if $A = \{0, 1, 5, 7\}$, $B = \{2, 3, 5, 8\}$, $C = \{5, 6, 9\}$, find

   (a) $A \cup B$ (e) $\bar{A} - \bar{B}$
   (b) $B \cap C$ (f) $A \cup (B \cap A)$
   (c) $\underline{A - B}$ (g) $(C - A) \cap (\bar{A})$
   (d) $\overline{A - B}$ (h) $(A - B) \cup (B - C)$

3. If U = universal set = $\{1, 2, 3, 4, 5\}$ and if $A = \{3, 5\}$, $B = \{1, 2, 3\}$, $C = \{2, 3, 4\}$, find

   (a) $\bar{A} - \bar{C}$ (e) $\overline{A - C}$
   (b) $(A \cup B) \cap C$ (f) $\overline{A \cup B}$

(c) $A \cup (B \cap C)$
(d) $(A \cup B) \cap (A \cup C)$
(g) $\bar{A} \cap \bar{B}$
(h) $(\bar{A})$

4. Find the most general condition on the sets A and B so that

(a) $A - B = B - A$
(b) $A \cap B = A \cup B$
(c) $A - B = A$
(d) $A - B = B$
(e) $A \cap B = B$

5. Find the most general conditions on a set X so that for a given nonempty set A, we have

(a) $A \cup X = X \cup A$
(b) $\bar{X} = A$
(c) $\bar{X} = U$
(d) $\bar{X} = \varnothing$
(e) $X \cap A = \varnothing$
(f) $X \cap A = U$

6. Prove that $A \subseteq A \cup B$ for any two sets A and B.

7. Prove that if $A \subseteq B$, then $A \cup B = B$.

8. If U = universal set $= \{1, 2, 3, 4, 5, 6\}$, $C = \{1, 3\}$ and A and B are nonempty, find all possibilities for A in each of the following:

(a) $A \cup B = U$ and $A \cap B = \{3\}$ and $B = \{2, 3, 4\}$
(b) $A \subseteq B$ and $A \cup B = \{2, 5, 6\}$
(c) $A \cap B = \varnothing$ and $A \cup B = \{1, 2, 3, 4, 5\}$ and $B \cup C = \{1, 2, 3\}$

9. If $A = \{1, 2, 3\}$, $B = \{3, 4, 5, 6\}$, $C = \{3, 5, 7\}$, find

(a) $A \cup B$
(b) $A \cup C$
(c) $(A \cup B) \cap C$
(d) $A \cap B$
(e) $A \cap C$
(f) $B \cap C$
(g) $(A \cap B) \cap C$
(h) $(A \cap B) \cup C$

10. If $A = \{1, 3, 5, 7\}$, $B = \{2, 4\}$, $C = \{1, 2, 3\}$, U = universal set $= \{1, 2, 3, 4, 5, 7\}$, find

(a) $\bar{A} \cup \bar{B}$
(b) $\overline{A \cap B}$
(c) $\bar{A} \cap \bar{B}$
(d) $\overline{A \cup B}$
(e) $\overline{U \cap B}$
(f) $\overline{C \cup B}$
(g) $\overline{C \cap A}$
(h) $C - A$
(i) $A - C$

11. Prove that if $A \subset B$, then $\bar{B} \subset \bar{A}$.

2.4 SOME RESULTS INVOLVING OPERATIONS WITH SETS

This section deals primarily with a discussion of some of the relationships which are obeyed by sets. For example, if

$$A = \{1, 2, 3\}, \qquad B = \{3, 4, 5\} \tag{2.4.1}$$

then

$$A \cap (A \cup B) = A \cap \{1, 2, 3, 4, 5\} = \{1, 2, 3\} = A \tag{2.4.2}$$

Thus, in this particular example, it turns out that

$$A \cap (A \cup B) = A \tag{2.4.3}$$

Later on in this section it is shown that no matter what two sets A and B are used, equation (2.4.3) is always true.

Although many relationships obeyed by sets will be discussed in this section, our primary purpose is to understand the meaning of these relationships. This does not mean they should be memorized. Our second goal is to be able to use and apply these laws or relationships to certain types of problems. Finally, the student, after he has mastered the first two goals, should strive to be able to reproduce the proofs of these laws.

The first law involves the unary operation of complement. It states that if this operation is performed twice on a set A, the resulting set is A itself.

Theorem 2.4.1

The operation of complementation of a set is *involutorial.* That is,

$$\bar{\bar{A}} = A, \qquad \text{for any set} \tag{2.4.4}$$

PROOF: If $A = \varnothing$, then $\overline{\varnothing} = U$ so that $(\bar{\bar{\varnothing}}) = \bar{U} = \varnothing$. If $(\bar{\bar{A}}) = \varnothing$, then $\bar{A} = U$ so that $A = \varnothing$. Henceforth, suppose $A \neq \varnothing$ and $(\bar{\bar{A}}) \neq \varnothing$. If $x \in A$, then $x \notin \bar{A}$ so that $x \in \bar{\bar{A}}$. Thus,

$$A \subseteq (\bar{\bar{A}}) \tag{2.4.5}$$

If $x \in (\bar{\bar{A}})$, then $x \notin \bar{A}$ so that $x \in A$. Thus,

$$\bar{\bar{A}} \subseteq A \tag{2.4.6}$$

Using (2.4.5), (2.4.6), and the anti-symmetric law, the theorem is proved.

The student should notice that Theorem 2.4.1 is very similar to the fact that $\sim(\sim p) \leftrightarrow p$ for any statement p. The next result for sets corresponds to the fact in logic that $p \wedge p \leftrightarrow p$, $p \vee p \leftrightarrow p$ are tautologies. It is only by using this fact from logic that an acceptable proof can be given.

Theorem 2.4.2

The Idempotent Laws: If A is any set, then

$$A \cup A = A, \qquad A \cap A = A \tag{2.4.7}$$

PROOF: Suppose $A \cup A = \varnothing$. Then $A = \varnothing$ by Theorem 2.3.2. If $A = \varnothing$, then $A \cup A = \varnothing \cup \varnothing = \varnothing$. Thus, if either $A \cup A$ or A is empty, the result is true. Henceforth, suppose neither $A \cup A$ nor A is empty.

Let $x \in A \cup A$. Then $x \in A$ or $x \in A$, so that $x \in A$. This says that any element in $A \cup A$ is also in A. Thus,

$$A \cup A \subseteq A \tag{2.4.8}$$

Let $x \in A$. Then $x \in A$ or $x \in A$ so that $x \in A \cup A$. Thus

$$A \subseteq A \cup A \tag{2.4.9}$$

The result follows by the anti-symmetric law.

In order to prove the next result, we use the fact from logic that $p \vee q \leftrightarrow q \vee p$ and $p \wedge q \leftrightarrow q \wedge p$ for any two statements p and q.

Theorem 2.4.3

The Commutative Laws: The operations of union and intersection relative to sets are *commutative*. That is, if A and B are two sets, then

$$A \cup B = B \cup A \tag{2.4.10}$$

and

$$A \cap B = B \cap A \tag{2.4.11}$$

PROOF: We shall only prove that the operation of union is commutative. The proof of the fact that the operation of intersection is commutative is left to the student.

First of all, if $A \cup B = \varnothing$, then by the remarks made above, it is seen that $A = \varnothing$ and $B = \varnothing$. Thus, $B \cup A = \varnothing \cup \varnothing = \varnothing$. Therefore, $A \cup B = B \cup A$. Similarly, if $B \cup A = \varnothing$, then $A \cup B = \varnothing$.

Henceforth, $A \cup B \neq \varnothing$ and $B \cup A \neq \varnothing$. Let $x \in A \cup B$. Then $x \in A$ or $x \in B$. Thus, $x \in B$ or $x \in A$ so that $x \in B \cup A$. Hence,

$$A \cup B \subseteq B \cup A \tag{2.4.12}$$

Also, if $x \in B \cup A$, then $x \in B$ or $x \in A$, which implies that $x \in A$ or $x \in B$ so that $x \in A \cup B$. Thus,

$$B \cup A \subseteq A \cup B \tag{2.4.13}$$

By the anti-symmetric law, we have

$$A \cup B = B \cup A \tag{2.4.14}$$

The commutative laws for the operations of union and intersection of sets tell us that the order in which we perform the operation will not affect the result. Thus, if we form the union of A and B, namely $A \cup B$, or if we form the union of B and A, namely, $B \cup A$, the result will always be the same.

Next, by using the fact from logic that $p \vee (q \vee r) \leftrightarrow (p \vee q) \vee r$ and $p \wedge (q \wedge r) \leftrightarrow (p \wedge q) \wedge r$, for any three statements p, q, r, we can prove that the operations of union and intersection of sets obey the following laws.

Theorem 2.4.4

The Associative Laws: The operations of union and intersection relative to sets are *associative*. That is, if A, B, C, are any three sets, then

$$A \cup (B \cup C) = (A \cup B) \cup C \tag{2.4.15}$$

and

$$A \cap (B \cap C) = (A \cap B) \cap C \tag{2.4.16}$$

PROOF: Again, we shall prove the first half and the proof that the operation of intersection is associative will be left to the reader.

Let us also assume that neither the set $A \cup (B \cup C)$ nor the set $(A \cup B) \cup C$ is empty. The student should verify for himself that the result is true if either of these is the case.

Let $x \in A \cup (B \cup C)$. Then $x \in A$ or $x \in (B \cup C)$. Thus, $x \in A$ or ($x \in B$ or $x \in C$) so that ($x \in A$ or $x \in B$) or $x \in C$. That is, $x \in A \cup B$ or $x \in C$. Hence $x \in (A \cup B) \cup C$. Thus

$$A \cup (B \cup C) \subseteq (A \cup B) \cup C \tag{2.4.17}$$

Next, if $x \in (A \cup B) \cup C$, then $x \in A \cup B$ or $x \in C$ so that ($x \in A$ or $x \in B$) or $x \in C$. This says that $x \in A$ or ($x \in B$ or $x \in C$). Thus, $x \in A$ or $x \in B \cup C$. Hence $x \in A \cup (B \cup C)$. Therefore,

$$(A \cup B) \cup C \subseteq A \cup (B \cup C) \tag{2.4.18}$$

By the anti-symmetric law, the associative law for the operation of union follows.

The associative laws for the operations of union and intersection of sets tell us that when performing the operation of union, say, on *three* sets, the way in which we group these three sets will not affect the result. Thus, for example, consider

$$A \cup B \cup C \tag{2.4.19}$$

Here, we have two choices. Either we compute $A \cup B$ and then unite this with C or else we compute $B \cup C$ and then find A united with $B \cup C$. That is,

$$\text{either} \quad (A \cup B) \cup C \quad \text{or} \quad A \cup (B \cup C) \tag{2.4.20}$$

The associative law guarantees that no matter which way we choose to group, the resulting set will be the same. *Note:* The associative law can be used only when *three* sets and the *same* binary operations are involved. The commutative law involved *two* sets and *one* binary operation.

Theorem 2.4.5

The Distributive Laws: The operation of union is distributive relative to the operation of intersection. That is, if A, B, C are three sets, then

$$A \cup (B \cap C) = (A \cup B) \cap (A \cup C) \tag{2.4.21}$$

Also, the operation of intersection is distributive relative to the operation of union. That is, if A, B, C are three sets, then

$$A \cap (B \cup C) = (A \cap B) \cup (A \cap C) \tag{2.4.22}$$

PROOF: The proof of these results is left to the student.

As a hint, use the facts from logic that $p \vee (q \wedge r) \leftrightarrow (p \vee q) \wedge (p \vee r)$ and $p \wedge (q \vee r) \leftrightarrow (p \wedge q) \vee (p \wedge r)$ for any *three* statements p, q, r.

Also, notice that the distributive laws involve *three* sets and two *different* binary operations.

Theorem 2.4.6

The Absorption Laws: Sets obey two absorption laws, namely,

$$A \cup (A \cap B) = A \tag{2.4.23}$$

and

$$A \cap (A \cup B) = A \tag{2.4.24}$$

for any two sets A and B.

The reason for the name given to the above laws is that the set B is "absorbed" by the set A.

PROOF: If $A \cup (A \cap B) = \varnothing$, then $A = \varnothing$ and $A \cap B = \varnothing$. That is, $A = \varnothing$. If $A = \varnothing$, then $A \cup (A \cap B) = \varnothing \cup (\varnothing \cap B) = \varnothing \cup \varnothing = \varnothing$. Thus, if either $A \cup (A \cap B)$ or A is empty, the first absorption law is valid.

Suppose neither A nor $A \cup (A \cap B)$ is empty.

Let $x \in A \cup (A \cap B)$. Then $x \in A$ or $x \in A \cap B$. That is, $x \in A$ or $x \in A$ and B. Hence, $x \in A$. That is,

$$A \cup (A \cap B) \subseteq A \tag{2.4.25}$$

Also, from equation (2.3.13) of Theorem 2.3.3, we have

$$A \subseteq A \cup (A \cap B) \tag{2.4.26}$$

By the anti-symmetric law,

$$A = A \cup (A \cap B) \tag{2.4.27}$$

Notice that in the proof of the absorption laws, we used the fact from logic that $p \vee (p \wedge q) \leftrightarrow p$ and $p \wedge (p \vee q) \leftrightarrow p$.

Theorem 2.4.7

De Morgan's Law: If A and B are any two sets, then

$$\overline{A \cup B} = \bar{A} \cap \bar{B} \tag{2.4.28}$$

and

$$\overline{A \cap B} = \bar{A} \cup \bar{B} \tag{2.4.29}$$

PROOF: The proof of these two results is left to the student. As a hint, refer to De Morgan's laws concerning statements in logic from Chapter 1, Section 1.3.

EXAMPLE 2.4.1

Using the laws given in Sections 2.3 and 2.4, prove that

$$(A - B) \cup (A \cap B) = A \tag{2.4.30}$$

Here, let us start with the left side and, using laws about sets, try to show the left side is the same as the right side.

The student should fill in a reason for each step. Now,

$$\begin{aligned}(A - B) \cup (A \cap B) &= (A \cap \bar{B}) \cup (A \cap B) \qquad (2.4.31)\\ &= A \cap (\bar{B} \cup B)\\ &= A \cap U\\ &= A\end{aligned}$$

Exercise 2.4

By using the laws given in Sections 2.3 and 2.4, prove the following. Give reasons for each step.

1. $A \cup (B \cup A) = A \cup B$
2. $A \cap (\bar{A} \cap B) = \varnothing$
3. $A \cup (\bar{A} \cup B) = U$
4. $A \cup (\bar{A} \cap B) = A \cup B$
5. $A \cap (\bar{A} \cup B) = A \cap B$
6. $A \cap (\bar{\bar{A}} \cup B) = A$
7. $A \cup (\bar{\bar{A}} \cap B) = A$
8. $(\overline{\bar{A} \cup \bar{B}}) = A \cap B$
9. $(A - B) \cup (B - A) = (A \cup B) - (A \cap B)$
10. $(\overline{A - B}) = \bar{A} \cup B$

2.5 VENN DIAGRAMS

A *Venn diagram* is a pictorial representation of relationships involving sets. Usually, sets are represented as interlocking circles, each of which is enclosed in a rectangle which represents the universal set U.

Venn diagrams are sometimes useful for *illustrating* theorems, but are *never* to be used as a replacement for a logical deductive proof of results concerning sets. They are to be used by the student only to gain insight into exactly what a given proposition says.

EXAMPLE 2.5.1

Illustrate $A \cup B$ and $\bar{A}$ by a Venn diagram.

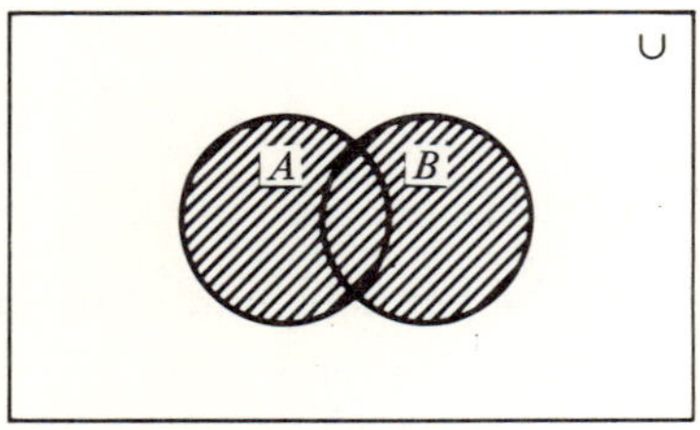

FIG. 2.1

Thus, $A \cup B$ consists of all elements in A or in B or in both A and B. We illustrate this by shading the region in A, the region in B, and the region in both A and B.

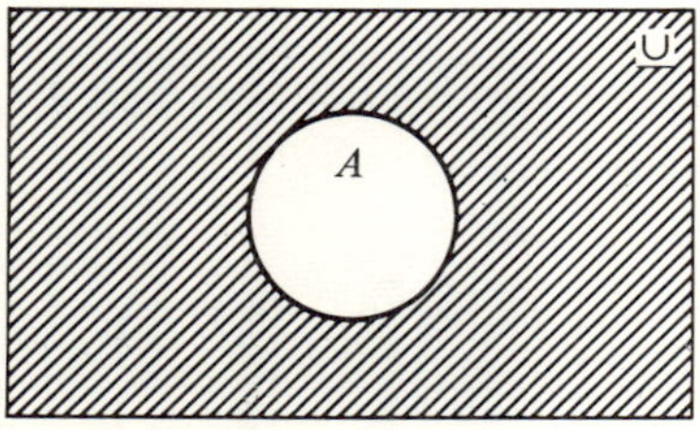

FIG. 2.2

Here $\bar{A}$ consists of all elements in U not in A. We illustrate this by shading everything except A.

EXAMPLE 2.5.2

Illustrate the absorption law by using a Venn diagram.

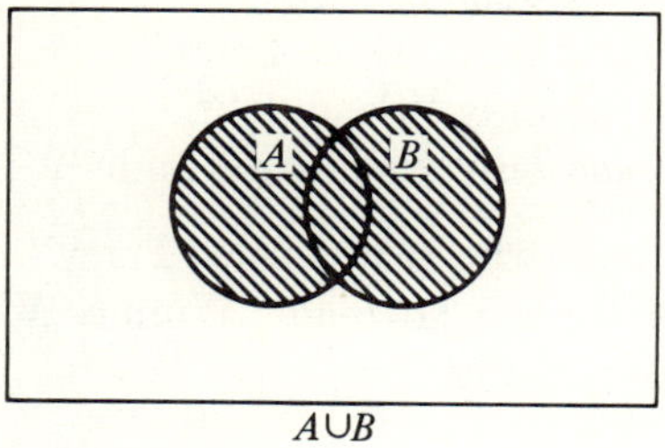

FIG. 2.3

Since the cross-hatched region in Fig. 2.4 matches the shaded region in Fig. 2.5, the fact that

$$A \cap (A \cup B) = A \tag{2.5.1}$$

has been illustrated.

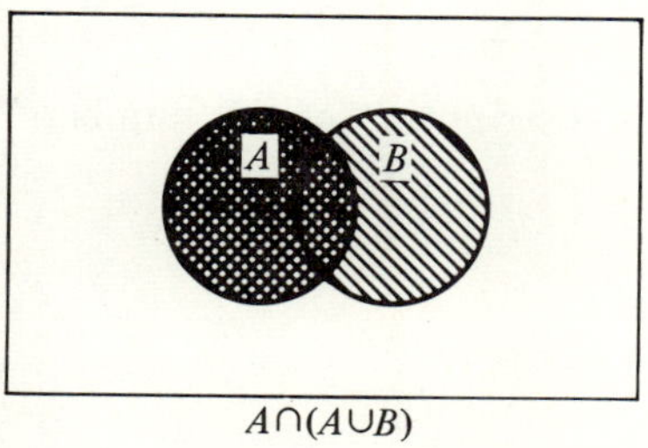

FIG. 2.4

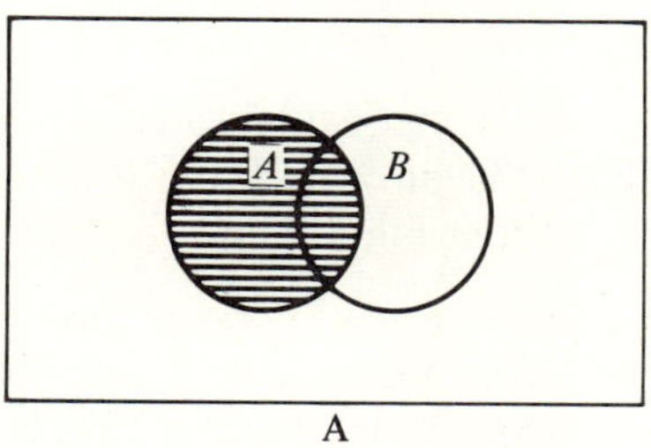

FIG. 2.5

EXAMPLE 2.5.3

In a survey of 50 school children, it was found that of the comic strips "Dick Tracy," "Peanuts," and "Winnie Winkle,"

20 read "Dick Tracy"

35 read "Peanuts"

9 read "Winnie Winkle"

15 read "Peanuts" and "Dick Tracy"

2 read "Dick Tracy" and "Winnie Winkle"

3 read "Peanuts" and "Winnie Winkle"

1 read all three comic strips

(a) How many read none of these?
(b) How many read only "Dick Tracy"?
(c) How many read only "Peanuts"?
(d) How many read only "Winnie Winkle"?

If we denote all those who read "Dick Tracy" by D, those who read "Peanuts" by P, and those who read "Winnie Winkle" by W, we can illustrate the above facts in Fig. 2.6.

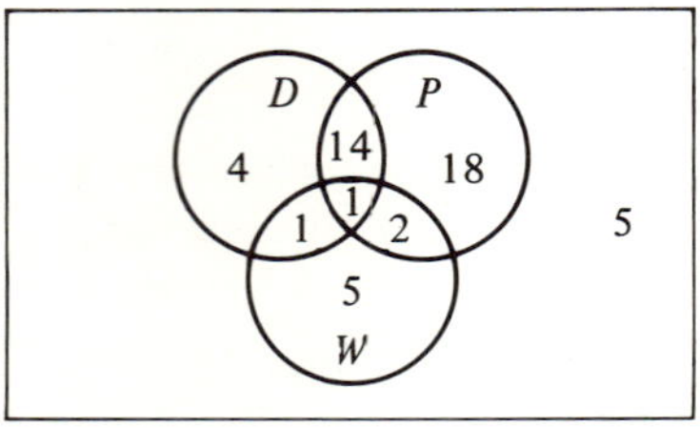

FIG. 2.6

Can you explain how we arrive at this distribution of the numbers? [HINT: begin with $D \cap P \cap W$.]

From Fig. 2.6, we see that

(a) 5 read none of these comic strips
(b) 4 read only "Dick Tracy"
(c) 18 read only "Peanuts"
(d) 5 read only "Winnie Winkle"

Exercise 2.5

1. Illustrate the following results by using Venn diagrams.

   (a) The commutative laws for union and intersection.
   (b) The associative laws for union and intersection.
   (c) The distributive laws for union and intersection.
   (d) The absorption laws.
   (e) De Morgan's laws.
   (f) $(A \cup B) \cap C$
   (g) $(\bar{A} \cup B) \cap (A \cup C)$

(h) $\bar{A} \cap \bar{B}) \cup C$
(i) $(\overline{A \cup B}) \cap (A \cup C)$

2. In a survey of 75 college students it was found that of the three weekly news magazines, *Time, Newsweek, U.S. News and World Report*

   23 read *Time*

   18 read *Newsweek*

   13 read *U.S. News and World Report*

   10 read *Time* and *Newsweek*

   9 read *Time* and *U.S. News and World Report*

   8 read *Newsweek* and *U.S. News and World Report*

   5 read all three magazines.

   (a) How many read none of the three magazines?
   (b) How many read *Time* as their only magazine?
   (c) How many read *Time* if and only if they read *Newsweek*?
   (d) How many read either *Time* or *Newsweek*?

3. A staff member at an engineering school was presenting data to show that the students there received a liberal education as well as a scientific one. "Look at the record," he said. "Out of our senior class of 500 students, 281 are taking English, 196 are taking English and history, 87 are taking history and a foreign language, 143 are taking a foreign language and English, and 36 are taking all of these." It sounded most impressive, but he was fired. Why?

2.6 THE CARDINAL NUMBER OF A SET

To ask *what is the cardinal number of this set* is essentially the same as asking *how many elements are in this set.* In other words, the concept of *cardinal number* is very closely related to the concept of *counting.*

The idea of counting is really one of comparison. That is to say, when one counts one always takes the objects to be counted and matches each of these objects exactly once to the numbers 1, 2, 3, etc., until *no* objects remain.

Because of the above, the following definition is given.

Definition 2.6.1

Two sets A and B are said to have the same *cardinal number* or the *same cardinality* or the *same potency* if and only if A and B are equivalent sets.

Thus, if there is a one-to-one correspondence between the elements of A and B, then A and B have the same cardinal number.

EXAMPLE 2.6.1

The two sets

$$A = \{1, 3, 5\}, \qquad B = \{2, 4, 6\} \tag{2.6.1}$$

are equivalent. Hence, they have the same cardinal number, in this case three.

DEFINITION 2.6.2

A *section* $[1, n]$ is defined as the set of counting numbers from 1 to n, inclusive. That is,

$$[1, n] = \{1, 2, 3, \cdots, n - 1, n\} \tag{2.6.2}$$

EXAMPLE 2.6.2

The section $[1, 6]$ is

$$[1, 6] = \{1, 2, 3, 4, 5, 6\} \tag{2.6.3}$$

DEFINITION 2.6.3

The empty set $\varnothing$ has the cardinal number 0. If a nonempty set A has the same cardinal number as the section $[1, n]$, then the set A is said to have the *cardinal number* n. In these situations, we write

$$c(\varnothing) = 0, \qquad c(A) = n \tag{2.6.4}$$

EXAMPLE 2.6.3

The set

$$A = \{a, b, c\} \tag{2.6.5}$$

has cardinal number 3, since

$$A \sim [1, 3] = \{1, 2, 3\} \tag{2.6.6}$$

DEFINITION 2.6.4

A set A is called *finite* if and only if either

$$A = \varnothing \tag{2.6.7}$$

or else, if $A \neq \varnothing$,

$$A \sim \{1, n\} \tag{2.6.8}$$

for some counting number n. If, for a set A, neither (2.6.7) nor (2.6.8) is true, then A is called *infinite*.

EXAMPLE 2.6.4

The set N of counting numbers, namely,

$$N = \{1, 2, 3, 4, \cdots, 635, \cdots\} \tag{2.6.9}$$

is an example of an infinite set.

DEFINITION 2.6.5

The set N of counting numbers is said to have the cardinal number *aleph null*, written as $\aleph_0$.

Because of the way in which cardinal number has been defined, some very strange results are found. For example, it turns out that two sets A and B can have the *same* cardinal number, and yet A can be a proper subset of B. (See problem 4 in the exercises below.)

Exercise 2.6

1. What are the cardinal numbers of the following sets?

   (a) Set D of digits $= \{0, 1, 2, 3, 4, 5, 6, 7, 8, 9\}$
   (b) $A = \{1, 3, 5, 7\}$
   (c) $A = \{0, 1, 2\}$
   (d) $W = \{0, 1, 2, 3, \cdots\}$

2. For the two sets $A = \{1, 2, 3, 5\}$, $B = \{4, 6, 8\}$, find the cardinal number of

   (a) A (d) $A \cup B$
   (b) B (e) $A - B$
   (c) $A \cap B$ (f) $B - A$

3. Show that the set N_e of even counting numbers, namely, $\{2, 4, 6, 8, \cdots\}$, has the same cardinal number as the set N_o of odd counting numbers, namely $\{1, 3, 5, \cdots\}$.

4. Show that the cardinal number of the set N_e of even counting numbers is $\aleph_0$.

5. For the sets $A = \{1, 3, 6, 8\}$, $B = \{8\}$, $C = \{8, 10\}$, find the cardinal number of the following:

   (a) $A \cup (B \cap C)$ (d) $A \cup B \cup C$
   (b) $A \cap (B \cap C)$ (e) $(A \cap B) \cup C$
   (c) $A - (B \cup C)$ (f) $A - (B \cap C)$

6. If $c(A) = m$ and $c(B) = n$ and $c(A \cup B) = m + n$, what is $c(A \cap B)$?

7. If $c(A) = m$, $c(B) = n$, and $c(A \cup B) = r$, where r is less than $m + n$, explain how to determine the cardinal number of $A \cap B$.

8. If $c(A) = 10$, $c(A \cup B) = 29$, and $c(A \cap B) = 5$, what is $c(B)$?

9. If $c(A) = c(B)$, $c(A \cup B) = 16$, and $c(A \cap B) = 6$, find $c(B)$.

10. Show that the union of two finite sets A and B is also a finite set. [HINT: Let $A = \{a_1, a_2, \cdots, a_n\}$ and $B = \{b_1, b_2, \cdots, b_m\}$. Then $A \cup B \sim C$, where $C \subseteq (1, m + n)$.]

2.7 PRODUCT SET

A set of great interest for the work to be done in Chapter 3 and in Chapter 9 is the *product set.*

DEFINITION 2.7.1

Let A and B be arbitrary sets.

If either $A = \varnothing$ or $B = \varnothing$ or $A = B = \varnothing$, the *product set* $A \times B$ is defined to be the empty set $\varnothing$.

If both A and B are nonempty sets, the *product set* $A \times B$ is defined to be a set whose elements are ordered pairs (a, b) for which $a \in A$ and $b \in B$.

Thus, in set notation, if $A \neq \varnothing$, $B \neq \varnothing$,

$$A \times B = \{(a, b) \mid a \in A, b \in B\} \tag{2.7.1}$$

It is important to note that in the above definition, the sets A and B need not be distinct. Also, the pair (a, b) is an *ordered* pair to indicate that the order in which a and b are placed makes a difference. That is, $(a, b) \neq (b, a)$, unless, of course, $a = b$. This fact is in contrast to what we learned with sets where $\{a, b\} = \{b, a\}$.

If A and B both consist of a single element; that is, if

$$A = \{a\}, \qquad B = \{b\} \tag{2.7.2}$$

then

$$A \times B = \{(a, b)\} \tag{2.7.3}$$

Thus, under these circumstances, $A \times B$ consists of an ordered pair, which is considered a single element.

The above definition of product set (sometimes called *cross product* or *Cartesian product*) can be generalized to more than two sets. Given A, B, C, arbitrary nonempty sets, the *cross product* $A \times B \times C$ is defined as follows.

$$A \times B \times C = \{(a, b, c) \mid a \in A, b \in B, c \in C\} \tag{2.7.4}$$

That is, $A \times B \times C$ is just the set of 3-tuples or ordered triplets. Of course, if any of the three sets A, B, or C is the empty set, then $A \times B \times C = \varnothing$.

EXAMPLE 2.7.1

If $A = \{1, 2, 3\}$ and $B = \{3, 4\}$, then

$$A \times B = \{(1,3)\ \ (1,4)\ \ (2,3)\ \ (2,4)\ \ (3,3)\ \ (3,4)\} \tag{2.7.5}$$

and

$$B \times A = \{(3,1)\ \ (3,2)\ \ (3,3)\ \ (4,1)\ \ (4,2)\ \ (4,3)\} \tag{2.7.6}$$

Also,

$$A \times A = \{(1,1)\ \ (1,2)\ \ (1,3)\ \ (2,1)\ \ (2,2)\ \ (2,3)\ \ (3,1)\ \ (3,2)\ \ (3,3)\} \tag{2.7.7}$$

and

$$B \times B = \{(3,3)\ \ (3,4)\ \ (4,3)\ \ (4,4)\} \tag{2.7.8}$$

Note that $A \times B \neq B \times A$.

From Example 2.7.1, we can conclude that ordinarily

$$A \times B \neq B \times A \tag{2.7.9}$$

EXAMPLE 2.7.2

Let $A = \{1, 2\}$ and $B = \{a, b\}$; then

$$A \times B = \{(1,a), (1,b), (2,a), (2,b)\} \tag{2.7.10}$$

and

$$B \times A = \{(a,1), (a,2), (b,1), (b,2)\} \tag{2.7.11}$$

EXAMPLE 2.7.3

Consider the three sets

$$A = \{1, 2\}, \qquad B = \{1, 3, 4\}, \qquad C = \{2, 3\} \tag{2.7.12}$$

We wish to find $A \times (B \cap C)$ and $(A \times B) \cap (A \times C)$.

First of all,

$$\begin{aligned} A \times (B \cap C) &= \{1, 2\} \times [\{1, 3, 4\} \cap \{2, 3\}] \\ &= \{1, 2\} \times \{3\} = \{(1,3), (2,3)\} \end{aligned} \tag{2.7.13}$$

Next, to find $(A \times B) \cap (A \times C)$, first

$$\begin{aligned} A \times B &= \{(1,1), (1,3), (1,4), (2,1), (2,3), (2,4)\} \\ A \times C &= \{(1,2), (1,3), (2,2), (2,3)\} \end{aligned} \tag{2.7.14}$$

Then

$$(A \times B) \cap (A \times C) = \{(1,3), (2,3)\} \tag{2.7.15}$$

Notice that

$$A \times (B \cap C) = (A \times B) \cap (A \times C) \tag{2.7.16}$$

The above, it turns out, is true no matter what three sets are involved. That is, $\times$ is distributive over intersection. Also, $\times$ is distributive over union. That is, for any three sets A, B, C, we have

$$A \times (B \cap C) = (A \times B) \cap (A \times C)$$

$$A \times (B \cup C) = (A \times B) \cup (A \times C) \tag{2.7.17}$$

Next, let us discuss the cardinality of the product set $A \times B$ of two sets A and B.

In Example 2.7.1 we see that the cardinality of $A \times B$ is 6. Also, the cardinal number of $B \times A$ is 6. From Example 2.7.2, the cardinal number of $A \times B$ is 4 and the cardinal number of $B \times A$ is also 4. From these examples, we can conjecture that even though $A \times B$ and $B \times A$ are usually unequal as sets, they are always equivalent as sets. That is,

$$A \times B \neq B \times A, \qquad A \times B \sim B \times A \tag{2.7.18}$$

Exercise 2.7

1. Find the product set $A \times E$ for the following sets.

   (a) $A = \{d, 8, 5\}$, $B = \{1, 2\}$
   (b) $A = \{1, a, x\}$, $B = \{c\}$
   (c) $A = \{x\}$, $B = \{x\}$
   (d) $A = \{1, a, x\}$, $B = \varnothing$
   (e) $A = \varnothing$, $B = \varnothing$
   (f) $A = \{1, 2, 3\}$, $B = \{2, 3\}$

2. If $A = \{1, 2, 3, 4\}$, $B = \{4, 5\}$, find

   (a) $A \times A$
   (b) $A \times B$
   (c) $B \times A$
   (d) $B \times B$

3. If $A = \{1, 2, 3\}$, $B = \{2, 3\}$, $C = \{1, 4\}$, find

   (a) $A \times (B \cap C)$
   (b) $(A \times B) \cup (C \times A)$
   (c) $(C \times A) \cap (C \times B)$
   (d) $(A \cup B) \times C$
   (e) $(A \cap B) \times C$
   (f) $(B \cap C) \times A$

4. Under what conditions on A and B will $A \times B = B \times A$?

5. Find the cardinal number of the sets in problem 1.

6. Find the cardinal number of the sets in problem 3.

3 The Set N of Natural Numbers and the Set W of Whole Numbers

3.1 INTRODUCTION

The set N of natural numbers, or counting numbers, consists of those numbers used for counting beginning with number one, namely,

$$N = \{1, 2, 3, \cdots\} \tag{3.1.1}$$

They are encountered early in life, when their recitation is learned for use in the process of counting. Yet when one is asked to give a definition of numbers, the task may appear impossible. The difficulty is that numbers are abstract concepts, and are therefore not easy to point to. For example, it is not difficult to find a set of four objects, but such a set is not the same as the number *four*. As a child, one learns to understand these abstract concepts from a large number of experiences. He is taught to recognize three horses, three dogs, three balls, and so on, until he understands the abstract notion of *three*.

There are many ways to introduce the natural numbers. However, at an elementary level, these many approaches boil down to two. In the first, the natural numbers are undefined objects which obey certain axioms or postulates. In the second, the natural numbers are defined in terms of sets, and the basic properties are proven from the corresponding properties of sets. It is the second method that is presented here.

The natural numbers were, no doubt, the first collection of numbers devised by man, and for a long time they were the only mathematical system used by him. However, other systems have been developed—integers, fractions, etc. The natural numbers remain as the basic system on which the others are constructed. Throughout the course, the student will realize that once the properties of natural numbers are known, he can pass along more easily to the invention and study of other kinds of number systems. It is for this reason that a study of the natural numbers is fundamental.

Consider the following collection of sets:

$$\{a, b, c, d\}, \qquad \{e, f, g, k\}, \qquad \{1, 2, 3, 4\}, \qquad \{*, \triangle, \checkmark, \#\} \tag{3.1.2}$$

All these sets have one property in common, namely, any one of them is in one-to-one correspondence to any other. That is, they are equivalent sets. We wish to abstract from this collection the property that all these sets have in common. In this example, the common property is that each set has *four* elements. That is, the number four is the idea which characterizes this particular collection of sets.

We can see from this that a natural number is what we described in Chapter 2 as a cardinal number. This leads us to the following definition.

Definition 3.1.1

A *natural number* is the cardinal number of any set in a collection of finite nonempty equivalent sets.

Thus, a natural number is the abstraction which the mind forms when it thinks of a collection of finite nonempty equivalent sets. To describe a natural number, we need only devise a symbol to represent this abstraction. This description can take any form, such as

$$//// \quad \text{or} \quad ---- \quad \text{or} \quad **** \tag{3.1.3}$$

In our particular system of writing, we use the symbol 4 to represent the abstraction *four*. It is to be noted that 4 is a symbol for the number four; it is *not* the number itself. Here four is the natural number associated with a set whose cardinal number is four.

Definition 3.1.2

The symbol used to represent a number is called a *numeral*.

By repeating the above discussion using collections of other finite, nonempty, equivalent sets, we obtain the set N of natural numbers. We shall denote the members of N by the numerals

$$1, 2, 3, 4, 5, 6, \cdots \tag{3.1.4}$$

3.2 EQUALITY, ADDITION, AND MULTIPLICATION OF NATURAL NUMBERS

Equality of natural numbers is described in terms of equivalent sets.

DEFINITION 3.2.1

Let a and b be two natural numbers associated with the finite, nonempty sets A and B. We shall say that *a equals b*, written as

$$a = b \tag{3.2.1}$$

if and only if

$$A \sim B \tag{3.2.2}$$

In the above definition, a and b are just two *different symbols* which represent the *same natural number*.

From now on, we shall use the symbol a to represent the natural number associated with a finite set A, whose cardinal number $c(A) = a$.

We are now able to state what the addition of two natural numbers means.

DEFINITION 3.2.2

Let a and b be two natural numbers associated with the finite nonempty sets A and B, respectively. Assume A and B are disjoint sets so that

$$A \cap B = \varnothing \tag{3.2.3}$$

The *sum of a and b*, written as

$$a + b \tag{3.2.4}$$

is defined as the natural number associated with the set $A \cup B$.

The above definition uses the fact that the union of two finite sets is also a finite set. Were this not a property of finite sets, Definition 3.2.2 would not make sense. See problem 10 in Exercise 2.6 of Chapter 2.

EXAMPLE 3.2.1

Let 2 and 3 be two natural numbers. We can associate 2 with the set

$$A = \{a, b\} \tag{3.2.5}$$

and 3 with the set

$$B = \{c, d, e\} \tag{3.2.6}$$

Since $A \cap B = \varnothing$, then

$$A \cup B = \{a, b, c, d, e\} \tag{3.2.7}$$

Here, 5 is the natural number associated with $A \cup B$. Thus,

$$2 + 3 = 5 \tag{3.2.8}$$

It is very important to remember that in Definition 3.2.2, the sets A and B must be disjoint. For example, if

$$A = \{a, b\}, \qquad B = \{b, c, d\} \tag{3.2.9}$$

Then,

$$A \cup B = \{a, b, c, d\} \tag{3.2.10}$$

Here 2 is associated with A, 3 is associated with B, and 4 is associated with $A \cup B$. The reason this apparent discrepancy occurs is that A and B are *not* disjoint.

Using Definition 3.2.2, we can prove the following useful properties.

Theorem 3.2.1

The Commutative Law for Addition: If a and b are any two natural numbers, then

$$a + b = b + a \tag{3.2.11}$$

PROOF: Let a be the natural number associated with set A and b be the natural number associated with set B. Assume $A \cap B = \varnothing$. Then $a + b$ is the natural number associated with $A \cup B$, and $b + a$ is the natural number associated with $B \cup A$.

In Chapter 2, we learned

$$A \cup B = B \cup A \tag{3.2.12}$$

Hence, by Definition 3.2.1, since equality of sets implies equivalence of sets,

$$a + b = b + a \tag{3.2.13}$$

EXAMPLE 3.2.2

Let 2 be associated with the set

$$A = \{l, m\} \tag{3.2.14}$$

and 3 be associated with

$$B = \{h, i, j\} \tag{3.2.15}$$

Since $A \cap B = \varnothing$, $2 + 3$ is associated with

$$A \cup B = \{l, m, h, i, j\}$$

and $3 + 2$ is associated with

$$B \cup A = \{l, m, h, i, j\}$$

But $A \cup B = B \cup A$. Thus,

$$2 + 3 = 3 + 2 \tag{3.2.16}$$

Because of Theorem 3.2.1, the order in which two natural numbers are added will not affect the sum.

Theorem 3.2.2

The Associative Law for Addition: If a, b, c are any three natural numbers, then

$$a + (b + c) = (a + b) + c \tag{3.2.17}$$

PROOF: Let a be the natural number associated with set A, b the natural number associated with B, and c the natural number associated with C. Here A, B, C have no elements in common. Then $a + (b + c)$ is the natural number associated with $A \cup (B \cup C)$, and $(a + b) + c$ is the natural number associated with $(A \cup B) \cup C$.

In Chapter 2 we have shown that

$$A \cup (B \cup C) = (A \cup B) \cup C \tag{3.2.18}$$

Since equality of sets implies equivalence of sets, by Definition 3.2.1,

$$a + (b + c) = (a + b) + c \tag{3.2.19}$$

EXAMPLE 3.2.3

Let 2 be associated with the set

$$A = \{l, m\} \tag{3.2.20}$$

and 3 be associated with

$$B = \{h, i, j\} \tag{3.2.21}$$

and 4 be associated with

$$C = \{a, b, c, d\} \tag{3.2.22}$$

Notice that $A \cap B = \varnothing$, $A \cap C = \varnothing$, and $C \cap B = \varnothing$. Also, $A \cap (B \cup C) = \varnothing$ and $(A \cup B) \cap C = \varnothing$. Now $2 + (3 + 4)$ is the natural number associated with $A \cup (B \cup C)$, and $(2 + 3) + 4$ is the natural number associated with $(A \cup B) \cup C$. But from Chapter 2 we know that

$$(A \cup B) \cup C = A \cup (B \cup C) \tag{3.2.23}$$

Since equality of sets implies equivalence of sets, we see that

$$(2 + 3) + 4 = 2 + (3 + 4) \tag{3.2.24}$$

In the expression $(a + b) + c$, the symbols ( ) are called *parentheses*. They give us a set of directions that must be followed. For example, in

$(2 + 3) + 4$, they tell us to add 2 and 3, obtaining 5, and then to add 5 to 4, obtaining 9. The associative law for addition tells us that if we are to add these natural numbers (in a given order), the placement of parentheses will not change the sum. Because of the associative law for addition, the expression $a + b + c$ now has meaning.

The reader should realize that at this point the concept of *subtraction* has not been introduced. Theorem 3.2.3 allows us to do subtraction problems without actually defining *subtraction*. In Chapter 4, where subtraction is defined, we will see that this property will not then be really necessary. The reason that subtraction cannot be introduced now is that an expression such as $3 - 5$ will have no meaning if we are only considering counting numbers.

Theorem 3.2.3

The Cancellation Law for Addition: Let a, b, c be any three natural numbers.

If

$$a + c = b + c \tag{3.2.25}$$

then

$$a = b \tag{3.2.26}$$

EXAMPLE 3.2.4

For example, if x is a natural number and

$$x + 2 = 5 + 2 \tag{3.2.27}$$

the cancellation law for addition tells us

$$x = 5 \tag{3.2.28}$$

Consider the following problem. A man has five shirts and three ties. What is the total number of possible combinations of a shirt with a tie?

To obtain the answer, suppose

$$S = \text{shirts} = \{\text{white, blue, yellow, green, red}\}$$

$$T = \text{ties} = \{\text{polka dot, striped, plain}\}$$

The possible combinations are

(white, polka dot), (white, striped), (white, plain),

(blue, polka dot), (blue, striped), (blue, plain),

(yellow, polka dot), (yellow, striped), (yellow, plain),

(green, polka dot), (green, striped), (green, plain),

(red, polka dot), (red, striped), (red, plain).

We see that there are 15 possible combinations.

If we take the cross product of the set S of shirts with the set T of ties, we obtain the set $S \times T$ whose elements are the above pairs. The cardinal number of $S \times T$ is 15—namely, the product of 5 and 3.

This leads us to make the following general definition for multiplication of natural numbers.

DEFINITION 3.2.3

Let a and b be two natural numbers associated with the finite, nonempty sets A and B. The *product of a and b*, written as

$$a \cdot b \qquad (3.2.29)$$

is the natural number associated with the product set $A \times B$.

EXAMPLE 3.2.5

Let 2 be associated with the set

$$A = \{a, b\} \qquad (3.2.30)$$

and let 3 be associated with the set

$$B = \{b, c, d\} \qquad (3.2.31)$$

Then

$$A \times B = \{(a, b), (a, c), (a, d), (b, b), (b, c), (b, d)\} \qquad (3.2.32)$$

The natural number associated with $A \times B$ is 6. Thus

$$2 \cdot 3 = 6 \qquad (3.2.33)$$

Of course, in practice one does not use the cross product of two sets in order to find the product of two natural numbers. However, in order to develop the properties of multiplication, it is necessary to use the concept of a cross product.

Before proceeding, let us make note of the following characteristic of cross products. In Chapter 2, we learned that

$$A \times B \neq B \times A \qquad (3.2.34)$$

Will this affect the commutativity of multiplication of natural numbers?

The answer is no. For example, consider the sets

$$A = \{h, i, j\}, \qquad B = \{l, m\} \qquad (3.2.35)$$

Then,

$$A \times B = \{(h, l), (h, m), (i, l), (i, m), (j, l), (j, m)\}$$

$$B \times A = \{(l, h), (m, h), (l, i), (m, i), (l, j), (m, j)\} \qquad (3.2.36)$$

Clearly, $A \times B$ and $B \times A$ are not equal. However, a one-to-one correspondence can be established between their elements. That is,

$$A \times B \sim B \times A \tag{3.2.37}$$

Because of this, we can prove the following result.

Theorem 3.2.4

The Commutative Law for Multiplication: If a and b are any two natural numbers, then

$$a \cdot b = b \cdot a \tag{3.2.38}$$

PROOF: Let a be the natural number associated with the set A and b the natural number associated with the set B. Then $a \cdot b$ is the natural number associated with $A \times B$ and $b \cdot a$ is the natural number associated with $B \times A$. But, as mentioned before,

$$A \times B \sim B \times A \tag{3.2.39}$$

Hence, by Definition 3.2.1,

$$a \cdot b = b \cdot a \tag{3.2.40}$$

Just as with addition, multiplication being commutative means that multiplying 3 by 2, namely $3 \cdot 2$, or 2 by 3, namely $2 \cdot 3$, will give the same result.

Theorem 3.2.5

The Associative Law for Multiplication: If a, b, c are any three natural numbers, then

$$a \cdot (b \cdot c) = (a \cdot b) \cdot c \tag{3.2.41}$$

The proof is left as an exercise.

Just as with the associative law for addition, the associative law for multiplication gives meaning to the expression $a \cdot b \cdot c$.

Together, the associative and commutative laws provide a great deal of freedom in grouping and ordering numbers in multiplication. The commutative law of multiplication permits one to change the order of numbers in multiplication, and the associative law permits grouping them at will. Thus, for the product $(a \cdot b) \cdot c$ there are eleven other arrangements, all of which look different and all of which, by the commutative and associative laws, are equal to this one.

At this point, the idea of *division* has not been introduced. The next theorem enables us to do division problems without actually defining *division*. The reason *division* cannot be introduced now is that an expression such as $5 \div 3$ has no meaning if only counting numbers are considered.

Theorem 3.2.6

The Cancellation Law for Multiplication: Let a, b, c be any three natural numbers.

If

$$a \cdot c = b \cdot c \tag{3.2.42}$$

then

$$a = b \tag{3.2.43}$$

As an example, suppose x is a natural number and

$$x \cdot 3 = 4 \cdot 3 \tag{3.2.44}$$

The cancellation law for multiplication tells us that

$$x = 4 \tag{3.2.45}$$

As another example of the cancellation law for multiplication, suppose it is asked to find a natural number x so that three times x equals 6. That is,

$$3 \cdot x = 6 \tag{3.2.46}$$

It is known that

$$6 = 3 \cdot 2 \tag{3.2.47}$$

Thus,

$$3 \cdot x = 3 \cdot 2 \tag{3.2.48}$$

Since multiplication is commutative,

$$x \cdot 3 = 2 \cdot 3 \tag{3.2.49}$$

The cancellation law for multiplication then tells us

$$x = 2 \tag{3.2.50}$$

Theorem 3.2.7

Multiplication is Distributive over Addition: Let a, b, c be any three natural numbers. Then multiplication is distributive over addition. That is

$$a \cdot (b + c) = (a \cdot b) + (a \cdot c) \tag{3.2.51}$$

PROOF: Let a, b, c be the natural numbers associated with the sets A, B, C, respectively. Assume $B \cap C = \varnothing$. Then $b + c$ is the natural number associated with $B \cup C$. Also, $a \cdot (b + c)$ is the natural number associated with $A \times (B \cup C)$.

Next, $a \cdot b$ and $a \cdot c$ are the natural numbers associated with $A \times B$ and $A \times C$, respectively. Since B and C are disjoint, then $A \times B$ and $A \times C$ are disjoint. Thus, $a \cdot b + a \cdot c$ is the natural number associated with

$(A \times B) \cup (A \times C)$. But from Chapter 2, we know that

$$A \times (B \cup C) = (A \times B) \cup (A \times C) \tag{3.2.52}$$

Hence,

$$a \cdot (b + c) = (a \cdot b) + (a \cdot c) \tag{3.2.53}$$

Note: Parentheses are always used to indicate a single quantity or to indicate an operation which must be performed first.

The natural number 1 plays a unique role in multiplication. When any natural number x is multiplied by 1, the product is always x.

For example

$$1 \cdot 5 = 5 \cdot 1 = 5, \qquad 1 \cdot 17 = 17 \cdot 1 = 17 \tag{3.2.54}$$

One is called the *identity for multiplication* or the *multiplicative identity.* This identity 1 has the property that for any natural number a,

$$a \cdot 1 = 1 \cdot a = a \tag{3.2.55}$$

The proof of this fact is left as an exercise.

Suppose we wish to find the value of

$$3 \cdot 4 + 5 \tag{3.2.56}$$

Do we

(a) multiply 3 by 4 and then add 5

or

(b) add 4 to 5 and multiply this result by 3?

Notice that a different answer is obtained depending on which choice (a) or (b) is made. Of course, only one of these can be the correct one. Henceforth, we employ the usual convention that, in the absence of parentheses, multiplication takes precedence over addition. Thus,

$$3 \cdot 4 + 5 = 12 + 5 = 17$$

$$3 \cdot 4 + 5 \neq 3 \cdot 9 = 27 \tag{3.2.57}$$

EXAMPLE 3.2.6

Find the value of

$$\text{(a)} \quad 5 + 7 \cdot 2 \qquad \text{and} \qquad \text{(b)} \quad (5 + 7) \cdot 2 \tag{3.2.58}$$

For (a), we have

$$5 + 7 \cdot 2 = 5 + 14 = 19 \tag{3.2.59}$$

since multiplication is performed before addition.

For (b), we have

$$(5 + 7) \cdot 2 = 12 \cdot 2 = 24 \tag{3.2.60}$$

since parentheses indicate that $5 + 7 = 12$ is to be treated as a single quantity.

Exercise 3.2

1. Give examples, using natural numbers, of the three properties of addition.

2. Give examples, using natural numbers, of the four properties of multiplication.

3. Using the definition of addition, show how to find

   (a) $6 + 4$ (c) $3 + 7$
   (b) $8 + 5$ (d) $9 + 2$

4. Using the definition of multiplication, show how to find

   (a) $8 \cdot 4$ (c) $9 \cdot 2$
   (b) $7 \cdot 8$ (d) $5 \cdot 4$

5. Using the properties of addition and multiplication, evaluate the following quantities:

   (a) $8 \cdot 3 + 4 \cdot 3$ (d) $(5 + 2) \cdot 7$
   (b) $8 \cdot (3 + 4) \cdot 3$ (e) $7 \cdot 2 + 6 \cdot 4$
   (c) $5 + 2 \cdot 7$ (f) $11 \cdot (2 + 1)$

6. Find the value of

   (a) $4 + 3 \cdot 2$ (d) $6 \cdot (2 + 7)$
   (b) $(4 + 3) \cdot 2$ (e) $5 + (2 + 3) \cdot 2$
   (c) $6 \cdot 2 + 7$ (f) $(5 + 2) + 3 \cdot 2$

7. (a) Write eleven different, but equal, arrangements for

   $$(2 + 3) + 5$$

   (b) Write eleven different, but equal, arrangements for

   $$(3 \cdot 5) \cdot 2$$

8. A traveler wants to go from New York to Los Angeles via Chicago. He has two possible routes from New York to Chicago and three possible routes from Chicago to Los Angeles. How many different routes can be selected? How did you obtain your answer?

9. By using the properties given in this section, the following problems can be performed mentally. Indicate which property you are using to make the computation easier.

   (a) $21 \cdot 14$ [HINT: $21 = 20 + 1$]
   (b) $3 + (7 + 21)$
   (c) $101 \cdot 24$
   (d) $66 + (77 + 34)$
   (e) $25 \cdot (22 \cdot 4)$

10. Prove Theorem 3.2.5.

11. Prove that for any natural number a, $a \cdot 1 = a$.

3.3 THE SET W OF WHOLE NUMBERS

Up to this point, we have concerned ourselves with the set N of natural numbers, namely,

$$N = \{1, 2, 3, \cdots\} \tag{3.3.1}$$

This set N was introduced by associating its members with nonempty finite sets.

In Chapter 2, we defined the cardinal number of the empty set $\varnothing$ to be zero, denoted by the symbol 0. The new set consisting of the elements

$$\{0, 1, 2, 3, \cdots\} \tag{3.3.2}$$

is called the *set W of whole numbers.*

By defining addition and multiplication of whole numbers in much the same way as already defined for natural numbers, the following properties are valid for whole numbers.

Theorem 3.3.1

If a is any whole number, then

$$a + 0 = 0 + a = a \tag{3.3.3}$$

PROOF: Let a be the whole number associated with the set A, and 0 be the whole number associated with the empty set $\varnothing$. Then, since $A \cap \varnothing = \varnothing$, the sum of $a + 0$ is the whole number associated with the set $A \cup \varnothing$. From Chapter 2,

$$A \cup \varnothing = \varnothing \cup A = A \tag{3.3.4}$$

Thus,

$$a + 0 = 0 + a = a \tag{3.3.5}$$

Because of the property of zero, we usually call zero the *additive identity* 0.

The following gives us the role for multiplying by zero.

Theorem 3.3.2

For any whole number a,

$$a \cdot 0 = 0 \cdot a = 0 \tag{3.3.6}$$

PROOF: Let a be the whole number associated with the set A, and 0 be the whole number associated with the empty set $\varnothing$. Then $a \cdot 0$ is the whole number associated with the set $A \times \varnothing$. From Chapter 2, we know that

$$A \times \varnothing = \varnothing \times A = \varnothing \tag{3.3.7}$$

Hence

$$a \cdot 0 = 0 \cdot a = 0 \tag{3.3.8}$$

Figure 3.1 summarizes some of the properties of the set W of whole numbers relative to addition and multiplication. In the table, a, b, c represent any three whole numbers.

Note: For the set W of whole numbers, it is necessary that $c \neq 0$ for the cancellation law for multiplication to be valid. For, consider the following arguments:

We know that

$$0 = 0 \tag{3.3.9}$$

Since for a, $a \cdot 0 = 0$, it is true that

$$2 \cdot 0 = 3 \cdot 0 \tag{3.3.10}$$

If the cancellation law for multiplication were valid for $c = 0$, this would mean

$$2 = 3 \tag{3.3.11}$$

This obvious impossibility tells us that c cannot be zero in the cancellation law for multiplication.

	Addition	Multiplication
Commutative Law	$a+b=b+a$	$a \cdot b = b \cdot a$
Associative Law	$a+(b+c)=(a+b)+c$	$a \cdot (b \cdot c) = (a \cdot b) \cdot c$
Cancellation Law	$a+c=b+c \rightarrow a=b$	$a \cdot c = b \cdot c,\ c \neq 0 \rightarrow a=b$
Identities	$a+0=0+a=a$	$a \cdot 1 = 1 \cdot a = a$
Distributive Law	$a \cdot (b+c) = a \cdot b + a \cdot c$	

FIG. 3.1

Exercise 3.3

1. Evaluate the following expressions:

(a) $[3 + 4\cdot 0] + 2$
(b) $3\cdot(2 + 0) + 6$
(c) $9\cdot 2 + 0\cdot 3$
(d) $9\cdot 2 + 0\cdot 2$
(e) $9\cdot(2 + 0)\cdot 2$
(f) $(9\cdot 2 + 0)\cdot 2$
(g) $9\cdot(2 + 0\cdot 2)$
(h) $(9 + 0)\cdot 2$
(i) $9\cdot 2\cdot 0 + 4$
(j) $9 + 4\cdot 2 + 0$

2. The following statements will be true if parentheses are placed properly. Place the parentheses so that the statement is true

(a) $2\cdot 6 + 4\cdot 6 = 120$
(b) $2 + 4\cdot 7 + 8 = 50$
(c) $2 + 4\cdot 7 + 8 = 62$
(d) $2 + 4\cdot 7 + 8 = 38$

3. Find whole number(s) x so that

(a) $2\cdot x = 8$
(b) $2\cdot x = 9$
(c) $x + 5 = 7$
(d) $(2\cdot x)\cdot 4 = 2\cdot(x\cdot 4)$

4. The operation *star*, denoted by $a * b$, where a, b are two whole numbers, is defined by

$$a * b = a\cdot b + a + b$$

For example,

$$2 * 4 = 2\cdot 4 + 2 + 4 = 14$$

$$4 * 2 = 4\cdot 2 + 4 + 2 = 14$$

(a) Is $*$ commutative?
(b) Is $*$ associative?
(c) Is there an identity, that is, a star identity?

3.4 ORDERING OF WHOLE NUMBERS

In this section, we shall discuss the concept of *ordering*. The student is already aware of the fact that 2 *is less than* 3 or 5 *is greater than* 1, etc. The terms *less than* or *greater than* are *order relations* for whole numbers.

The following statement, called the trichotomy law, can be proved.

Trichotomy Law

If x and y are any two whole numbers, one and only one of the following three possibilities can occur.

(I) There is a unique natural number u such that

$$y = x + u \tag{3.4.1}$$

In this case, we say that x *is less than* y, written as

$$x < y \tag{3.4.2}$$

(II) x *equals* y, written as

$$x = y \tag{3.4.3}$$

(III) There is a unique natural number v such that

$$x = y + v \tag{3.4.4}$$

In this case, we say that x *is greater than* y, written as

$$x > y \tag{3.4.5}$$

For example, we say that 5 is less than 7, $5 < 7$, since $7 = 5 + 2$. Of course, we can also say that 7 is greater than 5, $7 > 5$, since $7 = 5 + 2$. Thus, $x < y$ and $y > x$ are equivalent statements.

As another example, consider 0 and 3. Here, 0 is less than 3, since $3 = 0 + 3$.

Definition 3.4.1

If x and y are two whole numbers, we say that x *is less than or equal to* y, written as

$$x \leqq y \tag{3.4.6}$$

if and only if either $x < y$ or $x = y$.

Similarly, we say that x *is greater than or equal to* y, written

$$x \geqq y \tag{3.4.7}$$

if and only if either $x > y$ or $x = y$.

The order relation for whole numbers given above is called the *natural ordering*.

Now that we have discussed the concept of ordering for the set W of whole numbers, we can display the set W on a line, called the *whole number line*. We do this as follows:

Draw a horizontal line and pick an arbitrary point O on this line. Associate the whole number zero to this point.

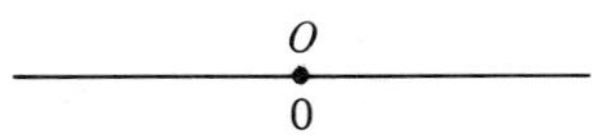

FIG. 3.2

Pick another point U on the line, different from the point associated to zero. Associate the whole number 1 to U. The length from O to U is called the *scale* or *unit of measurement*.

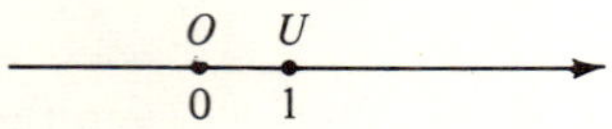

FIG. 3.3

Now we mark off this unit on the line. The whole numbers are then displayed according to the $<$ order relation.

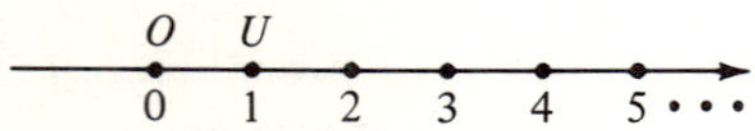

FIG. 3.4

Notice in Figure 3.5 that $1 < 3$ since 3 is two units to the right of 1.

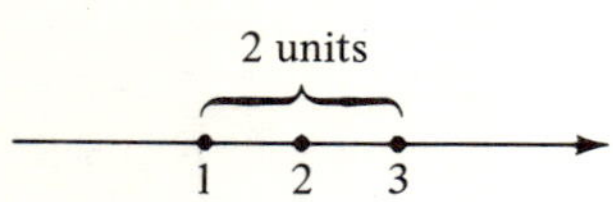

FIG. 3.5

Next, we have the following result.

Theorem 3.4.1

The set W of whole numbers relative to the natural ordering is a partially ordered set. That is, the following three properties are true:

(I) *Reflexive Law*

If x is any whole number, then

$$x \leqq x \tag{3.4.8}$$

(II) *Anti-Symmetric Law*

If x and y are two whole numbers and if

$$x \leqq y \quad \text{and} \quad y \leqq x \tag{3.4.9}$$

then

$$x = y \tag{3.4.10}$$

(III) *Transitive Law*

If x, y, z are three whole numbers and if

$$x \leqq y \quad \text{and} \quad y \leqq z \tag{3.4.11}$$

then

$$x \leqq z \tag{3.4.12}$$

PROOF: (I) $x \leqq x$ because of equation (3.4.3) and Definition 3.4.1.

(II) If $x \leqq y$, then there is a whole number c such that

$$x + c = y \tag{3.4.13}$$

If $y \leqq x$, there is a whole number d such that

$$y + d = x \tag{3.4.14}$$

Thus,

$$(y + d) + c = y \tag{3.4.15}$$

This is impossible, unless $d + c = 0$ which means $c = d = 0$. Thus,

$$x = y \tag{3.4.16}$$

(III) If $x \leqq y$, then there is a c such that

$$x + c = y \tag{3.4.17}$$

If $y \leqq z$, then there is a d such that

$$y + d = z \tag{3.4.18}$$

Thus, by combining (3 4.17) and (3.4.18), we have

$$(x + c) + d = z \tag{3.4.19}$$

By the associative law for addition,

$$x + (c + d) = z \tag{3.4.20}$$

This means

$$x \leqq z \tag{3.4.21}$$

DEFINITION 3.4.2

Suppose a and b are two whole numbers so that $a < b$. The notation

$$a \leqq x \leqq b \tag{3.4.22}$$

is used to denote all whole numbers x for which

$$a \leqq x \quad \text{and} \quad x \leqq b \tag{3.4.23}$$

The notation

$$a < x < b \tag{3.4.24}$$

is used to denote all whole numbers x for which

$$a < x \quad \text{and} \quad x < b \tag{3.4.25}$$

For example, if

$$3 \leqq x \leqq 5 \tag{3.4.26}$$

then this denotes the set $\{3, 4, 5\}$ of whole numbers since each of the elements 3, 4, and 5 satisfies the *two* conditions that

$$3 \leqq x \quad \text{and} \quad x \leqq 5 \tag{3.4.27}$$

See Fig. 3.6.

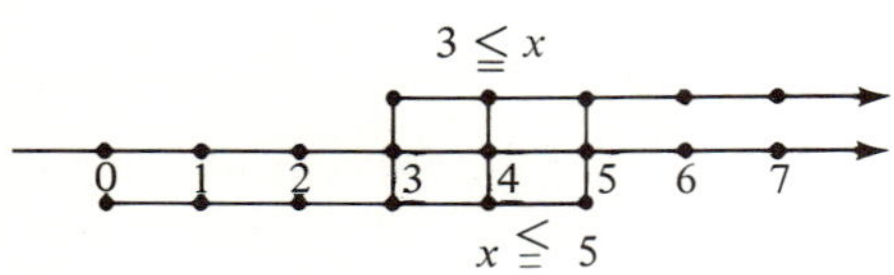

FIG. 3.6

There is also the notation

$$a \leqq x < b \tag{3.4.28}$$

which means

$$a \leqq x \quad \text{and} \quad x < b \tag{3.4.29}$$

Finally,

$$a < x \leqq b \tag{3.4.30}$$

means

$$a < x \quad \text{and} \quad x \leqq b \tag{3.4.31}$$

The set W of whole numbers, relative to the natural ordering, is a *simply ordered set*. That is, W obeys the reflexive law, anti-symmetric law, transitive law, and trichotomy law.

(IV) *Trichotomy Law*

If x and y are any two whole numbers, then

$$\text{either } x < y, \text{ or } x = y, \text{ or } y < x \tag{3.4.32}$$

This is merely a restatement of the definition of the relation $<$.

Thus, given *any* two whole numbers, we can *always* compare them by using the natural ordering. For example, the whole numbers 5 and 8 can be compared by writing that

$$5 < 8 \tag{3.4.33}$$

Recall that the ordering *is a subset of* for sets did *not* have this very desirable property.

Exercise 3.4

1. Write down the whole numbers x for which

(a)	$2 < x \leqq 5$	(e)	$4 < x < 5$
(b)	$3 \leqq x \leqq 7$	(f)	$1 \leqq x < 2$
(c)	$0 < x < 8$	(g)	$0 \leqq x < 1$
(d)	$6 < x \leqq 8$	(h)	$0 < x < 1$

Label your answer by using a whole number line.

2. Compare the following whole numbers by using the natural ordering:

(a)	3, 5	(e)	8, 10
(b)	6, 2	(f)	7, 8
(c)	5, 5	(g)	0, 3
(d)	4, 3	(h)	0, 1

3.5 DIVISION RELATION FOR THE SET N OF NATURAL NUMBERS

In this section, we shall discuss a way of relating or comparing two natural numbers.

DEFINITION 3.5.1

Let a and b be two natural numbers. We shall say that a *divides* b, written as

$$a \mid b \tag{3.5.1}$$

if and only if there is a natural number c so that

$$a \cdot c = b \tag{3.5.2}$$

For example,

$$3 \mid 12 \tag{3.5.3}$$

since $3 \cdot 4 = 12$.

This relation defined above is called *division relation*. The concept of division relation will be used in Chapter 5 and in Chapter 9.

Theorem 3.5.1

The set N of natural numbers, relative to division relation is a partially ordered set.

PROOF: (I) First of all, the reflexive law is obeyed. For, if a is any natural number, then

$$a \mid a \tag{3.5.4}$$

since $a \cdot 1 = a$.

(II) The anti-symmetric law is valid since if $a \mid b$ and if $b \mid a$, then there are two natural numbers c and d so that

$$a \cdot c = b \quad \text{and} \quad b \cdot d = a \tag{3.5.5}$$

Thus

$$(b \cdot d) \cdot c = b \tag{3.5.6}$$

or

$$b \cdot (d \cdot c) = b \cdot 1 \tag{3.5.7}$$

By the cancellation law for multiplication,

$$d \cdot c = 1 \tag{3.5.8}$$

Now, the product of two natural numbers can equal one if and only if the numbers themselves are one. Thus, from equation (3.5.8),

$$c = d = 1 \tag{3.5.9}$$

From equation (3.5.5), we have

$$a = b \tag{3.5.10}$$

(III) Finally, the transitive law is satisfied since if $a \mid b$ and if $b \mid c$, then there are natural numbers x and y such that

$$a \cdot x = b, \qquad b \cdot y = c \tag{3.5.11}$$

Then

$$(a \cdot x) \cdot y = c \tag{3.5.12}$$

or

$$a \cdot (x \cdot y) = c \tag{3.5.13}$$

Since $x \cdot y$ is a natural number, then from equation (3.5.13) it follows that

$$a \mid c \tag{3.5.14}$$

Now, let us look at the natural numbers, 2 and 3. We already know that

$$2 \neq 3 \tag{3.5.15}$$

Also, 2 does *not* divide 3 since there is *no* natural number which when multiplied by 2 gives 3. For the same reason 3 does *not* divide 2. Hence, relative to division ordering, the set N of natural numbers does not obey the trichotomy law. Thus, it is *not* always possible to compare two natural numbers when division ordering is used.

Exercise 3.5

1. If $|$ is used to denote the division relation and if $a \mid b$, let us define $a \mid x \mid b$ to mean that a divides x *and* x divides b. Which natural numbers x are given by

   (a) $2 \mid x \mid 4$ (c) $1 \mid x \mid 4$
   (b) $3 \mid x \mid 6$ (d) $2 \mid x \mid 5$

2. Which of the following pairs of natural numbers are comparable relative to division ordering? Why?

   (a) 3, 5 (c) 2, 4
   (b) 6, 10 (d) 8, 4

4 The Set J of Integers

4.1 INTRODUCTION

In Chapter 3 the concepts of natural number and whole number were introduced. Also, rules for adding and multiplying natural numbers and whole numbers were given and various properties of the set N of natural numbers and the set W of whole numbers were discussed.

However, the set W of whole numbers possesses certain deficiencies; that is to say, it lacks certain desirable properties. One of the properties missing for the set W of whole numbers is that equations of the type

$$a + x = b \tag{4.1.1}$$

for a, b given whole numbers, do not always have solutions for x in W.

For example, consider the equation

$$6 + x = 2 \tag{4.1.2}$$

There is *no* whole number x which when added to 6 gives 2.

Historically, the concept of numbers other than whole numbers was called "absurd" by the famous mathematician Diophontous of Alexandria. In fact, in 1225 the mathematician Fibonacci stated that the equation (4.1.1) with a larger than b has no solution unless it expresses that a man is in debt. As late as the seventeenth century numbers other than whole numbers were referred to as "false" or "fictitious." Finally, as a result of

Descarte's *La Geometrie,* published in 1637, numbers other than whole numbers were given concrete meaning.

In real life, the need for numbers other than whole numbers can be demonstrated by considering the following example. A housewife has a 55 dollar balance in her checking account and proceeds to write a check for 73 dollars. This happens every day. The physical situation is that her account is overdrawn by 18 dollars. However, without additional numbers, we cannot conveniently express this condition.

In this chapter we extend the set W of whole numbers so that the deficiency of equation (4.1.1) is removed.

Intuitively, we can accomplish this by demanding that for every whole number x there must be a number ^-x called the *additive inverse of* x with the property that

$$x + (^-x) = {}^-x + x = 0 \tag{4.1.3}$$

For example, the additive inverse of 2 is $^-2$, since when 2 is added to $^-2$ the result is 0.

This new set, consisting of the elements,

$$\{\cdots, {}^-3, {}^-2, {}^-1, 0, 1, 2, 3, \cdots\} \tag{4.1.4}$$

is called the *set* J *of integers.*

Notice that the set W of whole numbers is a proper subset of this set J of integers. Thus, the set J of integers is an *extension* of the set W of whole numbers. As will be seen shortly, this *new* set J of integers does *not* possess the aforementioned deficiency of the set W of whole numbers.

4.2 A DESCRIPTION OF THE SET J OF INTEGERS

The set J of integers, namely

$$J = \{\cdots, {}^-3, {}^-2, {}^-1, 0, 1, 2, \cdots\} \tag{4.2.1}$$

can be described as a set with two binary operations, called *addition* and *multiplication,* obeying certain laws.

The binary operation of addition of integers obeys the following laws:

(I) *Closure Law for Addition*

If x and y are two integers, their sum $x + y$ is also an integer. Moreover, this sum is *unique.* That is, if $x = x'$ and if $y = y'$, then $x + y = x' + y'$.

This property of the integers states that whenever two integers are added, the result is an integer. It also says that whenever two integers are added only *one* sum can result: that is, it is *not* possible to add the *same* two integers and obtain two *different* answers.

(II) *Commutative Law for Addition*

Addition is *commutative*. That is, if x and y are any two integers, then

$$x + y = y + x \tag{4.2.2}$$

This property of addition of integers is similar to many already discussed. It merely states that the order in which one adds will *not* affect the result. For example, this law allows one to add a column of two numbers either from the top down or from the bottom up. Since the commutative law guarantees the result will be the same in either case, this is usually a good way to check addition for accuracy.

(III) *Associative Law for Addition*

Addition is *associative*. That is, if x, y, z are any three integers, then

$$x + (y + z) = (x + y) + z \tag{4.2.3}$$

The associative law for addition states that when adding three integers, it makes no difference how they are *grouped*—the result will be the same. For example, if we wanted to add the integers 5, 4, and 7, we could either add 5 and 4, getting 9, and then add 7, getting 16; or else we could add 4 to 7, getting 11, and then add 5 to the 11, giving 16. That is,

$$5 + 4 + 7 = (5 + 4) + 7 = 9 + 7 + 16 \tag{4.2.4}$$

or

$$5 + 4 + 7 = 5 + (4 + 7) = 5 + 11 = 16 \tag{4.2.5}$$

(IV) *Existence of an Additive Identity*

In the set J of integers, there is an element 0, called the *additive identity zero*, with the property that

$$x + 0 = 0 + x = x \tag{4.2.6}$$

for any integer x.

This law tells us that any integer, say $^-3$, when added to zero, yields the same integer, $^-3$.

(V) *Existence of An Additive Inverse*

For every integer x, there is an integer ^-x, called the *additive inverse of* x, with the property that

$$x + (^-x) = {}^-x + x = 0 \tag{4.2.7}$$

To find the additive inverse ^-x of an element x, one must ask the question, "What should I add to x in order to obtain 0?" Thus, the additive inverse of $^-4$ is 4, since when I add 4 and $^-4$, I obtain 0.

It is very important that the student not give any additional meanings to the dash placed before a number, other than the meaning of additive inverse. At no time should ^-a be interpreted or read as *negative a*. As will be seen later in Chapter 5, the additive inverse symbol $^-$ does *not* tell us whether the given number is positive or negative.

In the set J of integers, the binary operation of multiplication has the following properties.

(VI) *Closure Law for Multiplication*

If x and y are two integers, their product $x \cdot y$ is also an integer. This product is unique. That is, if $x = x'$, $y = y'$, then $x \cdot y = x' \cdot y'$.

This law says that whenever you multiply two integers, the answer will always be an integer. Furthermore, it says that there can only be one answer whenever two integers are multiplied.

(VII) *Commutative Law for Multiplication*

Multiplication is *commutative*. That is, for any two integers x and y we have

$$x \cdot y = y \cdot x \tag{4.2.8}$$

Again, this states that the order in which integers are multiplied will not affect the result.

(VIII) *Associative Law for Multiplication*

Multiplication is *associative*. That is, for any three integers x, y, z, we have

$$x \cdot (y \cdot z) = (x \cdot y) \cdot z \tag{4.2.9}$$

Thus when multiplying three integers, one is free to group them in any way without changing the result. For example,

$$3 \cdot 4 \cdot 8 = (3 \cdot 4) \cdot 8 = 12 \cdot 8 = 96 \tag{4.2.10}$$

or

$$3 \cdot 4 \cdot 8 = 3 \cdot (4 \cdot 8) = 3 \cdot 32 = 96 \tag{4.2.11}$$

The student should compare the associative law for multiplication with the associative law for addition, noting that these two laws are identical except for the operation involved.

(IX) *Distributive Law*

Multiplication is distributive over addition. That is, for any three integers x, y, z, we have

$$x \cdot (y + z) = x \cdot y + x \cdot z \tag{4.2.12}$$

Since addition and multiplication are binary operations, they are operations performed between *two* elements. Thus, if one is asked to evaluate the quantity

$$2 \cdot 3 + 4 \tag{4.2.13}$$

it is seen that two choices are available: either multiply $2 \cdot 3$ and then add 4 or else add $3 + 4$ and multiply by 2. Again, we agree that, unless otherwise specified, the operation of multiplication is always performed before that of addition.

Thus,

$$2 \cdot 3 + 4 = 6 + 4 = 10, \qquad 2 \cdot 3 + 4 \neq 2 \cdot 7 = 14 \qquad (4.2.14)$$

(X) *Existence of a Multiplicative Identity*

In the set J of integers, there is an element 1, *not* the same as zero, called the *multiplicative identity one*, with the property that

$$x \cdot 1 = 1 \cdot x = x \qquad (4.2.15)$$

for *all* integers x.

This law states that the product of any integer, say $^-2$, by one, yields the same integer, $^-2$. Thus,

$$^-2 \cdot 1 = {}^-2 \qquad (4.2.16)$$

(XI) *The Product Law*

If x and y are two integers whose product is zero, that is, if

$$x \cdot y = 0 \qquad (4.2.17)$$

then

$$\text{either } x = 0 \text{ or } y = 0 \text{ or } x = y = 0 \qquad (4.2.18)$$

Thus, the only time a product of two integers will give zero is when one or both of the integers is itself zero. In other words, it *cannot* happen that the product of two nonzero integers will give the result zero.

The above eleven laws or properties of the set J of integers will be used extensively throughout this chapter and the next in two ways. First, they tell us what are correct and legal manipulations of integers. Second, they can be used to develop more properties of integers. By applying these laws and further developments of these laws in the right way, the student will gain confidence in his work and will not wonder whether a given manipulation is "legal" or "illegal."

Now, let us discuss a method for adding integers. First of all, display the set of integers in the following order:

$$\cdots, {}^-6, {}^-5, {}^-4, {}^-3, {}^-2, {}^-1, 0, 1, 2, 3, 4, \cdots \qquad (4.2.19)$$

For example to find $3 + ({}^-4)$, we proceed as follows: Start with 3 and count off *four* places to the *left* (because it is additive inverse of 4 which is to be added to 3), the result, namely $^-1$, is the desired answer.

EXAMPLE 4.2.1

Find $^-5 + 2$. Here start at $^-5$ and count off *two* places to the right.

$$\ldots, {}^-8, {}^-7, {}^-6, {}^-5, {}^-4, {}^-3, {}^-2, {}^-1, 0, 1, \ldots$$

1 2

$^-5$ = the starting point

Thus,

$$^-5 + 2 = {}^-3 \tag{4.2.20}$$

EXAMPLE 4.2.2

Find $^-6 + (^-3)$. Here we start at $^-6$ and count off 3 places to the left (to the *left* because we are adding the *additive inverse* of 3 to $^-6$). Thus,

$$\ldots, {}^-10, {}^-9, {}^-8, {}^-7, {}^-6, {}^-5, {}^-4, \ldots$$

3 2 1

$^-6$ = the starting point

Hence

$$^-6 + (^-3) = {}^-9 \tag{4.2.21}$$

EXAMPLE 4.2.3

Pick, at random, any six integers, say $^-3$, 2, 5, $^-4$, 6, and 8. Find the sum of each pair of these integers by using the table in Fig. 4.1.

+	5	7	$^-4$
$^-3$	2	4	$^-7$
2	7	9	$^-2$
6	11	13	2

FIG. 4.1

Circle any one of the integers in the box, say 4; and then cross out all the other numerals in the same row and column as 4. See Fig. 4.2.

+	5	7	$^-4$
$^-3$	2	4	$^-7$
2	7	9	$^-2$
6	11	13	2

FIG. 4.2

+	5	7	$^-4$
$^-3$	2	4	$^-7$
2	7	9	$^-2$
6	11	13	2

FIG. 4.3

Next, circle any one of the remaining integers, say $^{-}2$, and repeat the process. See Fig. 4.3. Circle the only remaining integer, namely, 11. The sum of the circled integers is

$$4 + (^{-}2) + 11 = 13 \tag{4.2.22}$$

What is interesting is that the sum of the three circled numbers will always equal 13, no matter which integer is circled first.

Can you explain why this works? [HINT: Notice that the sum of the original six integers is also 13.]

Exercise 4.2

1. Evaluate the following expressions:

   (a) $^{-}3 + (^{-}4)$ (c) $^{-}6 + (^{-}5)$

   (b) $5 + (^{-}7)$ (d) $^{-}6 + 5$

2. Find an integer x for which

   $$7 + x = 2$$

   and give a reason for each step. [HINT: Begin by adding the additive inverse of 7 to both sides.]

3. In problem 2, can there be more than one answer?

4. Work Example 4.2.3 using multiplication instead of addition. Is the same interesting fact true now? Why?

5. Make up a table for both addition and multiplication like Example 4.2.3, choosing eight integers so that sixteen entries are obtained. Is the same conclusion valid here as in Example 4.2.3? Why?

6. Pick any integer. Double this integer. Add two. Multiply this result by 4. Add $^{-}8$. Your result is then eight times the original integer. Why? It this always true? Can you make up a similar trick?

7. A magic square is an array of numbers such that the sum of the numbers in any row, or any column, or along the diagonals is always the same.

   In Fig. 4.4, there is a magic square with sixteen entries such that the sum is 34. Fill in the blank entries.

1		7	
			11
10		16	
15		9	

FIG. 4.4

8. Find a magic square with nine entries such that the sum of any row or any column or a diagonal is 15.

4.3 PROPERTIES OF INTEGERS

Just as in Chapter 2, when properties of sets were given, this section will deal with some properties of integers. These properties are not entirely theoretical in nature and the student will find them very useful for working many of the problems we shall encounter.

Theorem 4.3.1

Let a and b be two integers. The equation

$$a + x = b \tag{4.3.1}$$

has one and only one solution for x in J. Moreover, this solution is given by the formula

$$x = b + (^-a) \tag{4.3.2}$$

The student should notice that this result says three things: first, that any equation of the form (4.3.1) has a solution; secondly, that such equations have only one solution, that is once one solution has been found, then all solutions have been found; and third, the theorem tells us how to find the solution, that is, a formula (4.3.2) is given to find the solution.

PROOF: First of all, we shall show that if a solution x to (4.3.1) exists, it is unique. To do this, suppose there are two solutions x_1 and x_2 to (4.3.1), with $x_1 \neq x_2$. Then

$$a + x_1 = b, \qquad a + x_2 = b \tag{4.3.3}$$

Hence,

$$a + x_1 = a + x_2 \tag{4.3.4}$$

Add the additive inverse ^-a to both sides. This is possible because of the closure law for addition and the existence of an additive inverse. Then

$$^-a + (a + x_1) = {^-a} + (a + x_2) \tag{4.3.5}$$

By the associative law for addition, the above equation can be written as

$$(^-a + a) + x_1 = (^-a + a) + x_2 \tag{4.3.6}$$

By the additive inverse property,

$$0 + x_1 = 0 + x_2 \tag{4.3.7}$$

By the additive identity property, we have

$$x_1 = x_2 \tag{4.3.8}$$

This contradicts the fact that $x_1 \neq x_2$. Thus, the equation (4.3.1) can have *at most one* solution, since if we assume it has more than one solution, a contradiction is reached.

Next, consider the equation

$$a + x = b \tag{4.3.9}$$

Add the additive inverse ^-a of a to both sides. Then

$$^-a + (a + x) = {}^-a + b \tag{4.3.10}$$

By the associative law for addition, we see that

$$(^-a + a) + x = {}^-a + b \tag{4.3.11}$$

By the additive inverse property,

$$0 + x = {}^-a + b \tag{4.3.12}$$

By the additive identity property,

$$x = {}^-a + b \tag{4.3.13}$$

By the commutative law for addition, we have the result that

$$x = b + (^-a) \tag{4.3.14}$$

This value for x is *actually* a solution to equation (4.3.1), since

$$\begin{aligned} a + x &= a + [b + (^-a)] \\ &= a + [(^-a) + b] \\ &= [a + (^-a)] + b \\ &= 0 + b = b \end{aligned} \tag{4.3.15}$$

This completes the proof of Theorem 4.3.1.

Note: As was indicated before, the student should concentrate first on understanding what the theorem means, second on learning to properly apply the theorem in working problems, and third on mastering an acceptable proof of the theorem. It is not intended for the student to memorize the theorem or its proof. Rather it is intended that a working knowledge of the theorem should be obtained.

EXAMPLE 4.3.1

A man has 1020 dollars and wishes to purchase a boat costing 3195 dollars. How much additional money will he need for the purchase?

If we let x denote the amount of additional dollars needed for the purchase, then we are required to solve

$$1020 + x = 3195 \tag{4.3.16}$$

Using Theorem 4.3.1

$$x = 3195 + (^{-}1020) = 2175 \tag{4.3.17}$$

That is, he needs 2175 dollars.

EXAMPLE 4.3.2

Find an integer x for which $x + 4 = 7$.

Since this is an equation of the type given in (4.3.1), its solution is

$$x = 7 + (^{-}4) = 3 \tag{4.3.18}$$

It should be pointed out that Theorem 4.3.1 tells us that the deficiency of the set N of natural numbers discussed earlier in this chapter is *not* present in the set J of integers.

Corollary

In the set J of integers, there is one and only one additive identity, namely zero. Also, every integer a has one and only one additive inverse, namely ^{-}a.

Although the above results should be intuitively obvious to the student, it is necessary to give a formal proof. The reason, of course, is that intuition, even though a tremendous asset, is sometimes misleading.

PROOF: The additive identity is the solution x of the equation

$$a + x = a \tag{4.3.19}$$

for any integer a. But, by Theorem 4.3.1 such equations have one and only one solution. Hence, the additive identity 0 is unique.

The second half of this proof, which is very similar to the above, is left to the student.

If we think of additive inverse as a *unary* operation, that is, an operation performed on a single element, then we have the following result.

Theorem 4.3.2

The operation of additive inverse is *involutorial*. That is, for any integer a,

$$^{-}(^{-}a) = a \tag{4.3.20}$$

PROOF: Let

$$x = {}^{-}(^{-}a) \tag{4.3.21}$$

This means that

$$x + (^{-}a) = 0 \tag{4.3.22}$$

But this says that

$$x = a \tag{4.3.23}$$

Thus,

$$a = {}^{-}({}^{-}a) \qquad (4.3.24)$$

The above result can be interpreted another way. It says that if a is an integer, the additive inverse of the additive inverse of a is a itself.

Thus, we have now seen two unary operations which are involutorial. In Chapter 2, the unary operation of complementation of sets was shown to be involutorial. Now, it is seen that the unary operation of additive inverse is likewise involutorial.

EXAMPLE 4.3.2

The additive inverse of 3 is $^{-}3$. But the additive inverse of $^{-}3$, that is $^{-}(^{-}3)$, is 3. Hence

$$^{-}(^{-}3) = 3 \qquad (4.3.25)$$

Next, let us introduce another binary operation which can be performed on two integers. This binary operation, *subtraction*, is defined as follows.

DEFINITION 4.3.1

To *subtract a from b*, written as

$$b - a \qquad (4.3.26)$$

in which a is called the *subtrahend* and b is the *minuend*, means to find x so that

$$a + x = b \qquad (4.3.27)$$

Thus, the binary operation $-$ of *subtraction* has been defined.

For example, suppose it is desired to subtract 6 from 9. This means to find an integer x which when added to 6 gives 9. Thus, since $6 + 3 = 9$, we know that

$$9 - 6 = 3 \qquad (4.3.28)$$

The following results give us a rule or law to use for performing the operation of subtraction.

Theorem 4.3.3

If a and b are two integers, then

$$b - a = b + ({}^{-}a) \qquad (4.3.29)$$

PROOF: This follows since

$$b - a \quad \text{and} \quad b + ({}^{-}a) \qquad (4.3.30)$$

are *both* solutions of the equation

$$a + x = b \tag{4.3.31}$$

Since this equation has exactly *one* solution, the two solutions given in (4.3.30) must be the *same*. Hence, the result follows.

Because of this result, it is only necessary to be able to *add* integers, since the problem of *subtracting* integers can always be expressed as an addition problem by using Theorem 4.3.3.

Theorem 4.3.3 also gives an important relationship between the binary operation of subtraction and the unary operation of additive inverse. It says that to *subtract* an integer a from an integer b, $b - a$, is equivalent to finding the sum of b and the *additive inverse of* a, $b + (^-a)$. It also enables us always to change a *subtraction* problem into an *addition* problem.

EXAMPLE 4.3.4

Perform the indicated operations

$$(4 - 3)\cdot 2 + (^-3) \tag{4.3.32}$$

Here we evaluate the quantity in the parentheses first. Thus, using Theorem 4.3.3

$$4 - 3 = 4 + (^-3) = 1 \tag{4.3.33}$$

Now,

$$(4 - 3)\cdot 2 + (^-3) = 1\cdot 2 + (^-3) \tag{4.3.34}$$

As we agreed earlier, we multiply first and then add. Thus,

$$1\cdot 2 + (^-3) = 2 + (^-3) = {}^-1 \tag{4.3.35}$$

Thus,

$$(4 - 3)\cdot 2 + (^-3) = {}^-1 \tag{4.3.36}$$

EXAMPLE 4.3.5

Evaluate

$$[5 - (^-7)] + (^-2) \tag{4.3.37}$$

Here, we look first at the quantity in the brackets. Now,

$$5 - (^-7) = 5 + [^-(^-7)] = 5 + 7 = 12 \tag{4.3.38}$$

Thus,

$$[5 - (^-7)] + (^-2) = 12 + (^-2) = 10 \tag{4.3.39}$$

Exercise 4.3

Perform the indicated operations.

1. (a) $(3 - 5) + 2 - 3$
   (b) $(^-2 - 5) + 7$
   (c) $^-2 - (5 + 7)$
   (d) $^-8 - (^-2)$
   (e) $(^-4 + 3) - 7$
   (f) $^-4 + (3 - 7)$

2. (a) $^-7 + 5 - 2$
   (b) $^-(7 + 5) - 2$
   (c) $^-(^-6)\cdot(^-3 + 5)$
   (d) $[^-6 - (^-7)] - 2$
   (e) $^-[^-(^-2 - 3)]$
   (f) $[^-6 + (^-8)] - 3$

3. Find x in the following:

   (a) $x - 3 = 5$
   (b) $x + 4 = {}^-2\cdot 5$
   (c) $3\cdot(x + 2) = (5 - 2) + 2\cdot 1 + 1$
   (d) $3\cdot x + 2 - 4 = (5 - 7)\cdot 1$
   (e) $x + 3 - 7 = {}^-(2 - 3) - 3$
   (f) $2 - x = 4\cdot 2 - 1$

4. In the proof of Theorem 4.3.1, fill in reasons for each step in (4.3.15).

5. Prove the second half of the Corollary to Theorem 4.3.1.

6. Is the operation of subtraction commutative? Associative? Why? Give examples.

7. Prove that $a - b = {}^-(b - a)$ for *all* integers a and b.

8. Use the product law, equation (4.2.18), to solve the following problems: Find x if

   (a) $(x + 3)(x - 3) = 0$
   (b) $(x + 2)(x + 7) = 0$
   (c) $(x - 3)(x + 5) = 0$
   (d) $(x + 3)(x + 2)(x - 4) = 0$
   (e) $(x + 5)(x - 1)(x + 2) = 0$
   (f) $(x - 2)(x + 4)(x - 3) = 0$

4.4 FURTHER PROPERTIES OF INTEGERS

In this section, more results concerning integers are given.

Theorem 4.4.1

If a is any integer, then

$$a\cdot 0 = 0 \tag{4.4.1}$$

Although this is an intuitively obvious result, it should be verified logically.

PROOF: By the additive identity property

$$0 + 0 = 0 \tag{4.4.2}$$

Multiply both numbers by a. Then

$$a\cdot(0 + 0) = a\cdot 0 \tag{4.4.3}$$

By the distributive law, we have

$$a \cdot 0 + a \cdot 0 = a \cdot 0 \tag{4.4.4}$$

But $a \cdot 0$ has an additive inverse $^-(a \cdot 0)$. Add this to both sides. Then,

$$^-(a \cdot 0) + [a \cdot 0 + a \cdot 0] = {}^-(a \cdot 0) + a \cdot 0 \tag{4.4.5}$$

By the associative law for addition,

$$[^-(a \cdot 0) + a \cdot 0] + a \cdot 0 = {}^-(a \cdot 0) + a \cdot 0 \tag{4.4.6}$$

By the property of additive inverses,

$$0 + a \cdot 0 = 0 \tag{4.4.7}$$

and by the additive identity property,

$$a \cdot 0 = 0 \tag{4.4.8}$$

Thus, the product of any integer by zero gives zero. For example, $^-3 \cdot 0 = 0$.

The following formulas are quite useful for multiplying integers.

Theorem 4.4.2

Let a and b be two integers. Then,

$$\text{(a)} \quad a \cdot (^-b) = {}^-(a \cdot b) \tag{4.4.9}$$

and

$$\text{(b)} \quad (^-a) \cdot b = {}^-(a \cdot b) \tag{4.4.10}$$

and

$$\text{(c)} \quad (^-a) \cdot (^-b) = a \cdot b \tag{4.4.11}$$

PROOF: (a) Consider the quantity

$$a \cdot (^-b) + a \cdot b \tag{4.4.12}$$

Now, using the distributive law

$$a \cdot (^-b) + a \cdot b = a \cdot [(^-b) + b] \tag{4.4.13}$$

Since $^-b + b = 0$, and the product of any integer by zero is zero

$$a \cdot [^-b + b] = a \cdot 0 = 0 \tag{4.4.14}$$

That is, the additive inverse of $a \cdot b$ is $a \cdot (^-b)$. But the additive inverse of $a \cdot b$ is $^-(a \cdot b)$. Since additive inverses are unique, it follows that

$$a \cdot (^-b) = {}^-(a \cdot b) \tag{4.4.15}$$

(b) The proof of (4.4.10) is left to the student.
(c) The proof of (4.4.11) is left to the student.

The student is encouraged to *read* Theorem 4.4.2 as follows:

(a) a times the additive inverse of b, $a \cdot (^-b)$, equals the additive inverse of $a \cdot b$, $^-(a \cdot b)$.
(b) The additive inverse of a times b, $(^-a) \cdot b$, equals the additive inverse of $a \cdot b$, $^-(a \cdot b)$.
(c) The additive inverse of a times the additive inverse of b, $(^-a) \cdot (^-b)$, equals $a \cdot b$.

These particular results are easier to remember and easier to apply when they are read.

EXAMPLE 4.4.1

(a) $4 \cdot (^-3) = {}^-(4 \cdot 3) = {}^-12$
(b) $^-5 \cdot 6 = {}^-(5 \cdot 6) = {}^-30$
(c) $(^-7) \cdot (^-3) = 7 \cdot 3 = 21$

Theorem 4.4.3

For any integer a,

$$(^-1) \cdot a = {}^-a \tag{4.4.16}$$

This formula gives a relationship between the binary operation of multiplication and the unary operation of additive inverse.

PROOF: By (b) of Theorem 4.4.1, we see that

$$(^-1) \cdot a = {}^-(1 \cdot a) = {}^-a \tag{4.4.17}$$

EXAMPLE 4.4.2

If $a = 5$, then

$$(^-1) \cdot 5 = {}^-5 \tag{4.4.18}$$

That is, if we multiply the additive inverse of 1, $(^-1)$, by any integer a, we obtain the additive inverse ^-a of that integer.

Theorem 4.4.4

Cancellation Law for Addition: Let a, b, c, be three integers. If

$$a + c = b + c \tag{4.4.19}$$

then

$$a = b \tag{4.4.20}$$

PROOF: From the assumption

$$a + c = b + c \tag{4.4.21}$$

Add the additive inverse ^-c of c to both sides. Then,

$$[a + c] + (^-c) = [b + c] + (^-c) \tag{4.4.22}$$

By the associative law for addition,

$$a + [c + (^-c)] = b + [c + (^-c)] \tag{4.4.23}$$

By the additive inverse property,

$$a + 0 = b + 0 \tag{4.4.24}$$

By the additive identity property,

$$a = b \tag{4.4.25}$$

The reason for the name given this particular result is apparent because it seems that the c is merely cancelled or dropped.

Theorem 4.4.5

Cancellation Law for Multiplication: Let a, b, c be three integers. If $c \neq 0$ and if

$$a \cdot c = b \cdot c \tag{4.4.26}$$

then

$$a = b \tag{4.4.27}$$

Warning: It is very important to remember that the integer c cannot be zero. When $c = 0$, the result is *not* always true. For example, we know that $2 \cdot 0 = 1 \cdot 0$, since any integer times zero is zero. If the theorem were true for $c = 0$, it would then turn out that $2 = 1$, which we know is false. Thus, caution should be taken that the quantity "cancelled" is not equal to zero.

PROOF: Since $a \cdot c = b \cdot c$, then

$$a \cdot c + {}^-(b \cdot c) = 0 \tag{4.4.28}$$

By (b) of Theorem 4.4.2, we have

$$a \cdot c + (^-b) \cdot c = 0 \tag{4.4.29}$$

Use the distributive law. Then,

$$[a + (^-b)] \cdot c = 0 \tag{4.4.30}$$

Since $c \neq 0$, by the product law we have

$$a + (^-b) = 0 \tag{4.4.31}$$

But

$$b + (^-b) = 0 \tag{4.4.32}$$

Thus,

$$a + (^-b) = b + (^-b) \tag{4.4.33}$$

By the cancellation law for addition

$$a = b \tag{4.4.34}$$

EXAMPLE 4.4.3
Find x if

$$2 \cdot x + 4 = 6 \tag{4.4.35}$$

Here,

$$2 \cdot x = 6 + (^{-}4) \tag{4.4.36}$$

or

$$2 \cdot x = 2 \tag{4.4.37}$$

But $2 \cdot 1 = 2$. Thus,

$$2 \cdot x = 2 \cdot 1 \tag{4.4.38}$$

Using the cancellation law for multiplication, we can "cancel" the 2's so that

$$x = 1 \tag{4.4.39}$$

Henceforth, we shall usually employ the universally accepted convenience that when two symbols are placed next to each other, such as ab, it is assumed that a is multiplied by b. Thus, $2x$ means $2 \cdot x$. This convention is *never* used for two *numerals* since the meaning would then be ambiguous.

Quite often one hears, "I have a proof that $1 = 2$." Below is such a "proof." Can you find the fallacy in the argument?

Suppose a and b are two integers and suppose $a = b$. Then, by multiplying by a, we have

$$aa = ab \tag{4.4.40}$$

Subtract $b \cdot b$. Then,

$$aa - bb = ab - bb \tag{4.4.41}$$

Notice that

$$(a - b)(a + b) = aa - bb \tag{4.4.42}$$

Thus,

$$(a - b)(a + b) = ab - bb \tag{4.4.43}$$

By the distributive law,

$$(a - b)(a + b) = (a - b) \cdot b \tag{4.4.44}$$

By the cancellation law for multiplication,

$$a + b = b \tag{4.4.45}$$

Since $a = b$, we have

$$2b = b \tag{4.4.46}$$

Thus, since $b = 1 \cdot b$, by the cancellation law,

$$2 = 1 \tag{4.4.47}$$

Exercise 4.4

1. (a) $[3 + 4\cdot(^-3)] - 2$
   (b) $^-6\cdot 2 - 3$
   (c) $^-4 - 2\cdot(^-3)$
   (d) $(^-3)(^-3) + 5$
   (e) $5 - 2\cdot(^-3)$
   (f) $^-4 + (2\cdot 3)(^-2)$

2. (a) $(^-2 - 3)\cdot(^-4 + 1)$
   (b) $(^-5 - 2) + 4\cdot(^-3)$
   (c) $^-4[7 - (^-8)]$
   (d) $^-8\cdot[^-6 - 1] - 4$
   (e) $^-4\cdot(^-5 + 3)$
   (f) $(4 - 6)\cdot(^-7 + 2)$

3. Solve the following equations for x:

   (a) $3x + 5 = 2$
   (b) $(x - 5) + 2 = (x + x) - x$
   (c) $2x - 3 = {^-5} + x$
   (d) $3x - 7 = {^-5} + 2(x - 2)$

4. Solve the following equations for integral values of x:

   (a) $^-2(x + 5) = {^-3x} - 2$
   (b) $^-4 - 6 = 2 - 3x$
   (c) $4x + 6 = 2(x - 8)$
   (d) $^-3(x - 2) = 6x - 9$

5. Prove (b) of Theorem 4.4.2.

6. Prove (c) of Theorem 4.4.2.

7. Fill in reasons for each step of (4.4.13).

8. Prove that multiplication is distributive over subtraction, that is, show that $a\cdot(b - c) = a\cdot b - a\cdot c$.

9. A hat and coat cost 45 dollars together. If the cost of the coat is four times that of the hat, what are the individual costs. [HINT: If x = cost of hat, then $4x$ = cost of coat.]

5 Order for the Set J of Integers

5.1 NATURAL ORDERING

In Chapter 3, we discussed a way or method of ordering the set N of natural numbers. The different properties of natural ordering were discussed. In this section, we introduce a way of ordering the set J of integers.

Since the set N of natural numbers, $\{1, 2, 3, \cdots\}$, is a proper subset of the set J of integers, $\{\cdots, {}^-3, {}^-2, {}^-1, 0, 1, 2, 3, \cdots\}$, the following definition can be used.

DEFINITION 5.1.1

Let a and b be two integers. We shall say that *a is less than b*, written as

$$a < b \tag{5.1.1}$$

if and only if the quantity $b - a$, called the *difference* is a natural number.

This particular way of ordering the set J of integers is called *natural ordering*. The reason for this is that if one displays the set J of integers from left to right (see Fig. 5.1) then any integer to *the left* of a given integer is *less than* the given integer.

$$\cdots, {}^-6, {}^-5, {}^-4, {}^-3, {}^-2, {}^-1, 0, 1, 2, 3, \cdots$$

FIG. 5.1

Consider a line with the property that certain points on the line can be associated in a one-to-one way with the integers. Such a line is termed the *integer line.* By using the natural ordering given above, we can set up the correspondence between points and integers so that the ordering previously introduced for the whole number line is preserved. See Fig. 5.2.

EXAMPLE 5.1.1

Consider the two integers 4 and 6. Since the difference $6 - 4$ is a natural number, namely 2, then 4 is less than 6. We denote this by using the symbolism $4 < 6$. We can see this by noting that on the integer line 4 is to the left of 6.

EXAMPLE 5.1.2

Consider $^-4$ and 2. Again, since

$$2 - (^-4) = 2 + 4 = 6 \tag{5.1.2}$$

is a natural number, then $^-4 < 2$. Here again, on the integer line, $^-4$ is to the left of 2.

Sometimes it is more convenient to use the concept of *greater than.* We say that *an integer a is greater than an integer b*, written as

$$a > b \tag{5.1.3}$$

if and only if the quantity $a - b$ is a natural number.

The student should notice that the two statements *a is greater than b* and *b is less than a* are equivalent; that is

$$a > b \quad \text{if and only if} \quad b < a \tag{5.1.4}$$

The above is equivalent to saying that on the integer line, b is to the left of a if and only if a is to the right of b.

DEFINITION 5.1.2

Let a and b be two integers. *We say that a is less than or equal to b*, written as

$$a \leqq b \quad \text{(sometimes written } a \leq b\text{)} \tag{5.1.5}$$

if and only if either $a < b$ or $a = b$.

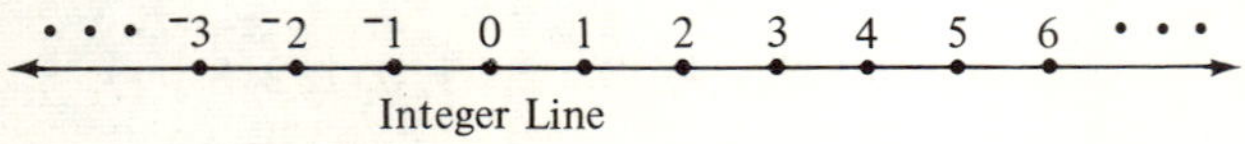

FIG. 5.2

As before, the statements *a is greater than or equal to b* and *b is less than or equal to a* are equivalent. In other words,

$$a \geqq b \quad \text{if and only if} \quad b \leqq a. \tag{5.1.6}$$

EXAMPLE 5.1.3

Since it is known that $^{-}4 < 2$, by the above definition, we know that $2 > {}^{-}4$.

Now, let us look at some of the properties of this relation $\leqq$ (less than or equal). Before doing so, however, notice that this relation is one between *two* integers; that is, the relation $\leqq$ is a *binary* relation.

Theorem 5.1.1

The binary relation $\leqq$ (less than or equal) between two integers has the following four properties.

(I) *Reflexive Law*

For any integer a,

$$a \leqq a$$

(II) *Anti-Symmetric Law*

For integers a and b,

$$\text{if } a \leqq b \text{ and } b \leqq a, \text{ then } a = b$$

(III) *Transitive Law*

Let a, b, c be three integers.

$$\text{If } a \leqq b \text{ and } b \leqq c, \text{ then } a \leqq c$$

(IV) *Trichotomy Law*

If a and b are any two integers,

$$\text{either } a < b \text{ or } a = b \text{ or } b < a$$

The three possibilities in (IV) are mutually exclusive.

Before proving these laws, let us review exactly what they mean. The reflexive law states that every integer is less than or equal to itself. To understand the validity of the statement, one must recall the meaning of the exclusive disjunctive *or*. If I am sleeping, the statement *I am sleeping or I am eating* is a *true* statement. Thus, to say $a \leqq a$ is a true statement since a is less than *or equal to* a. Even though $a < a$ is *not* true, the statement $a \leqq a$ is a true law for the relation $\leqq$. Recall, too, that the relation $\subseteq$ (is a subset of) is also reflexive.

The anti-symmetric law merely states that if a is less than or equal to b and simultaneously b is less than or equal to a, then we must have a equal to b. This law is also true for the relation $\subseteq$ (is a subset of) between sets.

The transitive law tells us that if a is less than or equal to b and if b, in turn, is less than or equal to c, then a must be less than or equal to c. Again, the relation $\subseteq$ between sets also obeys the transitive law.

Finally, the trichotomy law states that for *any* choice of two integers, say a and b, either a is less than b, or a equals b, or b is less than a. Moreover, one and only one of these possibilities can occur, and no two of them can occur at the same time. Recall that *no* such law is true for the relation $\subseteq$ between sets.

Now, let us give a formal proof of these laws.

PROOF: Before proceeding, review Definition 5.1.2, since it is used extensively.

(I) Since $a = a$, then $a \leqq a$.

(II) If $a \leqq b$, then either $a = b$ or $a < b$. If $b \leqq a$, then either $a = b$ or $b < a$. Suppose $a < b$ and $b < a$. Then $b - a$ and $a - b$ are both natural numbers. But

$$b - a = {}^{-}(a - b) \tag{5.1.7}$$

Thus, both $b - a$ and $a - b$ cannot be natural numbers. The only possibility remaining is that $a = b$.

(III) If $a \leqq b$, either $a = b$ or $a < b$. If $b \leqq c$, either $b = c$ or $b < c$. If $a = b$ and $b = c$, then $a = c$. If $a < b$ and $b < c$, then $b - a$ and $c - b$ are natural numbers. Their sum

$$(b - a) + (c - b) = c - a \tag{5.1.8}$$

must also be a natural number. This means that $a < c$. Hence, either $a = c$ or $a < c$. Thus, $a \leqq c$.

(IV) The proof of (IV) is left to the student.

Although Theorem 5.1.1 is primarily theoretical in nature, it is very useful for proving some practical results, as shown in the following sections.

The student should compare the concepts of *less than, less than or equal to, greater than,* and *greater than or equal to,* for integers, with the concepts of *subset, proper subset, contains,* and *properly contains* for sets from Chapter 2.

Exercise 5.1

1. For the following pairs of integers, replace $*$ by as many of the relations $<, \leqq, >, \geqq$, that are valid and show their position on the integer line.

(a) $3 * {}^{-}5$ (f) ${}^{-}4 * {}^{-}5$

(b) $4 * 2$ (g) ${}^{-}3 * {}^{-}3$

(c) $6 * 7$ (h) $2 * 0$
(d) $^{-}5 * ^{-}2$ (i) $0 * ^{-}1$
(e) $^{-}3 * 2$

2. Does the relation $<$ (is less than) obey the reflexive law? The transitive law? Why?

3. Prove part (IV) of Theorem 5.1.1.

5.2 POSITIVE INTEGERS

In Chapter 4, the set J of integers, namely

$$J = \{\cdots, ^{-}3, ^{-}2, ^{-}1, 0, 1, 2, 3, \cdots\} \tag{5.2.1}$$

was introduced. In the last section, an order relation of *less than* was defined for J. That is, if a and b are two integers, we say that *a is less than b*, written as

$$a < b \tag{5.2.2}$$

if and only if the quantity $b - a$ is a natural number. These concepts lead us to introduce the notion of a *positive* integer.

DEFINITION 5.2.1

An integer a is said to be a *positive integer* or merely *positive*, if and only if

$$0 < a \quad \text{or} \quad a > 0 \tag{5.2.3}$$

For example, since $0 < 3$, then 3 is a positive integer. Since $^{-}2 < 0$, then $^{-}2$ is *not* a positive integer. Notice that on the integer line a positive integer is always to the right of 0.

If J_+ represents the set of positive integers, it is easy to see that

$$J_+ = \{1, 2, 3, 4, \cdots\} \tag{5.2.4}$$

Notice that the set J_+ of positive integers and the set N of natural numbers are one and the same set. Because of this, the following result about positive integers is true.

Theorem 5.2.1

The set J of integers has the following three properties:

(I) In J, there is a nonempty subset J_+ consisting of positive integers.

(II) If a and b are two positive integers, their sum $a + b$ and their product $a \cdot b$ are also positive. In other words, the set J_+ of positive integers is closed relative to the operations of addition and multiplication.

(III) The set J obeys a trichotomy law. That is, if a is any integer, then

$$\text{either } a \text{ is positive, or } a \text{ is zero, or } {}^{-}a \text{ is positive} \tag{5.2.5}$$

These three possibilities in (5.2.5) are mutually exclusive.

PROOF: (I) It has already been demonstrated that a subset J_+ of J exists. This subset J_+ is the same as the subset N of natural numbers of J.

(II) Let a and b be two positive integers. Then we know that

$$a > 0, \quad \text{and} \quad b > 0 \tag{5.2.6}$$

Thus, a and b are both natural numbers. Hence, both their sum $a + b$ and their product $a \cdot b$ must be natural numbers. In other words,

$$a + b > 0, \qquad a \cdot b > 0 \tag{5.2.7}$$

(III) If in (IV) of Theorem 5.1.1 of Section 1, we replace b by 0, then for any integer a, we have

$$a < 0, \quad \text{or} \quad a = 0, \quad \text{or} \quad 0 < a. \tag{5.2.8}$$

If $a < 0$, then $0 - a = {}^{-}a$ is a natural number. Thus, (5.2.8) says that either ${}^{-}a$ is positive or $a = 0$ or a is positive.

DEFINITION 5.2.2

Let a be an integer. We shall say that a is a *negative integer*, or merely *negative*, if and only if a is neither positive nor zero. Thus, on the integer line, negative integers lie to the left of 0.

The student should be aware that Definition 5.2.2 has meaning only because of the trichotomy law of Theorem 5.2.1.

Also, it is clear that

$$a \text{ is negative if and only if } a < 0. \tag{5.2.9}$$

Of course, the set J_- of negative integers is

$$J_- = \{{}^{-}1, {}^{-}2, {}^{-}3, {}^{-}4, \cdots\}. \tag{5.2.10}$$

Finally, because of Definitions 5.2.1 and 5.2.2, we see that

$$a \text{ is positive if and only if } {}^{-}a \text{ is negative, and}$$

$$a \text{ is negative if and only if } {}^{-}a \text{ is positive} \tag{5.2.11}$$

Warning: For the first time, we can now distinguish between positive and negative integers. However, remember that the raised dash (${}^{-}$) before a number indicates the operation of additive inverse and *not* the concept of a negative number.

Thus, to indicate that 3 is positive we will write $3 > 0$. To indicate that $^{-}4$ is negative, we write $^{-}4 < 0$. If we write ^{-}a, we mean additive inverse of a. If we want to indicate a is negative, we write $a < 0$.

By remembering this important distinction, many of the difficulties of computation can be overcome.

The following theorem is very well known, although the reason for its validity is very seldom understood. It is hoped that the previous discussion, together with the proof given, will help to overcome this circumstance.

Theorem 5.2.2

Let a and b be two integers.

(1) If a is positive and if b is negative, their product $a \cdot b$ is negative. That is,

$$\text{if } a > 0,\ b < 0, \text{ then } a \cdot b < 0 \tag{5.2.12}$$

(2) If a is negative and b is negative, their product is positive. That is,

$$\text{if } a < 0,\ b < 0, \text{ then } a \cdot b > 0 \tag{5.2.13}$$

PROOF: (1) Suppose a is positive and b is negative. Recall that, if a and b are *any* two integers,

$$a \cdot (^{-}b) = {}^{-}(a \cdot b) \tag{5.2.14}$$

Now since b is negative, we know that ^{-}b is positive. Thus, the product of a and ^{-}b, namely, $a \cdot (^{-}b)$ must be positive. But $a \cdot (^{-}b)$ is the same as $^{-}(a \cdot b)$. Hence, $^{-}(a \cdot b)$ is positive. This says that $a \cdot b$ is negative.

(2) Let a and b be two negative integers. Then, ^{-}a and ^{-}b are two positive integers. Hence, their product $(^{-}a) \cdot (^{-}b)$ is positive. But

$$(^{-}a) \cdot (^{-}b) = a \cdot b \tag{5.2.15}$$

Thus, $a \cdot b$ is positive.

Corollary

The square $a^2 = a \cdot a$ of any integer a is *positive definite*. That is,

$$a^2 \geqq 0, \quad \text{and} \quad a^2 = 0 \quad \text{if and only if} \quad a = 0 \tag{5.2.16}$$

PROOF: By the trichotomy law, any integer a is positive, zero, or negative. If a is positive, then $a \cdot a = a^2$ is positive. If a is zero, then $a \cdot a = a^2 = 0 \cdot 0 = 0$. If a is negative, then by (2) of Theorem 5.2.2, a^2 is positive. Thus, for all possibilities, a^2 is zero or positive. Next, if $a^2 = a \cdot a = 0$, by the product law, we have that $a = 0$. Thus,

$$a^2 = 0 \quad \text{if and only if} \quad a = 0 \tag{5.2.17}$$

The notation $a^2 \geqq 0$ can be read as a^2 *is nonnegative*, meaning, of course, that either a^2 is positive or a^2 is zero.

5.3 THE ALGEBRA OF INEQUALITIES

Expressions involving the relations of $<$, $\leqq$, $>$, or $\geqq$ are called *inequalities*, just as an expression involving $=$ is called an equality or equation. To work with equations, one must first be aware of the laws or rules obeyed. So too, in order to handle inequalities, one must know and understand the laws and rules obeyed. This section deals with the "algebra" of inequalities; that is, with the rules which inequalities obey.

Theorem 5.3.1

The sense of an inequality is preserved under addition. That is, if a, b, c are three integers, and if

$$a \leqq b \tag{5.3.1}$$

then

$$a + c \leqq b + c \tag{5.3.2}$$

In other words, addition will retain the sense of an inequality.

PROOF: Here, we know that

$$(b + c) - (a + c) = b - a \tag{5.3.3}$$

Since $a \leqq b$, then $b - a$ is positive or zero. Hence, $(b + c) - (a + c)$ is positive or zero. Thus,

$$a + c \leqq b + c \tag{5.3.4}$$

Intuitively, the above result says that if a is to the left of b, then $(a + c)$ is to the left of $(b + c)$ on the integer line.

EXAMPLE 5.3.1

It is known that $^-3 < 5$. If we add $^-2$ to both sides of the relation $<$, Theorem 5.3.1 tells us we can retain the relation $<$ for the result. Thus,

$$^-3 + (^-2) < 5 + (^-2) \quad \text{or} \quad ^-5 < 3 \tag{5.3.5}$$

EXAMPLE 5.3.2

Find all x for which

$$x + 3 \leqq 5 \tag{5.3.6}$$

If this were an equality, we would add $(^-3)$ to both sides in order to isolate the unknown x. Here, for the inequality, we do the same. By Theorem 5.3.1 we know that adding $^-3$ will retain the sense of the inequality. Thus,

$$(x + 3) + (^-3) \leqq 5 + (^-3) \tag{5.3.7}$$

Using the associative law for addition,

$$x + [3 + (^-3)] \leqq 2 \tag{5.3.8}$$

By the additive inverse property

$$x + 0 \leqq 2 \tag{5.3.9}$$

By the additive identity property

$$x \leqq 2 \tag{5.3.10}$$

Thus, any integer less than or equal to 2 will make the original statement (5.3.6) a *true* statement. Figure 5.3 illustrates the set of solutions on the integer line

FIG. 5.3

Warning: Solutions to inequalities are usually *not* unique; rather there are ordinarily many solutions.

Thus, in Example 5.3.2, the solutions are given by the set X

$$X = \{2, 1, 0, {}^{-}1, {}^{-}2, {}^{-}3, \cdots\} = \{x \mid x \leqq 2\} \tag{5.3.11}$$

Theorem 5.3.2

The sense of an inequality is preserved under multiplication by positive integers. That is, if a and b are two integers with c a positive integer, and if

$$a \leqq b, \qquad c > 0 \tag{5.3.12}$$

then

$$a \cdot c \leqq b \cdot c \tag{5.3.13}$$

PROOF: It is known that

$$b \cdot c - a \cdot c = (b - a) \cdot c \tag{5.3.14}$$

Since $a \leqq b$, then $b - a$ is positive or zero. Since $c > 0$, then $(b - a) \cdot c$ is positive or zero. Thus,

$$a \cdot c \leqq b \cdot c \tag{5.3.15}$$

EXAMPLE 5.3.3

It is known that

$$^{-}3 < 4 \tag{5.3.16}$$

Thus, upon multiplying by 3, which is positive, we obtain

$$3 \cdot ({}^{-}3) < 3 \cdot 4 \tag{5.3.17}$$

Or

$$^{-}9 < 12 \tag{5.3.18}$$

Theorem 5.3.3

The sense of an inequality is reversed under multiplication by negative integers. That is, if a and b are two integers, and c is a negative integer, and if

$$a \leqq b, \qquad c < 0 \tag{5.3.19}$$

then

$$a \cdot c \geqq b \cdot c \tag{5.3.20}$$

PROOF: The proof of this result is left to the student.

EXAMPLE 5.3.4

We know that

$$^{-}2 < 5 \tag{5.3.21}$$

Multiply by $^{-}3$. Then

$$(^{-}3) \cdot (^{-}2) > (^{-}3) \cdot 5 \tag{5.3.22}$$

or

$$6 > {}^{-}15 \tag{5.3.23}$$

Theorem 5.3.4

Let a and b be two nonnegative integers. If

$$a \leqq b \tag{5.3.24}$$

then

$$a^2 \leqq b^2 \tag{5.3.25}$$

PROOF: Consider the fact that

$$(b - a) \cdot (b + a) = b^2 - a^2 \tag{5.3.26}$$

Since a and b are both nonnegative, then $a + b$ must be nonnegative. Also, since $a \leqq b$, then $b - a$ is positive or zero. Hence, their product $(b - a) \cdot (b + a)$ is positive or zero. Thus, $b^2 - a^2$ is positive or zero. That is,

$$a^2 \leqq b^2 \tag{5.3.27}$$

EXAMPLE 5.3.5

It is known that

$$3 < 5 \tag{5.3.28}$$

Theorem 5.3.4 tells us that if we square the left-hand side and the right-hand side, the *less than* symbol is retained. Thus, because of this,

$$9 < 25 \tag{5.3.29}$$

It is very important to remember that a and b must be nonnegative in order for Theorem 5.3.4 to be true. To see that this is the case, consider the following example.

EXAMPLE 5.3.6

We know that

$$-3 < 2 \tag{5.3.30}$$

Thus, if the Theorem were true for negative integers, then we would have

$$(-3)\cdot(-3) < 2\cdot 2 \tag{5.3.31}$$

or

$$9 < 4 \tag{5.3.32}$$

This is obviously incorrect.

The above example shows that it is just as important to know the hypothesis of a theorem as it is the conclusion.

The next result shows that the converse of Theorem 5.3.4 is also true.

Theorem 5.3.5

Let a and b be two nonnegative integers. If

$$a^2 \leqq b^2 \tag{5.3.33}$$

then

$$a \leqq b \tag{5.3.34}$$

PROOF: Assume the contrary. That is, suppose that a is not less than or equal to b. In other words, suppose

$$b < a \tag{5.3.35}$$

Since a and b are nonnegative, then by Theorem 5.3.4, it follows that

$$b^2 < a^2 \tag{5.3.36}$$

This is impossible since the hypothesis (5.3.33) of Theorem 5.3.5 states $b^2 < a^2$ is false. Thus,

$$a \leqq b \tag{5.3.37}$$

EXAMPLE 5.3.7

Find all nonnegative integers x for which

$$x^2 \leqq 16 \tag{5.3.38}$$

By Theorem 5.3.5, we have

$$x \leqq 4 \tag{5.3.39}$$

Thus, the set X of solutions to the inequality (5.3.38) is

$$X = \{x \mid x \geqq 0 \text{ and } x \leqq 4\} = \{0, 1, 2, 3, 4\} \tag{5.3.40}$$

Figure 5.4 illustrates the solution set on the integer line.

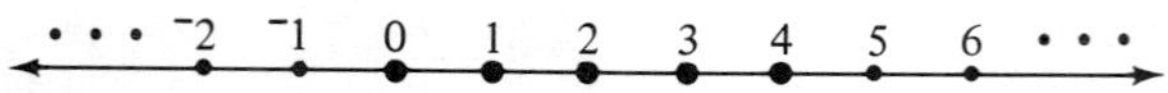

FIG. 5.4

Next, we shall introduce a shorthand convention that is useful for expressing certain sets in an easy way.

If a and b are two integers and if

$$a < b \tag{5.3.41}$$

the set X consisting of all integers x for which

$$x \geqq a \quad \text{and} \quad x \leqq b \tag{5.3.42}$$

is written as

$$X = \{x \mid a \leqq x \leqq b\} \tag{5.3.43}$$

Here we read X *is the set of all x for which a is less than or equal to x and x is less than or equal to b. It is understood that x is an integer.*

EXAMPLE 5.3.8

The set

$$X = \{x \mid 2 \leqq x \leqq 5\} \tag{5.3.44}$$

is

$$X = \{2, 3, 4, 5\} \tag{5.3.45}$$

since 2, 3, 4, 5 are the only integers both greater than or equal to 2 and less than or equal to 5. See Fig. 5.5.

Some variations of this notation are given below. Suppose a and b are two integers for which $a < b$.

The set X consisting of all integers x for which

$$x > a \quad \text{and} \quad x < b \tag{5.3.46}$$

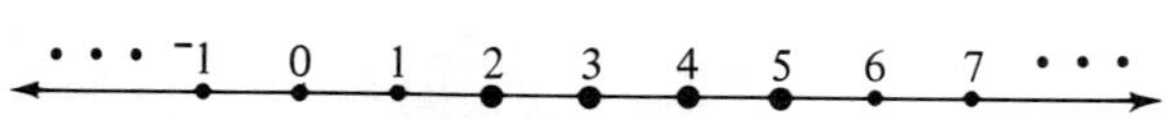

FIG. 5.5

is written as

$$X = \{x \mid a < x < b\} \tag{5.3.47}$$

The other two possibilities

$$X = \{x \mid a \leqq x < b\}, \qquad X = \{x \mid a < x \leqq b\} \tag{5.3.48}$$

are defined in a similar way.

EXAMPLE 5.3.9

$$3 \leqq x + 5 \leqq 8 \tag{5.3.49}$$

Here we want to isolate the x. To do this, remember that (5.3.49) means

$$3 \leqq x + 5 \quad \text{and} \quad x + 5 \leqq 8 \tag{5.3.50}$$

Treating these separately, we proceed as before by adding $(^{-}5)$ to both sides of the inequality. Thus,

$$^{-}2 \leqq x \quad \text{and} \quad x \leqq 3 \tag{5.3.51}$$

This means

$$^{-}2 \leqq x \leqq 3 \tag{5.3.52}$$

Thus, the set X of solutions x is

$$X = \{x \mid {}^{-}2 \leqq x \leqq 3\} = \{^{-}2, {}^{-}1, 0, 1, 2, 3\} \tag{5.3.53}$$

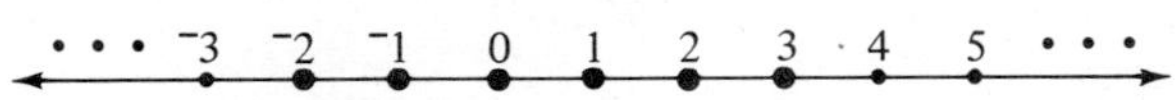

FIG. 5.6

Exercise 5.3

1. In the following problems, express each statement by means of symbols.

(a) 7 is greater than 3
(b) $^{-}4$ is less than $^{-}3$
(c) x is between 1 and $^{-}1$, inclusive
(d) x is positive
(e) x is nonnegative
(f) 2 is less than 5
(g) $^{-}4$ is greater than $^{-}7$
(h) x is negative
(i) x is nonpositive

2. In the following problems, each statement is justifiable by one of the theorems in this chapter. Cite the appropriate justification.

(a) If $y < 8$, then $3y < 24$
(b) If $x < 7$, then $x + 2 < 9$

(c) If $x + 1 < 0$, then $x < {}^-1$
(d) If $y < 4$, then $y + 2 < 6$
(e) If $x < 9$, then ${}^-2x > {}^-18$
(f) If $x > 0$ and $x^2 \leqq 9$, then $x \leqq 3$

3. Find all integers x for which
(a) $x - 3 \leqq 7$
(b) $x - 2 \leqq 7 + 5 - 3$
(c) ${}^-(x + 2) + 5 \leqq 4$
(d) $3 - x + 2 \leqq {}^-2x + 1$
(e) ${}^-2(x + 3) \geqq 2(3 - x) + x$

Show the solution set on the integer line.

4. Find all integers x for which

(a) $x^2 - 5 \leqq (3 + 4) - 3$ and $x > 0$
(b) $x + 5 \geqq 2$ and $x - 2 \leqq 6$
(c) ${}^-(x - 2) + 4 \leqq 3 - 2x$

Show the solution set on the integer line.

5. List the integers in the following sets:

(a) $X = \{x \mid 3 \leqq x + 1 \leqq 6\}$
(b) $A = \{a \mid 3 \leqq a + 4 \leqq 9\}$
(c) $B = \{b \mid 3 + b \leqq 2b < 7 + b\}$
(d) $C = \{c \mid 7 + 2c < 3c < 11 + 2c\}$
(e) $D = \{d \mid 3 \leqq d + 7 \leqq 20\}$

Show the solution set on the integer line.

6. Find all positive integers x for which

(a) $x^2 + 3 < 12$
(b) $x^2 + (3 - 4) + 7 > (2 + 7) + (10 - 3) + 6$
(c) $7 - (8 - 2) + x^2 < (11 - 4) + 10$
(d) ${}^-2 + 8 + 9 < x^2 - 10$
(e) $5 + x - 2 \leqq x^2 + 2 + x$
(f) $x(x + 1) \geqq x + 36$
(g) $(x + 11) - 71 + 3 < 2x + 7$

Show the solution set on the integer line.

7. Marmaduke has six coins of equal value and he also has a quarter. If the total value of the seven coins is less than 75¢, what are all the possible values of the coins?

8. Verify equation (5.3.26).

9. Prove Theorem 5.3.3.

5.4 ABSOLUTE VALUE

In this section, we shall introduce another concept, *absolute value,* directly related to the ordering properties of the set J of integers. The student should notice that the following definition can be given only because of the trichotomy property.

DEFINITION 5.4.1

The *absolute value* $|a|$ of an integer a is defined as

$$|a| = \begin{cases} a & \text{if} \quad a > 0 \\ 0 & \text{if} \quad a = 0 \\ {}^{-}a & \text{if} \quad a < 0 \end{cases} \tag{5.4.1}$$

EXAMPLE 5.4.1

Consider the three integers, 2, 0, $^{-}3$. Then,

$$|2| = 2, \qquad |0| = 0, \qquad |{}^{-}3| = {}^{-}({}^{-}3) = 3 \tag{5.4.2}$$

It is important for the student to realize that for a negative integer a, the absolute value of a is the additive inverse of a, *not negative* a. Thus, the absolute value of $^{-}5$ is the additive inverse of $^{-}5$ or 5.

Next, we shall develop some of the properties of absolute value.

Theorem 5.4.1

Let a be an integer. Then, the absolute value of a is positive definite. That is,

$$|a| \geqq 0 \quad \text{and} \quad |a| = 0 \quad \text{if and only if} \quad a = 0 \tag{5.4.3}$$

PROOF: By the trichotomy law, there are three possibilities for a. If a is positive, $a > 0$, then $|a| = a$ and so $|a| > 0$. If $a = 0$, then $|a| = |0| = 0$. If a is negative, $a < 0$, then $|a| = {}^{-}a$. Since $a < 0$, then ${}^{-}a > 0$. Then, if a is negative, $|a| > 0$. Finally, if $|a| = 0$, we know that $a = 0$. Thus, $|a|$ is positive definite.

The above result tells us that the absolute value $|a|$ of any integer a is nonnegative, that is, is *never* negative. Thus, the absolute value of a negative integer is a positive integer.

Note that for any integer a, the integer a is always less than or equal to its absolute value. That is,

$$a \leqq |a| \tag{5.4.4}$$

The equality sign will hold if $a > 0$ or $a = 0$.

EXAMPLE 5.4.2

Consider the integer $^-3$. Then, $|\,^-3\,| = 3$. Clearly, in this case, $^-3 < 3$. For the integer 4, we have $|\,4\,| = 4$.

In general, we see that $a \leqq |\,a\,|$. That is, an integer is *never* larger than its absolute value.

Theorem 5.4.2

Let $a > 0$ be a positive integer. If

$$|\,x\,| = a \tag{5.4.5}$$

then

$$\text{either } x = a \text{ or } x = {^-a} \tag{5.4.6}$$

PROOF: Since $a > 0$, it is clear that $x \neq 0$. Thus, either $x > 0$ or $x < 0$ by the trichotomy law. If $x > 0$, then $|\,x\,| = x = a$. If $x < 0$, then $|\,x\,| = {^-x} = a$. Thus, $x = {^-a}$. Hence if $|\,x\,| = a$, either $x = a$ or $x = {^-a}$.

EXAMPLE 5.4.3

Find all integers x for which

$$|\,x\,| = 3 \tag{5.4.7}$$

By Theorem 5.4.2, we know that

$$\text{either } x = 3 \text{ or } x = {^-3} \tag{5.4.8}$$

Let us check both these solutions. If $x = 3$, then $|\,3\,| = 3$ and if $x = {^-3}$, then $|\,x\,| = |\,^-3\,| = 3$. Thus, both solutions for x satisfy the given equation.

Theorem 5.4.3

Let $a > 0$ be a positive integer. If

$$|\,x\,| \leqq a \tag{5.4.9}$$

then

$$^-a \leqq x \leqq a \tag{5.4.10}$$

This says that all solutions of (5.4.9) lie between the additive inverse of a and a itself.

PROOF: Here we know that either $x \geqq 0$ or $x < 0$. If $x \geqq 0$, then $|\,x\,| = x$.

$$x \geqq 0 \quad \text{and} \quad x \leqq a \quad \text{or} \quad 0 \leqq x \leqq a \tag{5.4.11}$$

If $x < 0$, then $|\,x\,| = {^-x} \leqq a$. Here $x \geqq {^-a}$. Thus,

$$x < 0 \quad \text{and} \quad x \geqq {^-a} \quad \text{or} \quad {^-a} \leqq x < 0 \tag{5.4.12}$$

The set union of the solutions given in (5.4.11) and (5.4.12) is

$$^-a \leqq x \leqq a \tag{5.4.13}$$

EXAMPLE 5.4.4

Find all integers x for which

$$|x| \leqq 3 \tag{5.4.14}$$

By Theorem 5.4.3, all such x are

$$^-3 \leqq x \leqq 3 \tag{5.4.15}$$

The set X for all such solutions x is

$$X = \{^-3, ^-2, ^-1, 0, 1, 2, 3\} \tag{5.4.16}$$

See Fig. 5.7.

FIG. 5.7

It is important for the student to realize the difference between Theorem 5.4.2 and Theorem 5.4.3. Theorem 5.4.2 gives us a technique for finding integers x if we know what the absolute value of x is *equal to*. Theorem 5.4.3 gives us a way for finding integers x if it is known what the absolute value of x is *less than or equal to*. The next result will give us a method for finding integers x when it is known what the absolute value of x is *greater than or equal to*.

Theorem 5.4.4

Let $a > 0$ be a positive integer. If

$$|x| \geqq a \tag{5.4.17}$$

then

$$\text{either } x \geqq a \text{ or } x \leqq {}^-a \tag{5.4.18}$$

PROOF: This is left to the student.

EXAMPLE 5.4.5

Find all integers x for which

$$|x| \geqq 4 \tag{5.4.19}$$

By Theorem 5.4.4, we know that

$$x \geqq 4 \quad \text{or} \quad x \leqq {}^-4 \tag{5.4.20}$$

See Fig. 5.8.

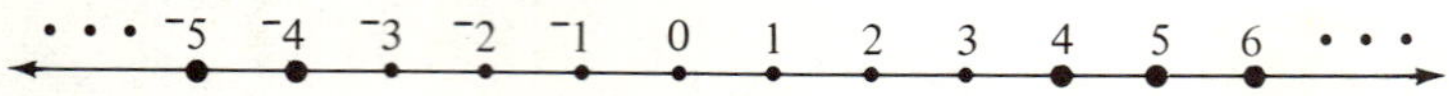

FIG. 5.8

The following two results will be very important for future applications of absolute value concepts.

Theorem 5.4.5

Let a and b be any two integers. The absolute value of their product is the same as the product of their absolute values. That is,

$$|a \cdot b| = |a| \cdot |b| \tag{5.4.21}$$

PROOF: To prove this result, we must consider the following four cases:

Case 1: $a = 0$

If $a = 0$, then

$$|a \cdot b| = |0 \cdot b| = |0| = 0 = |0 \cdot b| = |0| \cdot |b| = |a| \cdot |b| \tag{5.4.22}$$

Thus, in case $a = 0$, the result (5.4.21) is valid.

Case 2: $a > 0, \quad b > 0$

If $a > 0$, $b > 0$, then $a \cdot b > 0$. Thus,

$$|a \cdot b| = a \cdot b = |a| \cdot |b| \tag{5.4.23}$$

Case 3: $a > 0, \quad b < 0$

Here since $a > 0$, $b < 0$, we have $a \cdot b < 0$. Thus,

$$|a| = a, \qquad |b| = {}^{-}b, \qquad |a \cdot b| = {}^{-}(a \cdot b) \tag{5.4.24}$$

Here

$$|a \cdot b| = {}^{-}(a \cdot b) = a \cdot ({}^{-}b) = |a| \cdot |b| \tag{5.4.25}$$

Case 4: $a < 0, \quad b < 0$

This is left to the student.

The reason the case $a < 0$, $b > 0$ is excluded is that because multiplication is commutative we can interchange a and b so that this possibility is the same as that in Case 3.

Theorem 5.4.6

The Triangular Inequality: Let a and b be any two integers. The absolute value of their sum does not exceed the sum of their absolute values. That is,

$$|a + b| \leqq |a| + |b| \tag{5.4.26}$$

PROOF: For,

$$\begin{aligned}|a+b|^2 &= (a+b)^2 = (a+b)(a+b)\\ &= a^2+2ab+b^2 = |a|^2+2ab+|b|^2\\ &\leqq |a|^2+2|a||b|+|b|^2 = [|a|+|b|]^2 \qquad (5.4.27)\end{aligned}$$

By Theorem 5.3.5, we have

$$|a+b| \leqq |a|+|b| \qquad (5.4.28)$$

The student should verify each step of (5.4.27).

EXAMPLE 5.4.6

Consider the two integers $a = 3$, $b = {}^-2$. Then,

$$|a+b| = |3+({}^-2)| = |1| = 1 \qquad (5.4.29)$$

and

$$|a|+|b| = |3|+|{}^-2| = 3+2 = 5 \qquad (5.4.30)$$

In this case, $|a+b| < |a|+|b|$.

EXAMPLE 5.4.7

Consider the two integers $a = {}^-4$, $b = {}^-3$. Then,

$$|a+b| = |({}^-4)+({}^-3)| = |{}^-7| = 7 \qquad (5.4.31)$$

and

$$|a|+|b| = |{}^-4|+|{}^-3| = 4+3 = 7 \qquad (5.4.32)$$

In this case, $|a+b| = |a|+|b|$. The triangular inequality states that for any choice of a and b,

$$\begin{aligned}&\text{either} \quad |a+b| = |a|+|b|\\ &\text{or} \quad |a+b| < |a|+|b| \qquad (5.4.33)\end{aligned}$$

Exercise 5.4

1. Find all integers x for which

   (a) $|x| = 2$ (c) $|x| = 5$
   (b) $|x| = 0$ (d) $|x| = {}^-3$

   Show the solution set on the integer line.

2. Find all integers x for which

   (a) $|x| \leqq 3$ (c) $|x| < 2$
   (b) $|x| \geqq 3$ (d) $|x| > 4$

   Show the solution set on the integer line.

3. Find all integers x for which

(a) $|3 - x| \leqq 7$ (c) $|x + 3| > 2$
(b) $|4 - x| > 2$ (d) $|4 - x| < {}^{-}1$

Show the solution set on the integer line.

4. Find all integers x for which

(a) $|4 + x| \leqq 2$ (d) $|x + 7 - 2x| \leqq 3$
(b) $|6 - x| \leqq {}^{-}1$ (e) $|11 + x| > 7$
(c) $|(7 - 3) + ({}^{-}2) - x| > {}^{-}1$

Show the solution set on the integer line.

5. Find all integers x for which

(a) ${}^{-}|10 - 3 + x| \leqq 3$ (d) ${}^{-}3 + |{}^{-}7| + |x + 1| > 2$
(b) ${}^{-}|{}^{-}x + 3 - 2| \leqq 4$ (e) $|4 - 3||x + 1| < {}^{-}1$
(c) ${}^{-}|{}^{-}6| + |2 - x| \leqq 2$

Show the solution set on the integer line.

6. Find all integers x for which

(a) ${}^{-}|{}^{-}({}^{-}5) + ({}^{-}2) + x| \geqq 1$ (d) $|{}^{-}9| + |4 + x| \leqq 3$
(b) $|{}^{-}3 - x| - 7 < 2$ (e) $|x - 3| + 11 \leqq 4$
(c) $|6 + ({}^{-}8)| + |x| \leqq 7 + |{}^{-}1|$

Show the solution set on the integer line.

7. Find all integers x for which

(a) $|4x - 12| = 0$ (c) $|7x + 49| = 0$
(b) $|11 - x| = 7 + 2 - 11$ (d) $|121 - 11x| = 0$

Show the solution set on the integer line.

8. Find all integers x for which

(a) $|2x^2 + x - 2x + 3 - 3 + 2 - 2x^2| \leqq 7$
(b) $|3x + 7 - 11 + 2x + 8 - 9 + 0 - 4x| \geqq {}^{-}1$

Show the solution set on the integer line.

9. Prove that for any signed integer a

$$^{-}|a| \leqq a \leqq |a|$$

10. Use the result in problem 9 to prove the *reverse triangular inequality,* namely, $|a - b| \geqq ||a| - |b||$, for all integers a and b.

11. Use numerical examples to illustrate $|a + b| \leqq |a| + |b|$ for integers a and b.
    (a) Under what conditions does the equality hold?
    (b) Under what conditions does the strict inequality hold?

12. Use numerical examples to illustrate $|a - b| \geqq ||a| - |b||$, for integers a and b.
    (a) Under what conditions does the equality hold?
    (b) Under what conditions does the strict inequality hold?

13. Prove that for any choice of a and b

$$|a - b| = |b - a|$$

14. Prove Theorem 5.4.4.

15. Prove Case 4 of Theorem 5.4.5.

16. Fill in reasons for each step in equation (5.4.27).

5.5 PRIME INTEGERS; COMPOSITE INTEGERS

In this section, the fundamental concepts of *prime integer* and *composite integer* are introduced. It will be seen shortly that most of the practical applications involving integers deal with ideas involving a knowledge of what is meant by a prime integer.

The following definition is quite similar to one already discussed in Chapter 3.

DEFINITION 5.5.1

Let a and b, with $b \neq 0$, be two integers. We say that *a is divisible by b* or that *b divides a* if and only if there exists some integer c so that

$$a = b \cdot c \qquad (5.5.1)$$

If b divides a, b is called a *factor* of a.

Thus, for example, 6 is divisible by 3 since an integer, in this case 2, exists so that $6 = 3 \cdot 2$. Also, 7 is not divisible by 3 since there is *no integer* whose product with 3 gives 7.

The student should notice that division by zero is *not* defined. That is, we do not even talk about division by zero—it is impossible.

Let us now discuss the set J of integers and try to classify integers into various similar groups. The above definition leads us to a division or decomposition of the set J of integers into the following four nonempty disjoint subsets:

(I) *This set consists of the integer zero alone.*

Here, zero is the additive identity 0 of the set J of integers. Of course, zero has the property that when it is added to any integer a, the result is a.

Note that zero is divisible by every integer b, $b \neq 0$. This is so since

$$0 = b \cdot 0 \tag{5.5.2}$$

That is, we have found an integer, namely 0, so that b times 0 gives 0 no matter what b is. Thus, 0 is always divisible by any nonzero integer.

It turns out that 0 is the only integer with this property.

(II) *This set consists of the integers 1 and* $^-1$.

Here, one is the multiplicative identity of J and $^-1$ is its additive inverse. Of course, 1 has the property that

$$1 \cdot a = a \tag{5.5.3}$$

for any integer a.

Note that both 1 and $^-1$ have the property that they divide every integer a. This follows from equation (5.5.3).

For example, 1 divides 5 since $5 = 1 \cdot 5$. Also, $^-1$ divides 8 since $8 = (^-1) \cdot (^-8)$.

(III) *This set consists of the prime integers of* J.

Definition 5.5.2

Let a be an integer not equal to 0, 1, or $^-1$. This integer a is a *prime integer* or is *prime* if and only if the only divisors of a are a, ^-a, 1, $^-1$.

In other words, a is prime if and only if there *cannot* be found two integers b and c, neither of which is equal to 1 or $^-1$, so that $a = b \cdot c$.

For example, the only integers which divide 2 are 2, $^-2$, 1, $^-1$. Hence, 2 is prime.

Also, 6 is *not* prime since there is an integer besides 6, $^-6$, 1, $^-1$ (for example, 3), which divides 6.

(IV) *This set consists of the composite integers of* J.

Definition 5.5.3

An integer a is a *composite integer* or is *composite* if and only if a is neither zero, nor 1, nor $^-1$, nor prime.

In other words, an integer a is composite if and only if it is possible to find two integers b and c, with $b \neq 1, {}^-1$, $c \neq 1, {}^-1$, so that $a = b \cdot c$.

For example, 8 is composite since $8 = 4 \cdot 2$.

Thus, a is composite if and only if it is divisible by an integer besides itself and 1 or $^-1$.

Thus, we have divided the set J of integers into four subsets, each one being nonempty and each pair being disjoint, as shown in Fig. 5.9.

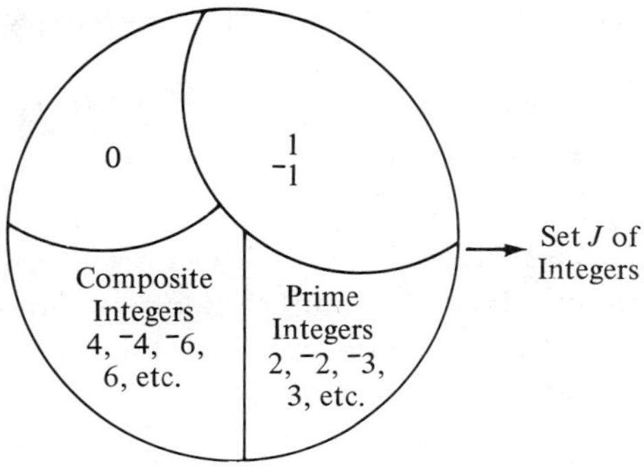

FIG. 5.9

Exercise 5.5.1

1. Write down the first 20 positive prime integers.
2. Write down the first 15 positive integers which are divisible by 3.
3. Do problem 2 for the integer 9.
4. If an integer a is divisible by 9, show that it is also divisible by 3.
5. If an integer a is divisible by 6, show that both 2 and 3 also divide a.
6. Write down all possible divisors of 8, of 12, of 13.
7. Two prime integers which differ by two, such as 5 and 7, 11 and 13, etc., are called *twin primes*. Find a pair of twin primes between

   (a) 25 and 35 (b) 55 and 65

   An unproved conjecture is that there are infinitely many twin primes.
8. Notice that 3, 5, 7 is a triplet of primes in the same sense as problem 7. Can you find another such triplet?
9. It has been proven that for $n \geqq 2$, there can be found a prime number between n and $2n$. Verify this for $n = 2, 3, 4$, and 20. From this fact, what do you think the cardinal number of the set of all prime numbers is?

We are now ready to discuss some further properties involving the concepts of prime and composite integers.

Henceforth, when we refer to prime integers, we shall mean the positive prime integers, namely, 2, 3, 5, 7, ···

The following theorem is of importance, but will be given without a formal proof.

Theorem 5.5.1

Unique Factorization Theorem: Any positive integer a, with $a \neq 0, 1, {}^{-}1$, can be written as the product of a finite number of prime integers. The product is unique, except for the order in which the factors appear.

For example, the integer 12 can be written as

$$12 = 2 \cdot 2 \cdot 3 \tag{5.5.4}$$

That is, we can express 12 as the product of the primes, 2, 2, and 3. The representation for 12 is unique, that is, there is no other way of expressing 12, except for changing the order of the factors, that is, to write 12 as $2 \cdot 3 \cdot 2$, for example.

This result for the set J of integers is important since there are some sets in which unique factorization is *not* possible.

Now, let us develop some methods or techniques for writing a positive integer as a product of a finite number of positive prime integers. There are two ways to write a positive integer as the product of prime integers.

(I) Test whether a prime integer is a factor of a by beginning with 2, and continuing until the integer has been written as a product of primes.

EXAMPLE 5.5.1

Factor 1080 into a product of prime integers. Using the above method, we try 2. Since 2 divides 1080, 2 is a factor of 1080 and

$$1080 = 2 \cdot 540 \tag{5.5.5}$$

Continue now with 540. Try 2. Then

$$1080 = 2 \cdot 2 \cdot 270 \tag{5.5.6}$$

With 270, we have that $270 = 2 \cdot 135$. Thus,

$$1080 = 2 \cdot 2 \cdot 2 \cdot 135 \tag{5.5.7}$$

Now with 135, try 2. Since 2 does not divide 135, we try the next prime, namely 3. Here $135 = 3 \cdot 45$. Thus,

$$1080 = 2 \cdot 2 \cdot 2 \cdot 3 \cdot 45 \tag{5.5.8}$$

With 45, we have $45 = 3 \cdot 15$. Thus,

$$1080 = 2 \cdot 2 \cdot 2 \cdot 3 \cdot 3 \cdot 15 \tag{5.5.9}$$

Finally, $15 = 3 \cdot 5$. Thus,

$$1080 = 2 \cdot 2 \cdot 2 \cdot 3 \cdot 3 \cdot 3 \cdot 5 \tag{5.5.10}$$

and so has been factored into a product of prime integers.

(II) The second method is to factor the given integer in any way possible and then to write each of these factors as a product of primes.

EXAMPLE 5.5.2

Write 1080 as a product of primes. Here we notice that 10 is a factor of 1080 so that

$$1080 = 108 \cdot 10 \tag{5.5.11}$$

Also, $108 = 9 \cdot 12$. Thus,

$$1080 = 9 \cdot 12 \cdot 10 \tag{5.5.12}$$

We have not factored 1080 into a product of primes, but we have factored it into integers which we can easily factor into products of primes. Thus, $9 = 3 \cdot 3$, $12 = 3 \cdot 2 \cdot 2$, $10 = 2 \cdot 5$. Then,

$$1080 = 3 \cdot 3 \cdot 2 \cdot 2 \cdot 3 \cdot 2 \cdot 5 \tag{5.5.13}$$

and we have factored 1080 into a product of primes.

We illustrate this method by using a "tree-diagram" in which the product of all numbers in a row equals the original number. See Fig. 5.10. *Note:* The second method will usually be faster than the first, provided we can easily recognize factors other than primes for the original number.

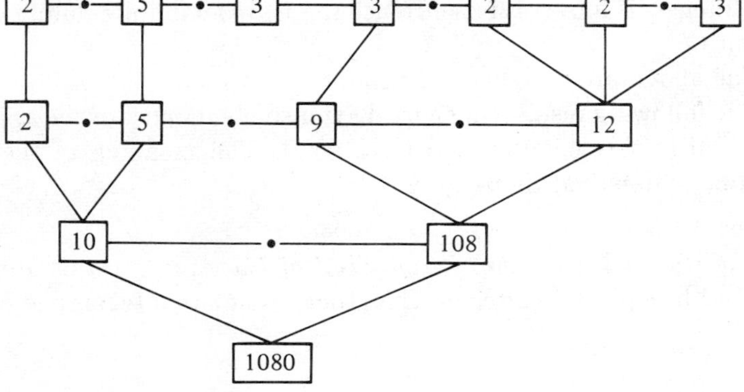

FIG. 5.10

DEFINITION 5.5.4

Let a be a positive integer, not 1. To *factor a completely* means to write a as a product of prime integers.

Note that the above definition is made possible because of the Unique Factorization Theorem. Without this theorem, the above definition could not be made.

Theorem 5.5.2

The set S consisting of all positive prime integers is infinite.

PROOF: Here we assume the contrary. That is, suppose that the set S is finite, not infinite. Then, this set S must have a largest element. Let p denote the largest prime integer. Suppose m is the product of all the prime integers in S.

Let $n = m + 1$. Now every positive integer larger than one is divisible by a prime integer. Thus, n is divisible by a prime integer, say q, and of course, q is in S. Since m is the product of all primes in S, then m is divisible by q, also. Thus, we have

$$n = q \cdot A, \qquad m = q \cdot B \tag{5.5.14}$$

for some integers A and B.

But $n = m + 1$. Thus,

$$q \cdot A = q \cdot B + 1 \tag{5.5.15}$$

or

$$q \cdot (A - B) = 1 \tag{5.5.16}$$

This says that 1 is divisible by q. This is impossible since no prime divides one—only 1 and $^-1$ divide 1.

This contradiction tells us the original assumption (that the set S of prime integers is finite) is false. Hence there are an infinite number of prime integers.

The above proof is due to Euclid.

The following result will be used extensively in most of the applications considered in this chapter and Chapter 9. The meaning of the theorem should be understood thoroughly.

Theorem 5.5.3

The Division Algorithm for the Set J of Integers: Let a be any integer. Let $b \geqq 1$ be a positive integer. Then there exist two integers q and r so that

$$a = b \cdot q + r \tag{5.5.17}$$

with

$$0 \leqq r < b \tag{5.5.18}$$

Moreover, the integers q and r are unique.

In the above result, a is called the *dividend*, b is the *divisor*, q is the *quotient*, and r is the *remainder*. Notice that the remainder r obtained upon dividing by b is always less than b, but greater than or equal to 0.

Of course, if upon dividing a by b, the remainder $r = 0$, then b divides a.

EXAMPLE 5.5.3

Find the quotient q and remainder r upon dividing 575 by 21. Here if we divide 575 by 21, we obtain

$$\begin{array}{r} 27 \\ 21\overline{)575} \\ \underline{42} \\ 155 \\ \underline{147} \\ 8 \end{array} \tag{5.5.19}$$

Thus,

$$575 = 21 \cdot 27 + 8 \tag{5.5.20}$$

The quotient is 27 and the remainder is 8.

The division algorithm can be explained as follows. Suppose we are *given* two integers a and b, where b is a positive integer. The division algorithm tells us that we can then *find* two other integers q and r where $r \geqq 0$ and $r < b$ so that when we multiply b by q and then add r, we obtain a.

It also tells us that the integers q and r that we find are unique. It is noted that there are many choices for q and r which make $q \cdot b + r = a$. However, there is only *one* choice of q and r, where $0 \leqq r < b$, that makes $q \cdot b + r = a$. It is the restriction on the remainder r which gives strength to the theorem.

Next, we have the following definitions.

DEFINITION 5.5.5

Let a and b be two positive integers. *The greatest common divisor of a and b*, abbreviated by

$$\gcd \{a, b\} \tag{5.5.21}$$

is defined to be the *largest* positive integer which *divides both a and b*.

EXAMPLE 5.5.4

Find the gcd {15, 18}. First of all, let us write down all the positive divisors of 15 and 18. The divisors of 15 are members of the set

$$\{1, 3, 5, 15\} \tag{5.5.22}$$

The divisors of 18 are members of the set

$$\{1, 2, 3, 6, 9, 18\} \tag{5.5.23}$$

The common divisors of 15 and 18 is the intersection of the two sets given above in (5.5.22) and (5.5.23). Then, the set of common divisors of 15 and 18 is

$$\{1, 3\} \tag{5.5.24}$$

Clearly, 3 is the largest member of this set. Thus, gcd $\{15, 18\} = 3$.

Note that if we factor the integers 15 and 18 into a product of prime integers so that

$$15 = 3 \cdot 5$$

$$18 = 2 \cdot 3 \cdot 3 \tag{5.5.25}$$

it is relatively easy to see that 3 is the largest integer dividing both 15 and 18, and so 3 is the gcd $\{15, 18\}$.

DEFINITION 5.5.6

Let a and b be two positive integers. If

$$\gcd \{a, b\} = 1 \tag{5.5.26}$$

then a and b are said to be *relatively prime integers*.

EXAMPLE 5.5.5

Find gcd $\{15, 28\}$. Here if we factor 15 and 28 completely, we obtain

$$15 = 3 \cdot 5$$

$$28 = 2 \cdot 2 \cdot 7 \tag{5.5.27}$$

Since 1 divides every integer, and since 15 and 28 possess no common prime factors, it follows that

$$\gcd \{15, 28\} = 1 \tag{5.5.28}$$

Hence, 15 and 28 are relatively prime integers.

DEFINITION 5.5.7

Let a and b be two positive integers. The *least common multiple of a and b*, abbreviated by

$$\text{lcm} \{a, b\} \tag{5.5.29}$$

is defined as the *smallest* positive integer *divisible by both a and b*.

EXAMPLE 5.5.6

Find lcm $\{15, 18\}$. Some of the integers divisible by 15 are

$$15, 30, 45, 60, 75, 90, 105, \cdots \tag{5.5.30}$$

Some of the integers divisible by 18 are

$$18, 36, 54, 72, 90, 108, \cdots \qquad (5.5.31)$$

The smallest positive integer divisible by both 15 and 18 is 90.
Thus,

$$\text{lcm}\ \{15, 18\} = 90 \qquad (5.5.32)$$

Again if we factor 15 and 18, we see

$$15 = 3 \cdot 5 \quad \text{and} \quad 18 = 2 \cdot 3 \cdot 3. \qquad (5.5.33)$$

Notice that if the one common 3 is crossed out, the product of the remaining prime integers 5, 2, 3, 3 is 90. This gives us a technique for finding the lcm of two integers.

EXAMPLE 5.5.7

Find the greatest common divisor and the least common multiple of 16 and 12.

Here, factoring these completely, we have

$$12 = 2 \cdot 2 \cdot 3 \quad \text{and} \quad 16 = 2 \cdot 2 \cdot 2 \cdot 2 \qquad (5.5.34)$$

Thus,

$$\text{gcd}\ \{12, 16\} = 2 \cdot 2 = 4$$

$$\text{lcm}\ \{12, 16\} = 3 \cdot 2 \cdot 2 \cdot 2 \cdot 2 = 48 \qquad (5.5.35)$$

In the above example, the lcm {12, 16} was obtained by eliminating the pair of 2's which leaves 3, 2, 2, 2, 2, and then finding their product.

EXAMPLE 5.5.8

Mike can throw a football exactly 36 yards and Danny can throw a football exactly 30 yards. If they start to throw their football from the same spot, what is the minimum number of throws for each of them so that the footballs will have been thrown the same distance? What is this distance?

Each time Mike throws the football, it travels a multiple of 36 yards and Danny's football travels a multiple of 30 yards. Figure 5.11 illustrates what happens.

Throws:	1st	2nd	3rd	4th	5th	6th	
Mike	36	72	108	144	180		
Danny	30	60	90	120	150	180	

FIG. 5.11

Thus, 180 yards is the distance covered in five throws by Mike and six throws by Danny.

But the shortest distance is $180 = \text{lcm}\,\{36, 30\}$. Also, the most throws needed is $6 = \text{gcd}\,\{36, 30\}$.

There is a third way to find the greatest common divisor of two integers a and b. This technique is called the Euclidean algorithm and is described below.

Consider the integers 1080 and 132. To find the gcd $\{1080, 132\}$, proceed as follows. Apply the division algorithm with 1080 as dividend and 132 as divisor. Then we have

$$1080 = 132 \cdot 8 + 24 \tag{5.5.36}$$

Repeat the above, using the divisor 132 and the remainder 24. Then

$$132 = 24 \cdot 5 + 12 \tag{5.5.37}$$

Again repeat the above, using the divisor 24 and the remainder 12. Then

$$24 = 12 \cdot 2 + 0 \tag{5.5.38}$$

Once a remainder of 0 is obtained, the divisor, here 12, is the gcd$\{1080, 132\}$.

Once the gcd of two integers a and b has been found, the lcm $\{a, b\}$ can be obtained by using the fact that

$$\text{lcm}\,\{a, b\} = \frac{a \cdot b}{\text{gcd}\,\{a \cdot b\}} \tag{5.5.39}$$

Thus, if $a = 1080$, $b = 132$, we know that

$$\text{gcd}\,\{1080, 132\} = 12 \tag{5.5.40}$$

Then,

$$\text{lcm}\,\{1080, 132\} = \frac{1080 \cdot 132}{12} = 1080 \cdot 11 = 11{,}880. \tag{5.5.41}$$

It is easy to see the pattern of the Euclidean algorithm. Look at Fig. 5.12.

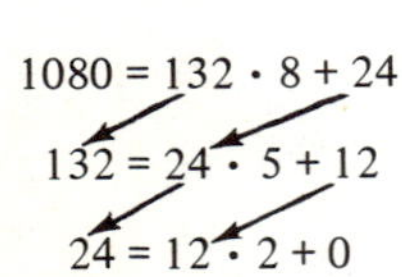

FIG. 5.12

This pattern is continued until a remainder of 0 is obtained. When this happens, the divisor appearing in this step is the gcd of the original two integers.

Thus, we have discussed *three* techniques for finding the greatest common divisor and least common multiple of two integers a and b. These three methods are reviewed in the following example.

EXAMPLE 5.5.9

Find gcd $\{24, 40\}$ and lcm $\{24, 40\}$.

METHOD I: *Definition Technique.* To find the gcd $\{24, 40\}$, we write down the set of all positive integers which divide both 24 and 40. Thus we obtain the two sets

$$\begin{aligned} &\{1, 2, 3, 4, 6, 8, 12, 24\} \text{ for } 24 \\ &\{1, 2, 4, 5, 8, 10, 20, 40\} \text{ for } 40 \end{aligned} \tag{5.5.42}$$

Those integers dividing both 24 and 40 are in the intersection of these two sets. Thus,

$$\{1, 2, 3, 4, 6, 8, 12, 24\} \cap \{1, 2, 4, 5, 8, 10, 20, 40\} = \{1, 2, 4, 8\} \tag{5.5.43}$$

The largest element in this last set is 8. Thus,

$$8 = \gcd \{24, 40\}. \tag{5.5.44}$$

To find the lcm $\{24, 40\}$, we write down the set of all positive integers divisible by both 24 and 40. Then we obtain

$$\begin{aligned} &\{24, 48, 72, 96, 120, 144, \cdots\} \text{ for } 24 \\ &\{40, 80, 120, 160, 200, \cdots\} \text{ for } 40 \end{aligned} \tag{5.5.45}$$

Those integers divisible by both 24 and 40 are in the intersection of these two sets. This is the set

$$\{120, 240, \cdots\} \tag{5.5.46}$$

The smallest element of this set is 120. Thus,

$$\text{lcm } \{24, 40\} = 120 \tag{5.5.47}$$

METHOD II: *Factoring Technique.* Here we factor both 24 and 40 completely. Then we obtain

$$\begin{aligned} 24 &= 2 \cdot 2 \cdot 2 \cdot 3 \\ 40 &= 2 \cdot 2 \cdot 2 \cdot 5 \end{aligned} \tag{5.5.48}$$

The common factors are $2 \cdot 2 \cdot 2 = 8$. Thus

$$8 = \gcd \{24, 40\} \tag{5.5.49}$$

Now, eliminating $2 \cdot 2 \cdot 2$ from the list of factors of 24 and 40 and finding the product of those remaining, we obtain

$$\operatorname{lcm} \{24, 40\} = 3 \cdot 2 \cdot 2 \cdot 2 \cdot 5 = 120 \tag{5.5.50}$$

METHOD III: *Euclidean Algorithm Technique.* Here we divide 40 by 24. Then

$$40 = 24 \cdot 1 + 16 \tag{5.5.51}$$

Continue the process:

$$24 = 16 \cdot 1 + 8$$

$$16 = 8 \cdot 2 + 0 \tag{5.5.52}$$

Thus

$$\gcd \{40, 24\} = 8 \tag{5.5.53}$$

To obtain lcm {24, 40}, we use (5.5.39):

$$\operatorname{lcm} \{24, 40\} = \frac{24 \cdot 40}{\gcd \{24, 40\}} = \frac{24 \cdot 40}{8} = 3 \cdot 40 = 120 \tag{5.5.54}$$

In some types of problems, one of these methods may prove superior to the others. However, it is only by knowing all three that one can take advantage of a shorter method.

Exercise 5.5.2

1. Fill in the blanks of the following "tree diagrams."

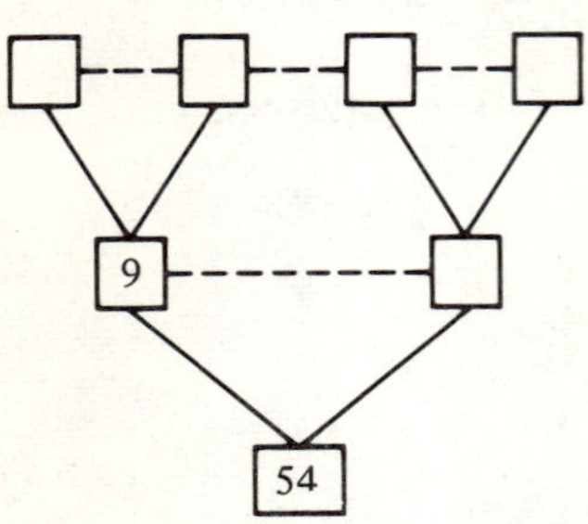

(a)

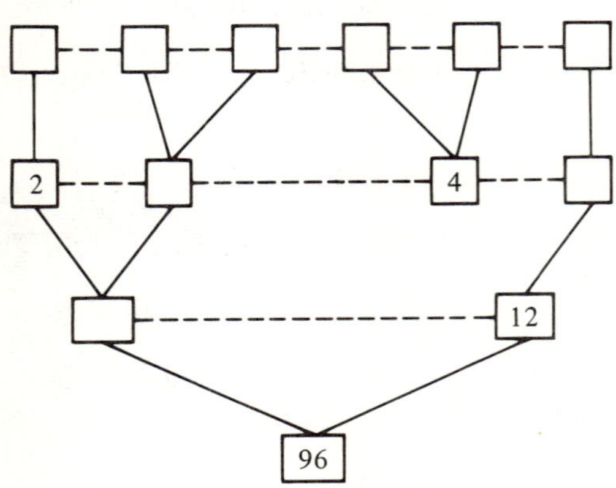

(b)

2. Write the following integers as a product of positive prime integers, using two different techniques.

(a)	24	(f)	72
(b)	18	(g)	90
(c)	35	(h)	152
(d)	42	(i)	19
(e)	65	(j)	28

3. Write the following integers as a product of positive prime integers, beginning with the smallest positive prime and ending with the largest positive prime.

(a)	48	(f)	1220
(b)	172	(g)	1480
(c)	156	(h)	760
(d)	2080	(i)	432
(e)	5842	(j)	648

4. Using the division algorithm, find the quotient q and remainder for the following.

	dividend	divisor		dividend	divisor
(a)	42	26	(g)	15	72
(b)	73	4	(h)	142	30
(c)	29	31	(i)	23	23
(d)	-23	4	(j)	550	25
(e)	471	22	(k)	0	3
(f)	-30	14			

5. Find the greatest common divisor and least common multiple of the following pairs of integers by using the factoring technique.

(a)	14, 35	(f)	1001, 7655
(b)	11, 15	(g)	142, 760
(c)	2873, 6643	(h)	1480, 648
(d)	4148, 7684	(i)	2080, 1220
(e)	180, 252	(j)	848, 152

6. Find the greatest common divisor and least common multiple of the pairs of integers of problem 5 by using the Euclidean algorithm technique.

7. Find the greatest common divisor and least common multiple of the pairs of integers of problem 5 by using the definition technique.

8. Find the greatest common divisor of

   (a) 198, 144, and 68
   (b) 363, 36, 42
   (c) 2080, 5642, 210
   (d) 66, 432, 648
   (e) 48, 24, 18, 28

Hint: The gcd $\{a, b, c\}$ is the largest integer that divides a, b, and c.

9. Find the least common multiple of the integers in problem 8.

10. Let a and b be two positive integers.

   (a) If gcd $\{a, b\} = 1$, find lcm $\{a, b\}$.
   (b) If gcd $\{a, b\} = b$, find lcm $\{a, b\}$.
   (c) If gcd $\{a, b\} = 5$ and lcm $\{a, b\} = 660$, find $a \cdot b$. What is a?
   (d) If gcd $\{a, b\} =$ lcm $\{a, b\}$, what if anything can be said about a and b.
   (e) The product of a and b is 480 and gcd $\{a, b\} = 12$. Find lcm $\{a, b\}$.

11. Mike can run around an oval track in 90 seconds and Dan can run around the same track in 132 seconds. If both of them start together, how long will it be before they cross the finish line together? How many laps will each of them have run? [HINT: Look at gcd $\{90, 132\}$ and lcm $\{90, 132\}$].

12. With a driver, a golfer can hit a golf ball 250 yards. With a seven iron, he can hit the ball 160 yards. Find the least number of strokes with the driver and the least number of strokes with the iron needed so that the ball is advanced the *same* distance.

13. A carpenter has two boards, one 72 inches long, the other 90 inches long. He wants to cut both boards into a number of pieces of equal lengths. What is the largest length he can obtain?

5.6 EVEN AND ODD INTEGERS

Suppose we are given an integer a and suppose we apply the division algorithm to a in which the divisor is 2. Then, using the theorem, we know that there are two unique integers q and r so that

$$a = 2 \cdot q + r \tag{5.6.1}$$

where

$$0 \leqq r < 2 \tag{5.6.2}$$

Examining (5.6.2), it is seen that the only possibilities for the remainder r are

$$r = 0 \quad \text{or} \quad r = 1 \tag{5.6.3}$$

Thus, whenever an integer is divided by 2, the remainder must be either 0 or 1.

This fact leads us to the following definition.

DEFINITION 5.6.1

An integer a is said to be an *even integer* if the remainder obtained upon dividing a by 2 is zero.

An integer a is said to be an *odd integer* if the remainder obtained upon dividing a by 2 is one.

For example, 5 is odd since when 5 is divided by 2, we obtain

$$5 = 2 \cdot 2 + 1 \tag{5.6.4}$$

a remainder of one.

Also, 8 is even since

$$8 = 2 \cdot 4 + 0 \tag{5.6.5}$$

Theorem 5.6.1

An integer a is even if and only if there can be found an integer k such that

$$a = 2 \cdot k \tag{5.6.6}$$

Similarly, an integer b is odd if and only if there can be found some integer p such that

$$b = 2 \cdot p + 1 \tag{5.6.7}$$

PROOF: This is an immediate consequence of the division algorithm.

Thus, any integer that can be written as twice some integer plus one is odd. If it can be written as twice some integer, it is even.

Theorem 5.6.2

(I) If a and b are two even integers, their product $a \cdot b$ is even.

(II) If a is an even integer and b is an odd integer, their product $a \cdot b$ is an odd integer.

(III) If a and b are both odd, their product $a \cdot b$ is an odd integer.

PROOF: (I) If a and b are even, then by Theorem 5.6.1 we can write a and b as

$$a = 2k, \qquad b = 2p \tag{5.6.8}$$

where k and p are some two integers. Then their product $a \cdot b$ is

$$a \cdot b = (2k) \cdot (2p) = 2 \cdot (2kp) \tag{5.6.9}$$

Since $2 \cdot k \cdot p$ is some integer, then $a \cdot b$ is twice some integer. Hence, $a \cdot b$ must be even.

The other two parts of Theorem 5.6.2 are left to the student.

Theorem 5.6.3

(I) If $a^2 = a \cdot a$ is even, then a itself must be even.
(II) If $a^2 = a \cdot a$ is odd, then a itself is odd.

PROOF: (I) Assume the contrary. That is, suppose a is not even. Then, a must be odd. By Theorem 5.6.2, the product of two odd integers is an odd integer. Hence, $a \cdot a$ is odd. But we know $a \cdot a$ is even. This is impossible. Thus, a is even.

(II) This is left to the reader.

Exercise 5.6

1. Are the following integers even or odd? Why?

(a)	3	(e)	$^-17$
(b)	$^-2$	(f)	$^-1$
(c)	0	(g)	1
(d)	16	(h)	$^-8$

2. Prove (II) of Theorem 5.6.2.

3. Prove (III) of Theorem 5.6.2.

4. Prove (II) of Theorem 5.6.3.

5. Are there any positive prime even integers larger than 2? Why?

6. A famous unsolved problem in mathematics is *Goldbach's conjecture,* which states that every even integer greater than 2 is expressible as the sum of two prime integers. Show that this conjecture is valid for each even integer from 4 through 30 inclusive.

7. Using the unique factorization theorem, we know that $12 = 2 \cdot 2 \cdot 3$. Notice that 12 has three prime factors. Thus, 12, an even integer, has an odd number (three) as one of the prime factors. However, $6 = 2 \cdot 3$ is even and has an even number of prime factors. Find an odd integer that has an even number of prime factors and one which has an odd number as a prime factor.

8. A *perfect number* is one that has the property that the sum of all positive integers which divide it, excluding the number itself,

equal the number itself. Which of the following are perfect numbers?

(a) 28 (c) 14
(b) 6 (d) 32

9. Pick any odd number (greater than one) and multiply it by itself and then subtract 1. Is the result divisible by 8? Make a conjecture. Try to prove it. [HINT: $(2k + 1)(2k + 1) - 1 = 4k(k + 1)$.]

10. In problem 8, Exercise 5.5.1, the question of triplets of primes was raised. Show that given any three consecutive odd numbers (say 17, 19, 21), one of them must be divisible by 3—thus proving that 3, 5, 7 is the *only* triplet of primes. [HINT: If the first is not divisible by 3, what is the remainder?]

6 The Set R of Rational Numbers

6.1 INTRODUCTION

In the last two chapters, the set J of integers

$$J = \{\cdots, {}^-3, {}^-2, {}^-1, 0, 1, 2, 3, \cdots\} \tag{6.1.1}$$

was introduced and various properties of the integers were discussed. It was noted, for example, that the set W of whole numbers

$$W = \{0, 1, 2, 3, \cdots\} \tag{6.1.2}$$

was a proper subset of J and that the integers possessed certain desirable properties which were lacking for the set W of whole numbers.

It became apparent as we studied the integers that they too have certain deficiencies. For example, suppose we want to measure the length of a fence with a stick and we find the length of the fence is more than 51 sticks and less than 52 sticks. What should we do? Without introducing *new* numbers, this situation is impossible to handle.

As everyone knows, tomatoes are sold in stores by the pound. Does this mean that Mrs. Murphy must buy 1 pound, or 2 pounds, or 3 pounds, etc., of tomatoes? If we only have integers with which to measure, the answer is yes.

In this chapter, the set J is extended to form a new set of numbers, for which such deficiencies no longer exist. This *new* set of numbers is called the *set R of rational numbers*. It will be seen that the set J of integers is a proper subset of R so that

$$J \subset R \tag{6.1.3}$$

6.2 DEFINITION OF A RATIONAL NUMBER

Suppose we have a pie which is cut into eight equal pieces. It is clear that one piece of pie represents one part of the whole pie (eight pieces). Three pieces of pie represent three parts of the whole pie (eight pieces). The word *ratio* is used to denote the relationship that exists between parts of the whole and the whole. We denote the ratio of 1 to 8, such as, one piece of a pie cut into eight equal pieces, by the symbol $\frac{1}{8}$. In general, if an object is divided into B equal parts, then A parts of this object of B parts is the *ratio of A to B*, denoted by $\frac{A}{B}$.

For example, if a plot of land is divided into ten equal parts, then one of these parts represents $\frac{1}{10}$ of the original plot. See Fig. 6.1.

Definition 6.2.1

A *fraction* is a name for a ratio $\frac{A}{B}$, in which A and $B \neq 0$ are integers.

Note: The restriction that B cannot be zero is discussed later.

For a fraction $\frac{A}{B}$, the integer A is called the *numerator*, and B is called the *denominator*.

Let us return to the plot of land in Fig. 6.1. Suppose we first divide this plot of land into two equal parts. Then each part can be represented by the fraction $\frac{1}{2}$. If we had divided this plot into four equal parts and also

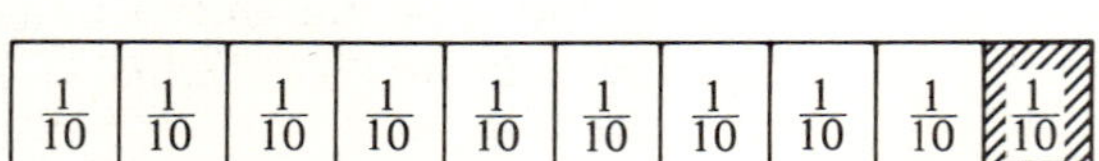

FIG. 6.1

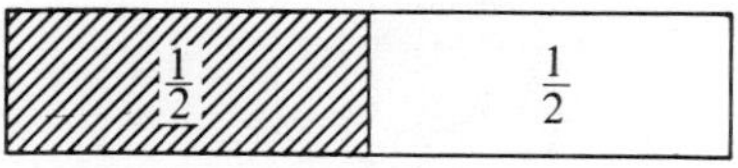

FIG. 6.2

into ten equal parts, we would obtain Figs. 6.3 and 6.4. Notice that the

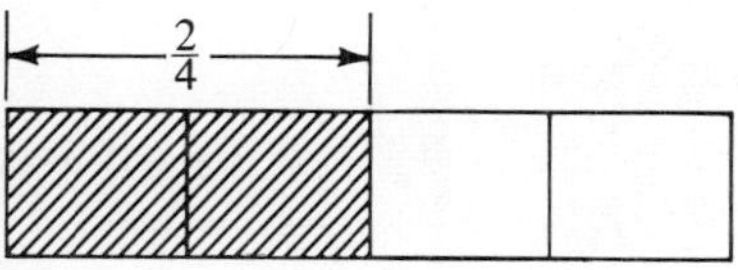

FIG. 6.3

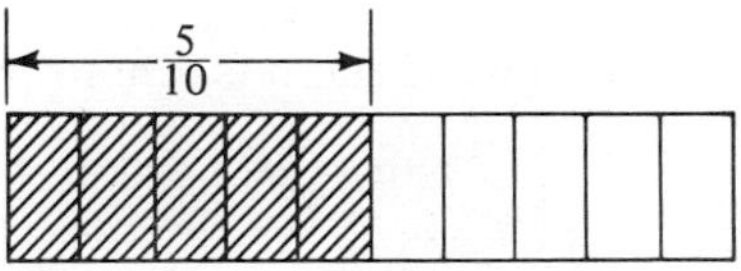

FIG. 6.4

fraction $\frac{1}{2}$ (see Fig. 6.2), the fraction $\frac{2}{4}$ (see Fig. 6.3), and the fraction $\frac{5}{10}$ (see Fig. 6.4) all name the same ratio, that is, all name the *same* part of the whole. In other words, we have found three seemingly different fractions which are, in reality, the same. That is, the fractions $\frac{1}{2}, \frac{2}{4}, \frac{5}{10}$ are *equal.*

Let us consider another example. Suppose Mrs. Murphy wants to buy some cans of peaches. The peaches are marked 3 cans for 42¢. Her cost for one can is 14¢. Her cost for two cans is 28¢. Her cost for 6 cans is 84¢. In general, the cost for n cans is $(14n)$ ¢. We can display these as the fractions

$$\frac{1}{14}, \frac{2}{28}, \frac{3}{42}, \ldots, \frac{6}{84}, \ldots, \frac{n}{14n}, \ldots \tag{6.2.1}$$

To convince yourself that each of these fractions represents the *same* ratio, consider Fig. 6.5.

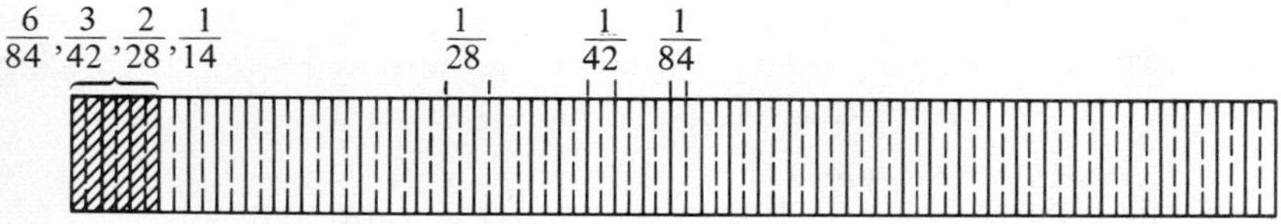

FIG. 6.5

We are thus led to make the following definition.

Definition 6.2.2

Let $\frac{a}{b}$, $\frac{c}{d}$ be two fractions. We shall say that $\frac{a}{b}$ *equals* $\frac{c}{d}$, written as

$$\frac{a}{b} = \frac{c}{d} \tag{6.2.2}$$

if and only if there is a nonzero integer n such that

$$n \cdot a = c, \qquad n \cdot b = d \tag{6.2.3}$$

For example,

$$\frac{2}{3} = \frac{8}{12} \tag{6.2.4}$$

since $2 \cdot 4 = 8$ and $3 \cdot 4 = 12$. Also,

$$\frac{4}{9} \neq \frac{1}{2} \tag{6.2.5}$$

since $1 \cdot 4 = 4$ and $2 \cdot 4 = 8 \neq 9$.

Of course, if $\frac{a}{b} = \frac{c}{d}$ and if $\frac{a}{b} = \frac{e}{f}$, then we will agree that $\frac{c}{d} = \frac{e}{f}$. Thus, for example,

$$\frac{2}{28} = \frac{3}{42} \tag{6.2.6}$$

since

$$\frac{2}{28} = \frac{1}{14} \quad \text{and} \quad \frac{3}{42} = \frac{1}{14} \tag{6.2.7}$$

Let us return to our physical example of a plot of land divided into ten equal parts. Each one of these parts can be represented by $\frac{1}{10}$. If we have five of these parts $\left(5 \cdot \frac{1}{10}\right)$, this quantity can be represented by $\frac{5}{10}$. Also, if we have three of these parts $\left(3 \cdot \frac{1}{10}\right)$, this amount is $\frac{3}{10}$. The question is asked, "If we have five parts and three parts, how much do we have all told?" It is clear that the answer is $\frac{8}{10}$ or $8 \cdot \frac{1}{10}$. In other words, we have an operation of addition of fractions. That is,

$$\frac{5}{10} + \frac{3}{10} = 5 \cdot \frac{1}{10} + 3 \cdot \frac{1}{10} = (5 + 3) \cdot \frac{1}{10} = 8 \cdot \frac{1}{10} = \frac{8}{10} \qquad (6.2.8)$$

Suppose we had instead wanted to find the sum of $\frac{3}{4}$ and $\frac{4}{5}$. In this case, we know that

$$\frac{3}{4} = \frac{15}{20}, \qquad \frac{4}{5} = \frac{16}{20} \qquad (6.2.9)$$

Thus,

$$\frac{3}{4} + \frac{4}{5} = \frac{3 \cdot 5}{4 \cdot 5} + \frac{4 \cdot 4}{5 \cdot 4} = \frac{15}{20} + \frac{16}{20} = \frac{31}{20} \qquad (6.2.10)$$

Figure 6.6 gives a physical interpretation of this.

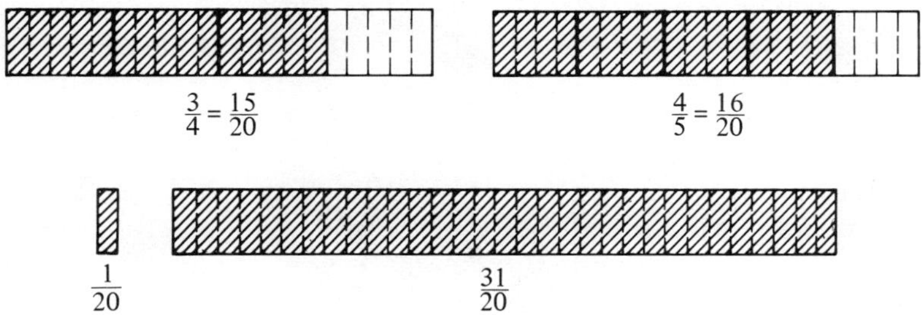

FIG. 6.6

Now we are ready to define the *sum* of two ratios or fractions.

Definition 6.2.3

If $\frac{a}{b}$ and $\frac{c}{d}$ are two fractions, their *sum* $\frac{a}{b} + \frac{c}{d}$ is the fraction

$$\frac{a}{b} + \frac{c}{d} = \frac{a \cdot d + c \cdot b}{b \cdot d} \qquad (6.2.11)$$

Notice that the definition makes sense since

$$\frac{a}{b} + \frac{c}{d} = \frac{a \cdot d}{b \cdot d} + \frac{c \cdot b}{b \cdot d} = \frac{a \cdot d + c \cdot b}{b \cdot d} \tag{6.2.12}$$

Note: Even though the physical examples given were for $a, b, c, d,$ positive integers, Definition 6.2.3 is valid for *all* integers with $b \neq 0$, $d \neq 0$.

EXAMPLE 6.2.1

Find the sum of $\frac{7}{4}$ and $\frac{^-8}{5}$. Here

$$\frac{7}{4} + \frac{^-8}{5} = \frac{7 \cdot 5}{4 \cdot 5} + \frac{(^-8) \cdot 4}{5 \cdot 4}$$

$$= \frac{35}{20} + \frac{(^-32)}{20} = \frac{35 + (^-32)}{20} = \frac{3}{20} \tag{6.2.13}$$

The addition of two fractions contains an application of the concept of least common multiple of two integers (see Chapter Five). Suppose it is desired to add two fractions $\frac{a}{b}$ and $\frac{c}{d}$. We know their sum is some third fraction. That is,

$$\frac{a}{b} + \frac{c}{d} = \frac{?}{?} \tag{6.2.14}$$

Instead of using $b \cdot d$ for the denominator of the sum (as Definition 6.2.3 indicates), suppose instead we use the smallest positive integer divisible by both b and d, that is, use lcm $\{b, d\}$. If $L = \text{lcm}\ \{b, d\}$, we know that integers x and y exist so that

$$L = x \cdot b, \qquad L = y \cdot d \tag{6.2.15}$$

Then

$$\frac{a}{b} + \frac{c}{d} = \frac{a \cdot x}{b \cdot x} + \frac{c \cdot y}{d \cdot y} = \frac{ax}{L} + \frac{cy}{L} = \frac{ax + cy}{L} \tag{6.2.16}$$

In such applications, $L = \text{lcm}\ \{b, c\}$ is usually called the *least common denominator* of $\frac{a}{b}$ and $\frac{c}{d}$.

EXAMPLE 6.2.2

Find

$$\frac{5}{12} + \frac{7}{18} \tag{6.2.17}$$

Here lcm $\{12, 18\} = 36$. Thus,

$$\frac{5}{12} + \frac{7}{18} = \frac{5 \cdot 3}{12 \cdot 3} + \frac{7 \cdot 2}{18 \cdot 2} = \frac{15}{36} + \frac{14}{36} = \frac{29}{36} \tag{6.2.18}$$

Had we not used this technique, we would obtain

$$\frac{5}{12} + \frac{7}{18} = \frac{5 \cdot 18 + 7 \cdot 12}{12 \cdot 18} = \frac{90 + 84}{216} = \frac{174}{216} \tag{6.2.19}$$

Of course, $\frac{174}{216}$ and $\frac{29}{36}$ are equal fractions so that either technique gives the same answer. The importance of using the least common denominator to add fractions will be seen at the end of this section.

Let us again return to our physical example of a plot of land. Suppose we divide this plot into three equal parts. Consider two parts or $\frac{2}{3}$. The question is asked, "If we want $\frac{5}{7}$ of $\frac{2}{3}$ of the plot, how much land do we have?" To see how we might obtain the answer, let us look at Fig. 6.7.

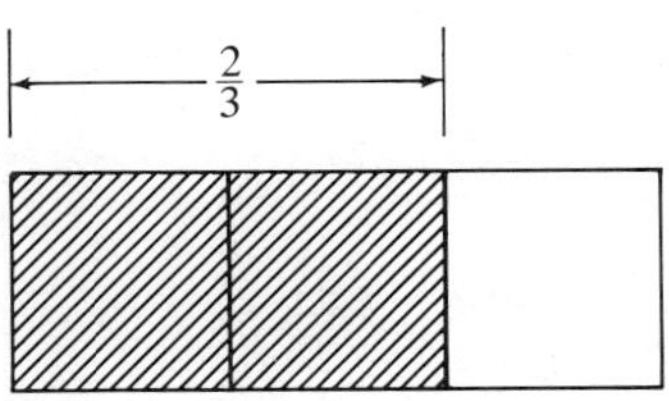

FIG. 6.7

Let us now divide the hatched region above into seven parts.

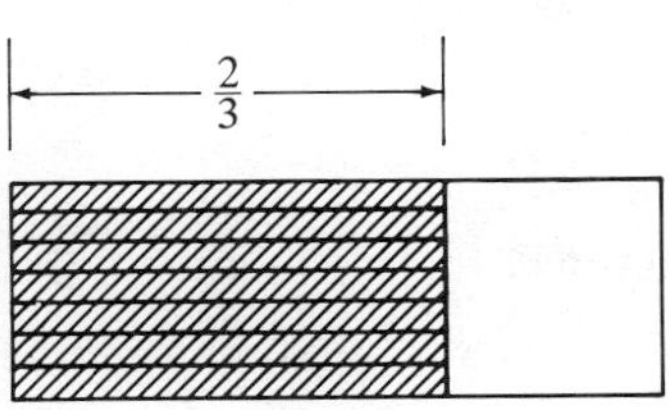

FIG. 6.8

Now, $\frac{5}{7}$ of this is depicted by the checkered region in Fig. 6.9.

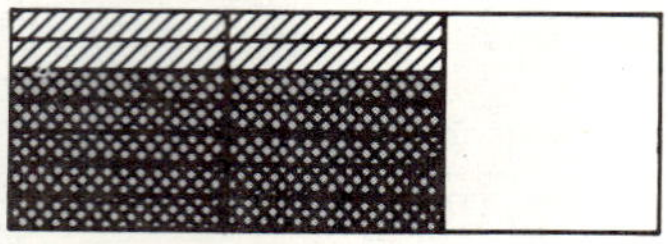

FIG. 6.9

To determine how much this checkered region represents, suppose we extend the division into seven parts to the uncheckered portion.

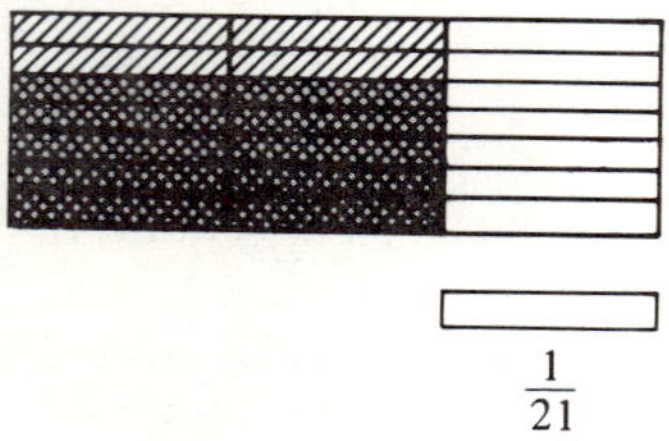

FIG. 6.10

This amounts to dividing the original plot into 21 parts. The checkered region accounts for 10 such parts or $\frac{10}{21}$ of the total. That is, $\frac{5}{7}$ of $\frac{2}{3}$ is $\frac{10}{21}$.

Such operations we shall call *products* and denote them by $\frac{5}{7}\cdot\frac{2}{3}$. Notice that

$$\frac{5}{7}\cdot\frac{2}{3} = \frac{10}{21} \quad \text{and} \quad \frac{10}{21} = \frac{5\cdot 2}{7\cdot 3} \tag{6.2.20}$$

This leads us to make the following definition.

Definition 6.2.4

If $\frac{a}{b}, \frac{c}{d}$ are two fractions, their *product* $\frac{a}{b}\cdot\frac{c}{d}$ is the fraction

$$\frac{a}{b}\cdot\frac{c}{d} = \frac{a\cdot c}{b\cdot d} \tag{6.2.21}$$

Note: Definition 6.2.4 is valid whether the integers a, b, c, d are positive or negative. Of course, $b \neq 0$, $d \neq 0$.

EXAMPLE 6.2.3

Find the product of $\frac{3}{^-5}$ and $\frac{7}{12}$. Here

$$\frac{3}{^-5}\cdot\frac{7}{12} = \frac{3\cdot 7}{^-5\cdot 12} = \frac{21}{^-60} \tag{6.2.22}$$

Consider a set of fractions in which each fraction in the set equals any other fraction in the set. For example

$$\left\{\cdots, \frac{^-4}{^-14}, \frac{^-2}{^-7}, \frac{2}{7}, \frac{4}{14}, \frac{6}{21}, \cdots\right\} \tag{6.2.23}$$

is such a set. Let us agree that any member of the set could be used to represent the set. For our purposes, the *best* representative of the set will be that ratio $\frac{a}{b}$ for which

$$(1)\quad b \geqq 1 \qquad (2)\quad \gcd\{a, b\} = 1 \tag{6.2.24}$$

That is, $\frac{a}{b}$ is the *best* representative provided the denominator b is positive and the numerator a and denominator b are relatively prime.

In our example, $\frac{2}{7}$ is the best representative. In other words,

$$\frac{2}{7} \quad \text{represents} \quad \left\{\cdots, \frac{^-4}{^-14}, \frac{^-2}{^-7}, \frac{2}{7}, \frac{4}{14}, \cdots\right\} \tag{6.2.25}$$

The set of all such "best" representatives is defined as the *set R of rational numbers.*

Once again, let us illustrate the distinction between a fraction and a rational number by the following example. The ratio $\frac{25}{35}$ is a fraction, but is *not* a rational number. The ratio $\frac{5}{7}$ is a rational number since it is the best representative of the set

$$\left\{\cdots, \frac{^-5}{^-7}, \frac{5}{7}, \frac{10}{14}, \frac{15}{21}, \frac{20}{28}, \frac{25}{35}, \cdots\right\} \tag{6.2.26}$$

Thus, we see that every rational number is a fraction (that is, a ratio of two integers), but not every fraction is a rational number.

A technique for finding that rational number which is the best representative uses the concept of greatest common divisor. For example, consider $\frac{12}{18}$. Since gcd $\{12, 18\} = 6$, this is not the best representative. But using the definition of equality of fractions,

$$\frac{12}{18} = \frac{2 \cdot 6}{3 \cdot 6} = \frac{2}{3} \tag{6.2.27}$$

Since $3 > 0$ and gcd $\{2, 3\} = 1$, the fraction $\frac{2}{3}$ is a rational number.

EXAMPLE 6.2.4

Write $\frac{30}{12}$ as a rational number. First of all we find gcd $\{12, 30\}$. Thus,

$$\text{gcd } \{30, 12\} = 6 \tag{6.2.28}$$

Then, we can write both 30 and 12 as

$$30 = 6 \cdot 5, \qquad 12 = 6 \cdot 2 \tag{6.2.29}$$

That is,

$$\frac{30}{12} = \frac{6 \cdot 5}{6 \cdot 2} = \frac{5}{2} \tag{6.2.30}$$

Thus, $\frac{30}{12}$ is best represented by $\frac{5}{2}$.

Of course, if for a fraction $\frac{a}{b}$, it turns out that $b \geqq 1$ and gcd $\{a, b\} = 1$, then $\frac{a}{b}$ is already the best representative. Sometimes we refer to the best representative by saying the fraction is *written in lowest terms.*

Note: A rigorous construction of the set R of rational numbers is given in Chapter 9.

Let us now find out if there is any relationship between the set J of integers and the set R of rational numbers. Recall that the rational number $\frac{3}{4}$ is a name for 3 parts of a unit divided into 4 parts. The rational number $\frac{2}{1}$

would then be a name for 2 units. That is, the rational number $\frac{2}{1}$ and the integer 2 have the *same* meaning. See Fig. 6.11.

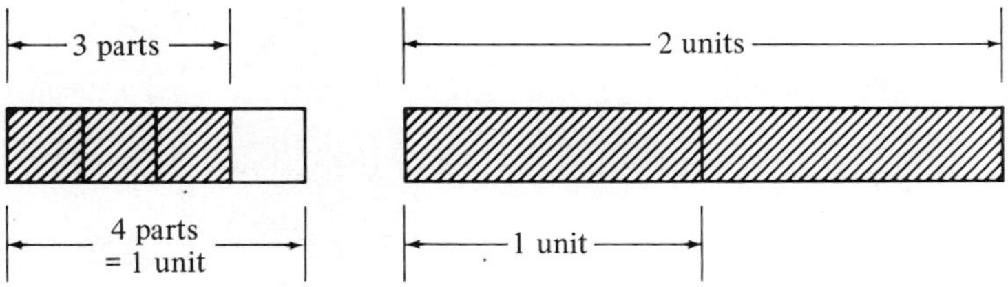

FIG. 6.11

In general, the rational number $\frac{a}{1}$ has the same meaning as the integer a.

Henceforth, we agree that the rational number $\frac{a}{1}$ can be written as a and that a can be written as $\frac{a}{1}$.

For example, the rational number $\frac{0}{1}$ can now be written as 0. The integer 1 can be written as the rational number $\frac{1}{1}$.

The set of all rational numbers of the form $\frac{a}{1}$, in other words, the set J of integers, is thus a proper subset of the set R of rational numbers. That is,

$$J \subset R \tag{6.2.31}$$

The student is probably familiar with the notation $2\frac{3}{4}$. This is merely a convention or shorthand for $2 + \frac{3}{4}$. Thus, whenever a positive integer appears next to a positive rational number, we agree that this means to add the integer and the rational number. That is,

$$N\frac{a}{b} = N + \frac{a}{b} \tag{6.2.32}$$

where N is a positive integer and $\frac{a}{b}$ is a positive rational number.

The rest of this chapter concerns itself with the properties of rational numbers.

Exercise 6.2

1. Find two fractions equal to each of the following fractions.

(a) $\frac{3}{5}$ (g) $\frac{14}{3}$

(b) $\frac{^-32}{8}$ (h) $\frac{^-6}{5}$

(c) $\frac{2}{70}$ (i) $\frac{^-2}{^-3}$

(d) $\frac{^-4}{3}$ (j) $\frac{3}{^-7}$

(e) $\frac{4}{3}$ (k) $\frac{^-1}{7}$

(f) $\frac{11}{10}$ (l) $\frac{30}{4}$

2. Find the sum of the following fractions by using first the Definition 6.2.3 and then the least common denominator method.

(a) $\frac{3}{5} + \frac{4}{3}$ (g) $\frac{^-2}{5} + \frac{6}{13}$

(b) $\frac{6}{5} + \frac{7}{2}$ (h) $\frac{9}{13} + \frac{4}{^-3}$

(c) $\frac{4}{3} + \frac{7}{11}$ (i) $\frac{4}{9} + \frac{^-3}{18}$

(d) $\frac{^-2}{7} + \frac{5}{2}$ (j) $\frac{5}{8} + \frac{^-2}{9}$

(e) $\frac{-1}{6} + \frac{5}{-2}$

(f) $\frac{5}{12} + \frac{2}{5}$

(k) $\frac{1}{12} + \frac{-2}{30}$

(l) $\frac{4}{15} + \frac{-3}{27}$

3. Find the product of the following fractions.

(a) $\frac{3}{5} \cdot \frac{-2}{1}$

(b) $\frac{4}{11} \cdot \frac{7}{12}$

(c) $\frac{-1}{3} \cdot \frac{13}{-2}$

(d) $\frac{4}{9} \cdot \frac{2}{5}$

(e) $\frac{15}{4} \cdot \frac{5}{8}$

(f) $\frac{4}{17} \cdot \frac{-3}{7}$

(g) $\frac{9}{8} \cdot \frac{-3}{2}$

(h) $\frac{-2}{3} \cdot \frac{8}{13}$

(i) $\frac{5}{9} \cdot \frac{16}{3}$

4. For the fractions of problem 1, find the best representative.

5. A bank charges \$2 for every \$400 worth of cashier's checks issued. If the charge for such checks was \$10, how many checks were purchased?

6. A car averages 14 miles for each gallon of gasoline. How many gallons will be used after the car has been driven 42 miles?

7. On a map, the scale indicates that 1 inch is equivalent to 100 miles. The distance between two cities on the map measures 3 inches. How far apart are the cities?

8. For your answers to problem 2, compare those obtained using Definition 6.2.3 with those obtained for the least common denominator method. Does this convince you that the least common denominator method will save time?

9. Write the following as a rational number.

(a) $3\frac{5}{8} + 7\frac{2}{3}$

(b) $4\frac{6}{7} + 2\frac{4}{9}$

(c) $4\frac{4}{5} + 6\frac{9}{10}$

(d) $3\frac{3}{4} + 4\frac{1}{2}$

10. A *proper* rational number $\frac{a}{b}$ is one for which $a < b$. An *improper* rational number $\frac{a}{b}$ is one for which $a \geqq b$. Which of the following are improper?

(a) $\frac{11}{6}$ (c) $\frac{17}{18}$

(b) $\frac{19}{5}$. (d) $\frac{16}{3}$

Write those that are improper in the notation given by (6.2.32).

11. A contractor divides a plot of land into 80 equal parcels and then divides each such parcel into 12 lots. If the original plot cost the contractor \$480,000 and if he sells 2 lots for \$2000, how much profit (or loss) did the contractor have on this sale?

6.3 SOME PROPERTIES OF THE SET R OF RATIONAL NUMBERS

In this section, we shall state, without proof, eleven fundamental properties which describe the set R of rational numbers. These results should be thoroughly understood by the student since it will be these properties that will be used later in the chapter to deduce further conclusions concerning rational numbers.

The set R of rational numbers is described as a set of elements, called *rational numbers*, together with *two binary* operations, *addition* and *multiplication*, satisfying the following eleven properties:

(I) *Closure Law for Addition*

If $\frac{a}{b}, \frac{c}{d}$ are two rational numbers, their *sum*

$$\frac{a}{b} + \frac{c}{d} \tag{6.3.1}$$

is also a rational number. Furthermore, this sum is unique. That is, if x, x', y, y' are four rational numbers for which $x = x'$, and $y = y'$, then

$$x + y = x' + y' \tag{6.3.2}$$

This is the same property which the set N of natural numbers, the set W of whole numbers, and the set J of integers possessed. It states that when two rational numbers are added, *only one* (never more than one) rational number is obtained as their sum.

(II) *Commutative Law for Addition*

If $\frac{a}{b}, \frac{c}{d}$ are two rational numbers, then

$$\frac{a}{b} + \frac{c}{d} = \frac{c}{d} + \frac{a}{b} \tag{6.3.3}$$

Thus, the order in which *two* rational numbers are added will not affect the result.

EXAMPLE 6.3.1

Consider the rational numbers $\frac{2}{3}$ and $\frac{4}{5}$. Using the rule for adding two rational numbers (see 6.2.11), we have

$$\frac{2}{3} + \frac{4}{5} = \frac{2\cdot 5 + 3\cdot 4}{3\cdot 5} = \frac{10 + 12}{15} = \frac{22}{15} \tag{6.3.4}$$

and

$$\frac{4}{5} + \frac{2}{3} = \frac{4\cdot 3 + 5\cdot 2}{5\cdot 3} = \frac{12 + 10}{15} = \frac{22}{15} \tag{6.3.5}$$

(III) *Associative Law for Addition*

If $\frac{a}{b}, \frac{c}{d}, \frac{e}{f}$ are *three* rational numbers, then

$$\frac{a}{b} + \left(\frac{c}{d} + \frac{e}{f}\right) = \left(\frac{a}{b} + \frac{c}{d}\right) + \frac{e}{f} \tag{6.3.6}$$

Thus, when adding *three* rational numbers, the way in which they are grouped together, provided their order is not disturbed, will make *no* difference in the result.

EXAMPLE 6.3.2

Consider $\frac{2}{3}, \frac{1}{2}$, and $\frac{3}{5}$. Then

$$\left(\frac{2}{3} + \frac{1}{2}\right) + \frac{3}{5} = \frac{2\cdot 2 + 3\cdot 1}{3\cdot 2} + \frac{3}{5} = \frac{4 + 3}{6} + \frac{3}{5} = \frac{7}{6} + \frac{3}{5}$$

$$= \frac{7\cdot 5 + 6\cdot 3}{6\cdot 5} = \frac{35 + 18}{30} = \frac{53}{30} \tag{6.3.7}$$

and

$$\frac{2}{3}+\left(\frac{1}{2}+\frac{3}{5}\right)=\frac{2}{3}+\frac{5\cdot 1+2\cdot 3}{2\cdot 5}=\frac{2}{3}+\frac{5+6}{10}=\frac{2}{3}+\frac{11}{10}$$

$$=\frac{2\cdot 10+3\cdot 11}{3\cdot 10}=\frac{20+33}{30}=\frac{53}{30} \tag{6.3.8}$$

This example illustrates the fact that addition of rational numbers is associative.

(IV) *The Existence of an Additive Identity*

In the set R of rational numbers, there is a rational number $\frac{0}{1}$ or 0, called the *additive identity zero,* with the property that

$$\frac{a}{b}+\frac{0}{1}=\frac{a}{b} \tag{6.3.9}$$

for any rational number $\frac{a}{b}$.

EXAMPLE 6.3.3

For example,

$$\frac{5}{6}+\frac{0}{1}=\frac{5\cdot 1+6\cdot 0}{6\cdot 1}=\frac{5+0}{6}=\frac{5}{6} \tag{6.3.10}$$

This illustrates the fact that $\frac{0}{1}$ or 0 is an additive identity for the set R of rational numbers.

(V) *Existence of an Additive Inverse*

If $\frac{a}{b}$ is a rational number, there is a rational number $-\frac{a}{b}$, called the *additive inverse of* $\frac{a}{b}$, with the property that

$$\frac{a}{b}+\left(-\frac{a}{b}\right)=0 \tag{6.3.11}$$

Thus, for every rational number in R, there is another rational number in R such that when these two rational numbers are added, the result is zero.

For example, the additive inverse of the rational number $\frac{2}{3}$ is $-\frac{2}{3}$. Since

$$\frac{2}{3}+\left(\frac{^{-}2}{3}\right)=\frac{2\cdot 3+3\cdot(^{-}2)}{3\cdot 3}=\frac{0}{9}=0 \tag{6.3.12}$$

it is seen that the additive inverse of $\frac{2}{3}$ is $\frac{^{-}2}{3}$. Thus, we have shown that

$$-\frac{2}{3}=\frac{^{-}2}{3} \tag{6.3.13}$$

That is, instead of writing the additive inverse of $\frac{2}{3}$ as $-\frac{2}{3}$, we can write it as $\frac{^{-}2}{3}$, the ratio of $^{-}2$ to 3. It will be shown later in the chapter that the fact shown in (6.3.13) is true for any two rational numbers.

Next, let us look at some properties of rational numbers, relative to the binary operation of multiplication.

(VI) *Closure Law for Multiplication*

If $\frac{a}{b}$, $\frac{c}{d}$ are two rational numbers, their product

$$\frac{a}{b}\cdot\frac{c}{d} \tag{6.3.14}$$

is also a rational number. Moreover, the product is unique. That is, if x, x', y, y' are four rational numbers so that $x = x'$, $y = y'$, then

$$x\cdot y = x'\cdot y' \tag{6.3.15}$$

Thus, when one multiplies two rational numbers, not only is the result a rational number, but there is also just one such product obtainable.

(VII) *Commutative Law for Multiplication*

If $\frac{a}{b}$, $\frac{c}{d}$ are any two rational numbers, then

$$\frac{a}{b}\cdot\frac{c}{d}=\frac{c}{d}\cdot\frac{a}{b} \tag{6.3.16}$$

This law tells us that the order in which two rational numbers are multiplied does not alter their product.

EXAMPLE 6.3.4

Consider the rational numbers $\frac{2}{5}$ and $\frac{-3}{7}$. Using the rule for multiplying two rational numbers [see (6.2.21)], we have

$$\frac{2}{5}\cdot\frac{-3}{7} = \frac{2\cdot(-3)}{5\cdot 7} = \frac{-6}{35} \tag{6.3.17}$$

and

$$\frac{-3}{7}\cdot\frac{2}{5} = \frac{-3\cdot 2}{7\cdot 5} = \frac{-6}{35} \tag{6.3.18}$$

(VIII) *Associative Law for Multiplication*

If $\frac{a}{b}, \frac{c}{d}, \frac{e}{f}$, are three rational numbers, then

$$\left(\frac{a}{b}\cdot\frac{c}{d}\right)\cdot\frac{e}{f} = \frac{a}{b}\cdot\left(\frac{c}{d}\cdot\frac{e}{f}\right) \tag{6.3.19}$$

Thus, when multiplying three rational numbers, the way in which they are grouped together, provided their order is not disturbed, will not affect the result.

EXAMPLE 6.3.5

Consider the three rational numbers $\frac{2}{3}, \frac{-4}{5}, \frac{1}{3}$. Then

$$\frac{2}{3}\cdot\left(\frac{-4}{5}\cdot\frac{1}{3}\right) = \frac{2}{3}\cdot\left(\frac{-4\cdot 1}{5\cdot 3}\right) = \frac{2}{3}\cdot\frac{-4}{15} = \frac{2\cdot(-4)}{3\cdot 15} = \frac{-8}{45} \tag{6.3.20}$$

and

$$\left(\frac{2}{3}\cdot\frac{-4}{5}\right)\cdot\frac{1}{3} = \left(\frac{2\cdot(-4)}{3\cdot 5}\right)\cdot\frac{1}{3} = \frac{-8}{15}\cdot\frac{1}{3} = \frac{-8\cdot 1}{15\cdot 3} = \frac{-8}{45} \tag{6.3.21}$$

(IX) *Distributive Law*

In the set R of rational numbers, the binary operation of multiplication distributes relative to the binary operation of addition. That is, if $\frac{a}{b}, \frac{c}{d}, \frac{e}{f}$, are three rational numbers, then

$$\frac{a}{b}\cdot\left(\frac{c}{d} + \frac{e}{f}\right) = \left(\frac{a}{b}\cdot\frac{c}{d}\right) + \left(\frac{a}{b}\cdot\frac{e}{f}\right) \tag{6.3.22}$$

EXAMPLE 6.3.6

Consider the rational numbers $\frac{2}{3}, \frac{-1}{5}, \frac{4}{7}$. Then

$$\frac{2}{3}\cdot\left(\frac{-1}{5}+\frac{4}{7}\right)=\frac{2}{3}\cdot\left(\frac{-1\cdot 7+5\cdot 4}{5\cdot 7}\right)=\frac{2}{3}\cdot\left(\frac{-7+20}{35}\right)$$

$$=\frac{2}{3}\cdot\frac{13}{35}=\frac{2\cdot 13}{3\cdot 35}=\frac{26}{105} \tag{6.3.23}$$

and

$$\left(\frac{2}{3}\cdot\frac{-1}{5}\right)+\left(\frac{2}{3}\cdot\frac{4}{7}\right)=\frac{2\cdot(-1)}{3\cdot 5}+\frac{2\cdot 4}{3\cdot 7}=\frac{-2}{15}+\frac{8}{21}$$

$$=\frac{-2\cdot 7}{15\cdot 7}+\frac{8\cdot 5}{21\cdot 5}=\frac{-14+40}{105}=\frac{26}{105} \tag{6.3.24}$$

Of course, just as with the integers, we agree to the convention that if a choice is to be made whether to perform the operation of multiplication or that of addition first, multiplication will always take precedence.

EXAMPLE 6.3.7

Consider the three rational numbers $\frac{2}{3}, \frac{-1}{5}, \frac{4}{7}$. Here

$$\frac{2}{3}\cdot\frac{-1}{5}+\frac{4}{7}=\frac{2\cdot(-1)}{3\cdot 5}+\frac{4}{7}=\frac{-2}{15}+\frac{4}{7}=\frac{(-2)\cdot 7+15\cdot 4}{15\cdot 7}$$

$$=\frac{-14+60}{105}=\frac{46}{105} \tag{6.3.25}$$

(X) *Existence of a Multiplicative Identity*

In the set R of rational numbers, there is a rational number $\frac{1}{1}$ or 1, called the *multiplicative identity one*, with the property that for any rational number $\frac{a}{b}$. we have

$$\frac{a}{b}\cdot\frac{1}{1}=\frac{a}{b} \tag{6.3.26}$$

For example, for the rational number $\frac{2}{3}$, we have

$$\frac{2}{3}\cdot\frac{1}{1} = \frac{2\cdot 1}{3\cdot 1} = \frac{2}{3} \tag{6.3.27}$$

It should be recognized that the properties listed thus far for the set R of rational numbers are identical with those listed in Chapter 4 for the set J of integers, with the exception of the Product Law.

The following property for rational numbers will be fundamental for the ensuing development of properties of rational numbers.

(XI) *Existence of a Multiplicative Inverse*

Let $a \neq 0$, $b \neq 0$ be two nonzero integers. For each rational number $\frac{a}{b}$ in R, there is a rational number $\frac{b}{a}$ in R, called the *multiplicative inverse of* $\frac{a}{b}$ such that

$$\frac{a}{b}\cdot\frac{b}{a} = 1 \tag{6.3.28}$$

Note that 0 is the only rational number that has *no* multiplicative inverse.

This property is very similar to the property stating the existence of an additive inverse. Recall that the *additive inverse* of a rational number $\frac{a}{b}$ is a rational number which when *added* to $\frac{a}{b}$ gives the *additive identity zero.* The *multiplicative inverse* of a rational number $\frac{a}{b}$ is a rational number which when *multiplied* by $\frac{a}{b}$ gives the *multiplicative identity one.*

EXAMPLE 6.3.8

Find the additive inverse and multiplicative inverse of the rational number $\frac{4}{3}$.

The additive inverse of $\frac{4}{3}$ is $\frac{^-4}{3}$ since

$$\frac{4}{3} + \frac{^-4}{3} = \frac{4 \cdot 3 + 3 \cdot (^-4)}{3 \cdot 3} = \frac{12 + (^-12)}{9} = \frac{0}{9} = 0 \qquad (6.3.29)$$

Since the sum of $\frac{4}{3}$ and $\frac{^-4}{3}$ is 0, then $\frac{^-4}{3}$ is the additive inverse of $\frac{4}{3}$.

The multiplicative inverse of $\frac{4}{3}$ is $\frac{3}{4}$ since

$$\frac{4}{3} \cdot \frac{3}{4} = \frac{4 \cdot 3}{3 \cdot 4} = \frac{12}{12} = 1 \qquad (6.3.30)$$

Since the product of $\frac{4}{3}$ and $\frac{3}{4}$ is the multiplicative identity one, then the multiplicative inverse of $\frac{4}{3}$ is $\frac{3}{4}$.

Exercise 6.3

1. Show that for the following pairs of rational numbers, the second one is the additive inverse of the first.

(a) $\frac{2}{3}, \frac{^-2}{3}$ (d) $\frac{3}{2}, \frac{^-3}{2}$

(b) $\frac{7}{6}, \frac{^-7}{6}$ (e) $\frac{5}{3}, \frac{^-5}{3}$

(c) $\frac{^-4}{5}, \frac{4}{5}$ (f) $\frac{^-6}{5}, \frac{6}{5}$

2. Show that for the following pairs of rational numbers, the second one is the multiplicative inverse of the first.

(a) $\frac{2}{3}, \frac{3}{2}$ (d) $\frac{6}{7}, \frac{7}{6}$

(b) $\frac{5}{3}, \frac{3}{5}$

(c) $\frac{-5}{6}, \frac{-6}{5}$

(e) $\frac{-4}{5}, \frac{-5}{4}$

(f) $\frac{4}{3}, \frac{3}{4}$

3. Find the additive inverse and multiplicative inverse of the following rational numbers. Verify your results.

(a) $\frac{7}{5}$

(b) $\frac{5}{3}$

(c) $\frac{-4}{5}$

(d) $\frac{-6}{11}$

(e) $\frac{15}{8}$

(f) $\frac{3}{7}$

(g) 0

(h) 1

4. Evaluate the following quantities as a single rational number.

(a) $\frac{2}{3}\cdot\left(\frac{4}{7}+\frac{4}{5}\right)$

(b) $\frac{2}{3}\cdot\frac{4}{7}+\frac{4}{5}$

(c) $\frac{2}{3}\cdot\frac{5}{6}\cdot\frac{3}{4}$

(d) $\frac{4}{3}+\frac{2}{5}\cdot\frac{1}{7}$

5. Evaluate the following quantities as a single rational number.

(a) $\frac{-3}{2}\cdot\left(\frac{-1}{2}+\frac{1}{2}\right)$

(b) $\frac{11}{7}\cdot\frac{3}{2}+\frac{11}{7}\cdot\frac{1}{2}$

(c) $\frac{-1}{1}+\frac{7}{2}\cdot\frac{2}{7}$

(d) $\frac{3}{2}\cdot\left(\frac{3}{4}+\frac{-5}{6}\right)$

6. Use numerical examples to illustrate the following:

(a) The closure law for addition of rational numbers.
(b) The closure law for multiplication of rational numbers.
(c) The commutative law for addition of rational numbers.
(d) The commutative law for multiplication of rational numbers.

(e) The associative law for addition of rational numbers.
(f) The associative law for multiplication of rational numbers.
(g) The distributive law.

7. What is wrong with

$$\frac{a}{b} + \frac{c}{d} = \frac{a+c}{b+d}?$$

Use examples to convince yourself.

8. Find the error:

$$1 = \frac{12}{12} = \frac{12}{6+6} = \frac{12}{6} + \frac{12}{6} = 2 + 2 = 4$$

6.4 FURTHER RESULTS FOR THE SET R OF RATIONAL NUMBERS

Now, let us use the eleven properties given in Section 6.3 to develop some further laws that are valid for rational numbers. Each of the following results will be proved by using the definition of a rational number and the eleven properties already discussed.

The first result we shall discuss is one we have already seen. It states that whenever the product of two rational numbers is zero, then either the first or the second or both rational numbers are zero. That is, the Product Law is valid for the set R of rational numbers.

Theorem 6.4.1

The Product Law: If $\frac{a}{b}$ and $\frac{c}{d}$ are two rational numbers whose product is zero, that is, if

$$\frac{a}{b} \cdot \frac{c}{d} = 0 \qquad (6.4.1)$$

then

$$\text{either } \frac{a}{b} = 0 \quad \text{or} \quad \frac{c}{d} = 0 \quad \text{or} \quad \frac{a}{b} = \frac{c}{d} = 0 \qquad (6.4.2)$$

PROOF: Suppose $\frac{a}{b}, \frac{c}{d}$ are two rational numbers for which

$$\frac{a}{b} \cdot \frac{c}{d} = 0 \qquad (6.4.3)$$

Then, we know that

$$\frac{a \cdot c}{b \cdot d} = \frac{0}{1} \tag{6.4.4}$$

But this is true if and only if

$$a \cdot c = 0 \tag{6.4.5}$$

By the Product Law for integers, this says

$$a = 0 \quad \text{or} \quad c = 0 \quad \text{or both} \tag{6.4.6}$$

Hence,

$$\frac{a}{b} = 0 \quad \text{or} \quad \frac{c}{d} = 0 \quad \text{or both} \tag{6.4.7}$$

Because of Theorem 6.4.1 it is deduced that the eleven properties given in Chapter 4, Section 4.2 for the set J of integers are also obeyed by the set R of rational numbers.

The student should recall that the Theorems derived in Sections 4.3 and 4.4 of Chapter 4 for the integers were proved by using the eleven properties of integers. Thus, it follows that the Theorems of Sections 4.3 and 4.4, Chapter 4, are valid for rational numbers as well since they obey these same laws. Because of this, it will not be necessary to prove these results for rational numbers.

To review some of these results, using the notation for rational numbers, we list them below.

The equation

$$\frac{a}{b} + x = \frac{c}{d} \tag{6.4.8}$$

has one and only one solution for x. This solution x is given by the formula

$$x = \frac{c}{d} + \left(\frac{{}^{-}a}{b}\right) \tag{6.4.9}$$

Also, the unary operation of additive inverse relative to rational numbers is involutorial. That is, for any rational number $\frac{a}{b}$, we have

$$-\left(-\frac{a}{b}\right) = \frac{a}{b} \tag{6.4.10}$$

The binary operation of subtraction is defined as follows. If $\frac{a}{b}$ and $\frac{c}{d}$ are two rational numbers, to subtract $\frac{a}{b}$ from $\frac{c}{d}$, written as

$$\frac{c}{d} - \frac{a}{b} \tag{6.4.11}$$

means to find the solution x of the equation

$$\frac{a}{b} + x = \frac{c}{d} \tag{6.4.12}$$

It then follows that for rational numbers $\frac{a}{b}$ and $\frac{c}{d}$ we have the following relationship between the two binary operations of addition and subtraction, namely,

$$\frac{c}{d} - \frac{a}{b} = \frac{c}{d} + \left(-\frac{a}{b}\right) \tag{6.4.13}$$

Next, if $\frac{a}{b}$ is any rational number, then

$$\frac{a}{b}\cdot 0 = 0 \tag{6.4.14}$$

That is, the product of any rational number and zero is zero.

Also, if $\frac{a}{b}$, $\frac{c}{d}$ are two rational numbers, then

$$\left(-\frac{a}{b}\right)\cdot\frac{c}{d} = -\left(\frac{a}{b}\cdot\frac{c}{d}\right) \tag{6.4.15}$$

$$\frac{a}{b}\cdot\left(-\frac{c}{d}\right) = -\left(\frac{a}{b}\cdot\frac{c}{d}\right) \tag{6.4.16}$$

$$\left(-\frac{a}{b}\right)\cdot\left(-\frac{c}{d}\right) = \frac{a}{b}\cdot\frac{c}{d} \tag{6.4.17}$$

EXAMPLE 6.4.1

Consider the rational numbers $-\frac{3}{4}$ and $\frac{5}{2}$. Then using (6.4.15), we have

$$\left(-\frac{3}{4}\right)\cdot\left(\frac{5}{2}\right) = -\left(\frac{3}{4}\cdot\frac{5}{2}\right) = -\frac{3\cdot 5}{4\cdot 2} = -\frac{15}{8} \tag{6.4.18}$$

EXAMPLE 6.4.2

Consider $-\frac{2}{3}$ and $-\frac{7}{5}$. Then using (6.4.17) we have

$$\left(-\frac{2}{3}\right)\cdot\left(-\frac{7}{5}\right) = \frac{2}{3}\cdot\frac{7}{5} = \frac{2\cdot 7}{3\cdot 5} = \frac{14}{15} \tag{6.4.19}$$

Next, if $^-1$ is used to denote the rational number which when added to 1 gives 0, and if $\frac{a}{b}$ is any rational number, then

$$({}^-1)\cdot\frac{a}{b} = -\frac{a}{b} \tag{6.4.20}$$

That is, multiplying a rational number by $^-1$ gives the additive inverse of that rational number.

Finally, we have the cancellation laws for rational numbers.

The cancellation law for addition states that if $\frac{a}{b}, \frac{c}{d}, \frac{e}{f}$, are three rational numbers for which

$$\frac{c}{d} + \frac{a}{b} = \frac{e}{f} + \frac{a}{b} \tag{6.4.21}$$

then

$$\frac{c}{d} = \frac{e}{f} \tag{6.4.22}$$

The cancellation law for multiplication says that if $\frac{a}{b} \neq 0, \frac{c}{d}, \frac{e}{f}$ are three rational numbers and if

$$\frac{a}{b}\cdot\frac{c}{d} = \frac{a}{b}\cdot\frac{e}{f} \tag{6.4.23}$$

then

$$\frac{c}{d} = \frac{e}{f} \tag{6.4.24}$$

Consider two integers a and $b \neq 0$. Suppose that we know

$$\frac{a}{b} = 0 \tag{6.4.25}$$

It follows that the numerator a must be zero. Thus, if $\frac{a}{b} = 0$, then $a = 0$.

The student should apply these laws and properties in the following exercise.

Exercise 6.4

1. Solve the following for x.

(a) $\frac{2}{21}\cdot x = \left(\frac{2}{3} + \frac{^-4}{7}\right)\cdot\frac{9}{8}$

(b) $\frac{5}{^-6} + x = \frac{23}{9} + \frac{5}{^-6}$

(c) $\frac{9}{8}\cdot x + \frac{7}{6} = \frac{9}{8}\cdot\frac{4}{3} + \frac{7}{6}$

(d) $\frac{5}{3} + \frac{2}{3}\cdot x = \frac{5}{3}$

2. Find all x for which

(a) $\frac{3}{2}\cdot x = \frac{3}{2}$

(b) $\left(x - \frac{3}{2}\right)\cdot\frac{4}{5} = 0$

(c) $\left(\frac{3}{4} - \frac{4}{3}\right)\cdot x = 0$

(d) $\left(\frac{6}{5} + \frac{4}{3}\right)\cdot x = \frac{7}{6} + \frac{^-7}{6}$

(e) $\left(\frac{17}{9} + \frac{^-8}{9}\right)\cdot x = \frac{4}{3}$

3. Show that the operation of subtraction is not commutative.

4. Show that the operation of subtraction is not associative.

5. Using the Product Law, find all x for which

(a) $\left(x + \frac{3}{2}\right)\left(x - \frac{3}{2}\right) = 0$

(b) $\left(x - \frac{7}{2}\right)\left(x + \frac{1}{2}\right)\left(x - \frac{3}{2}\right) = 0$

(c) $\left(x + \frac{3}{4}\right)\left(x - \frac{11}{7}\right) = 0$

(d) $(x + 1)\left(x - \frac{1}{2}\right)\left(x + \frac{5}{4}\right) = 0$

(e) $\left(x - \frac{3}{5}\right)\left(x - \frac{7}{5}\right) = 0$

6. Solve the following for x.

(a) $\frac{4}{3}\left(x + \frac{1}{2}\right) + \frac{3}{4} = \frac{3}{4}$

(b) $\frac{7}{8}\cdot\frac{3}{4} + \frac{1283}{18}\cdot x - 1 = 1 + \frac{7}{3}\cdot\frac{4}{3}$

(c) $\frac{4}{7}\cdot x = \frac{4}{7}\cdot\frac{^-3}{4} + \frac{4}{7}\cdot\frac{^-1}{3}$

(d) $1\cdot x - \frac{5}{11} = \frac{3}{7}\cdot\frac{1}{8} + \frac{^-3}{7} + \frac{1}{8}$

7. Evaluate the following:

(a) $\frac{3}{4} + \left(\frac{^-6}{5}\right)\left(\frac{^-7}{11}\right)$

(b) $\frac{^-7}{8} + \left(\frac{^-5}{4}\right)\left(\frac{4}{5}\right)$

(c) $\frac{6}{7} + \left(\frac{^-5}{7}\right)$

(d) $\left(\frac{5}{2}\right)\left(\frac{4}{3}\right) + \frac{7}{11} + \frac{3}{4}$

(e) $\left(\frac{3}{4} + \frac{^-7}{8}\right) + \frac{5}{4}$

(f) $\frac{^-3}{4} + \left(\frac{^-7}{8}\right)\left(\frac{5}{4}\right) + \frac{1}{2}$

8. Find the value of x of the following equations.

(a) $\frac{3x - 5}{7x + 14} = 0$

(b) $\dfrac{\frac{1}{2}x + 8}{\frac{3}{4}x + 6} = 0$

(c) $\dfrac{-\frac{3}{4}x + 17}{2x - 5} = 0$

Warning: The value of x must be checked in the denominator to assure that the denominator is *not* zero.

6.5 FURTHER PROPERTIES OF THE SET R OF RATIONAL NUMBERS

In this section we shall state several more properties of the set R of rational numbers. These results may be proven by using the eleven properties given in Section 6.3 and those properties discussed in Section 6.4.

Theorem 6.5.1

For any rational number $\frac{a}{b}$, we have

$$-\frac{a}{b} = \frac{{}^{-}a}{b} \tag{6.5.1}$$

That is, the additive inverse of the ratio of a to b is the same as the ratio of the additive inverse of a to b.

First of all,

$$\frac{a}{b} + \frac{{}^{-}a}{b} = \frac{a \cdot b + b \cdot ({}^{-}a)}{b \cdot b} = \frac{a \cdot b + [{}^{-}(a \cdot b)]}{b \cdot b} = \frac{0}{b \cdot b} = 0 \tag{6.5.2}$$

That is, when we add $\frac{{}^{-}a}{b}$ to $\frac{a}{b}$, we get zero. This means that $\frac{{}^{-}a}{b}$ is the additive inverse of $\frac{a}{b}$. But $-\frac{a}{b}$ is the additive inverse of $\frac{a}{b}$. Since additive inverses are unique, we have

$$-\frac{a}{b} = \frac{{}^{-}a}{b} \tag{6.5.3}$$

Because of Theorem 6.5.1, we can write the additive inverse of any rational number in two ways. For example, the additive inverse of $\frac{2}{3}$ can be written as

$$-\frac{2}{3} \quad \text{or} \quad \frac{^-2}{3} \tag{6.5.4}$$

EXAMPLE 6.5.1
Find the value of

$$\frac{3}{5} - \frac{1}{2} \tag{6.5.5}$$

Now,

$$\frac{3}{5} - \frac{1}{2} = \frac{3}{5} + \left(-\frac{1}{2}\right) = \frac{3}{5} + \frac{^-1}{2} = \frac{3 \cdot 2 + 5 \cdot (^-1)}{5 \cdot 2}$$

$$= \frac{6 + (^-5)}{10} = \frac{1}{10} \tag{6.5.6}$$

EXAMPLE 6.5.2
Find the value of

$$\frac{4}{7} - \frac{3}{2} \cdot \frac{5}{2} \tag{6.5.7}$$

Here,

$$\frac{4}{7} - \frac{3}{2} \cdot \frac{5}{2} = \frac{4}{7} - \frac{3 \cdot 5}{2 \cdot 2} = \frac{4}{7} - \frac{15}{4} = \frac{4}{7} + \left(\frac{^-15}{4}\right) = \frac{4}{7} + \frac{^-15}{4}$$

$$= \frac{4 \cdot 4 + 7(^-15)}{7 \cdot 4} = \frac{16 + (^-105)}{28} = \frac{^-89}{28} \tag{6.5.8}$$

With the set J of integers, we learned that equations of the form

$$a + x = b, \qquad a, b \in J \tag{6.5.9}$$

always have one and only one solution for x in J. It was also discovered that equations of the type

$$2 \cdot x = 3 \tag{6.5.10}$$

do *not always* have integral solutions for x. The next theorem guarantees that this deficiency of the set J of integers is *not* present in the set R of rational numbers.

Theorem 6.5.2

Let $\frac{a}{b} \neq 0$ and $\frac{c}{d}$ be two rational numbers. The equation

$$\frac{a}{b} \cdot x = \frac{c}{d} \tag{6.5.11}$$

has one and only one solution for x. This solution x is the rational number

$$x = \frac{b}{a} \cdot \frac{c}{d} \tag{6.5.12}$$

PROOF: The fact that if a solution to (6.5.11) exists, it must be unique, is left to the reader. (Use the proof of Theorem 4.3.1 in Chapter 4 as a guide.)

Next, consider the equation

$$\frac{a}{b} \cdot x = \frac{c}{d} \tag{6.5.13}$$

Since $\frac{a}{b} \neq 0$, then by Property XI for the rational numbers, there is a rational number $\frac{b}{a}$ such that $\frac{a}{b} \cdot \frac{b}{a} = 1$, the multiplicative identity. Multiply both sides of (6.5.13) by $\frac{b}{a}$. Then we have

$$\frac{b}{a} \cdot \left(\frac{a}{b} \cdot x\right) = \frac{b}{a} \cdot \frac{c}{d} \tag{6.5.14}$$

By the associative law for multiplication,

$$\left(\frac{b}{a} \cdot \frac{a}{b}\right) \cdot x = \frac{b}{a} \cdot \frac{c}{d} \tag{6.5.15}$$

Using Property XI, the multiplicative inverse property, we have

$$1 \cdot x = \frac{b}{a} \cdot \frac{c}{d} \quad \text{or} \quad x = \frac{b}{a} \cdot \frac{c}{d} \tag{6.5.16}$$

Now, the expression (6.5.16) is *really* a solution since

$$\frac{a}{b}\cdot x = \frac{a}{b}\cdot\left[\frac{b}{a}\cdot\frac{c}{d}\right] = \left[\frac{a}{b}\cdot\frac{b}{a}\right]\cdot\frac{c}{d} = 1\cdot\frac{c}{d} = \frac{c}{d} \tag{6.5.17}$$

Thus, the theorem is proved.

The student should compare this proof with the proof given for Theorem 4.3.1 in Chapter 4.

EXAMPLE 6.5.3

Solve for x if

$$\frac{3}{2}\cdot x = \frac{5}{6} \tag{6.5.18}$$

Here, the multiplicative inverse of $\frac{3}{2}$ is $\frac{2}{3}$. Thus,

$$\frac{2}{3}\cdot\left(\frac{3}{2}\cdot x\right) = \frac{2}{3}\cdot\frac{5}{6} \tag{6.5.19}$$

or

$$\left(\frac{2}{3}\cdot\frac{3}{2}\right)\cdot x = \frac{2\cdot 5}{3\cdot 6} \tag{6.5.20}$$

or

$$1\cdot x = \frac{10}{18} \tag{6.5.21}$$

or

$$x = \frac{5}{9} \tag{6.5.22}$$

Thus, $\frac{5}{9}$ is the desired answer.

EXAMPLE 6.5.4

Find all x for which

$$\frac{3}{2} - \frac{4}{3}\cdot x = \frac{6}{5} + \frac{1}{2} \tag{6.5.23}$$

Here, we first find the sum of the right-hand side. Then,

$$\frac{6}{5} + \frac{1}{2} = \frac{6\cdot 2 + 5\cdot 1}{5\cdot 2} = \frac{12+5}{10} = \frac{17}{10} \tag{6.5.24}$$

Now, the original problem can be replaced by

$$\frac{3}{2} - \frac{4}{3}\cdot x = \frac{17}{10} \tag{6.5.25}$$

Add the additive inverse of $\frac{3}{2}$, namely $-\frac{3}{2}$, to both sides. Then, also using a property of subtraction,

$$-\frac{3}{2} + \left[\frac{3}{2} + \left(-\frac{4}{3}\cdot x\right)\right] = -\frac{3}{2} + \frac{17}{10} \tag{6.5.26}$$

or

$$\left[-\frac{3}{2} + \frac{3}{2}\right] + \left(-\frac{4}{3}\cdot x\right) = -\frac{3}{2} + \frac{17}{10} \tag{6.5.27}$$

or

$$0 + \left(-\frac{4}{3}\right)\cdot x = -\frac{3}{2} + \frac{17}{10} \tag{6.5.28}$$

or

$$-\frac{4}{3}\cdot x = -\frac{3}{2} + \frac{17}{10} \tag{6.5.29}$$

Now, $-\frac{4}{3} = \frac{^-4}{3}$ and $-\frac{3}{2} = \frac{^-3}{2}$. Thus,

$$\frac{^-4}{3}\cdot x = \frac{^-3}{2} + \frac{17}{10} \tag{6.5.30}$$

Thus, since $\frac{^-3}{2} + \frac{17}{10} = \frac{1}{5}$, we have

$$\frac{^-4}{3}\cdot x = \frac{1}{5} \tag{6.5.31}$$

The multiplicative inverse of $\frac{^-4}{3}$ is $\frac{^-3}{4}$, since their product is 1. Then,

$$\frac{^-3}{4}\cdot\left(\frac{^-4}{3}\cdot x\right) = \frac{^-3}{4}\cdot\frac{1}{5} \tag{6.5.32}$$

or

$$\frac{^-3}{4}\cdot\frac{^-4}{3}\cdot x = \frac{^-3\cdot 1}{4\cdot 5} \tag{6.5.33}$$

or

$$1\cdot x = x = \frac{^-3}{20} \tag{6.5.34}$$

Because of Theorem 6.5.2 we can think of a rational number $\frac{a}{b}$ as being the solution of an equation of the form

$$b\cdot x = a \tag{6.5.35}$$

where $a, b \geqq 1$ are two integers with gcd $\{a, b\} = 1$. For example, the rational number $\frac{5}{3}$ can now be thought of as the solution x of the equation

$$3\cdot x = 5 \tag{6.5.36}$$

Sometimes this interpretation of a rational number is useful for solving problems.

We are now in a position to explain why we have imposed the restriction that $b \neq 0$ for a ratio $\frac{a}{b}$. If we think of $\frac{a}{b}$ as the solution of the equation

$$b\cdot x = a \tag{6.5.37}$$

and if we allow $b = 0$, this means

$$0\cdot x = a \tag{6.5.38}$$

This equation has *no* solution for $a \neq 0$, and *any* solution is valid for $a = 0$. This ambiguity is one reason for not allowing a denominator to be zero.

Thus far, we have discussed *three* binary operations—addition, multiplication, and subtraction. We are now ready to introduce our fourth binary operation—*division.*

Definition 6.5.1

Let $\frac{a}{b} \neq 0$, $\frac{c}{d}$ be two rational numbers. To *divide* $\frac{c}{d}$ *by* $\frac{a}{b}$, written as

$$\frac{c}{d} \div \frac{a}{b} \tag{6.5.39}$$

means to find that rational number x for which

$$\frac{a}{b}\cdot x = \frac{c}{d} \tag{6.5.40}$$

In the expression (6.5.39), the rational number $\frac{c}{d}$ is called the *dividend* and $\frac{a}{b}$ is termed the *divisor.*

Notice that without Theorem 6.5.2, we cannot even talk of an operation of division. Also, because of Theorem 6.5.2 and Definition 6.5.1, we have the following relationship between the operations of multiplication and division.

Theorem 6.5.3

Let $\frac{a}{b} \neq 0$, $\frac{c}{d}$ be two rational numbers. Then

$$\frac{c}{d} \div \frac{a}{b} = \frac{c}{d}\cdot\frac{b}{a} \tag{6.5.41}$$

Thus, to divide $\frac{c}{d}$ by $\frac{a}{b}$ is the same as multiplying $\frac{c}{d}$ by $\frac{b}{a}$.

EXAMPLE 6.5.5

Find

$$\frac{2}{3} \div \frac{5}{7} \tag{6.5.42}$$

Here,

$$\frac{2}{3} \div \frac{5}{7} = \frac{2}{3}\cdot\frac{7}{5} = \frac{2\cdot 7}{3\cdot 5} = \frac{14}{15} \tag{6.5.43}$$

Of course, Theorem 6.5.3 gives us a relationship between the two binary operations of multiplication and division. The student should compare this result with Theorem 4.3.3 of Chapter 4.

Theorem 4.3.3 in Chapter 4 states that *to subtract b from a is the same as adding the additive inverse of b to a.* Theorem 6.5.3 of this chapter states that *to divide* $\frac{c}{d}$ *by* $\frac{a}{b}$ *is the same as multiplying the multiplicative inverse of* $\frac{a}{b}$ *by* $\frac{c}{d}$.

It is noted that the rational number $\frac{2}{3}$ is the solution of the equation

$$3 \cdot x = 2 \tag{6.5.44}$$

Because of this fact, the ratio of two rational numbers, say $\frac{a}{b}$ and $\frac{c}{d}$, namely

$$\frac{\frac{a}{b}}{\frac{c}{d}} \tag{6.5.45}$$

is the solution of the equation

$$\frac{c}{d} \cdot x = \frac{a}{b} \tag{6.5.46}$$

But the solution of this equation is

$$x = \frac{a}{b} \cdot \frac{d}{c} \tag{6.5.47}$$

Thus, we have the formula that

$$\frac{\frac{a}{b}}{\frac{c}{d}} = \frac{a}{b} \div \frac{c}{d} = \frac{a}{b} \cdot \frac{d}{c} \tag{6.5.48}$$

This is the rule which states *to divide two rational numbers, multiply the numerator* $\frac{a}{b}$ *by the multiplicative inverse* $\frac{d}{c}$ *of the denominator* $\frac{c}{d}$.

EXAMPLE 6.5.6

Evaluate

$$\frac{\frac{3}{5}}{\frac{4}{7}} \tag{6.5.49}$$

Here,

$$\frac{\frac{3}{5}}{\frac{4}{7}} = \frac{3}{5}\cdot\frac{7}{4} = \frac{3\cdot 7}{5\cdot 4} = \frac{21}{20} \tag{6.5.50}$$

With the set J of integers, we discovered two cancellation laws, one for addition and one for multiplication. The set R of rational numbers has not only these, but also a third, the Cancellation Law for Division.

Theorem 6.5.4

Cancellation Law for Division: If $\frac{a}{b}$, and $\frac{c}{d}\,\frac{e}{f} \neq 0$ are three rational numbers, then

$$\frac{\frac{c}{d}\cdot\frac{a}{b}}{\frac{c}{d}\cdot\frac{e}{f}} = \frac{\frac{a}{b}}{\frac{e}{f}} \tag{6.5.51}$$

PROOF: The quantity

$$\frac{\frac{c}{d}\cdot\frac{a}{b}}{\frac{c}{d}\cdot\frac{e}{f}} \tag{6.5.52}$$

is the solution x of the equation

$$\left(\frac{c}{d}\cdot\frac{e}{f}\right)\cdot x = \frac{c}{d}\cdot\frac{a}{b} \tag{6.5.53}$$

or

$$\frac{c}{d}\cdot\left(\frac{e}{f}\cdot x\right) = \frac{c}{d}\cdot\frac{a}{b} \tag{6.5.54}$$

Using the cancellation law for multiplication, we get

$$\frac{e}{f}\cdot x = \frac{a}{b} \tag{6.5.55}$$

The solution x of this is

$$x = \frac{\frac{a}{b}}{\frac{e}{f}} \tag{6.5.56}$$

Since such solutions are unique, the result follows.

Warning: A very common error made by students is to attempt to apply a cancellation law when the operation involved is addition. For example, in

$$\frac{\frac{1}{2}+\frac{1}{3}}{\frac{1}{2}} \tag{6.5.57}$$

one cannot "cancel" the $\frac{1}{2}$ to obtain $\frac{1}{3}$. Rather, one must find the sum of $\frac{1}{2}$ and $\frac{1}{3}$ in the numerator first. Thus,

$$\frac{\frac{1}{2}+\frac{1}{3}}{\frac{1}{2}} = \frac{\frac{5}{6}}{\frac{1}{2}} = \frac{5}{6}\cdot\frac{2}{1} = \frac{5}{3} \tag{6.5.58}$$

Exercise 6.5

1. Write the following fractions as rational numbers.

(a) $\frac{72}{144}$

(b) $\frac{727}{1441}$

(c) $\frac{163{,}264}{2{,}754{,}108}$

(d) $\frac{29+27}{52+26}$

(e) $\frac{228\cdot 19}{204\cdot 19}$

(f) $\frac{^-54}{81}$

(g) $\frac{72}{38}+\frac{16}{19}$

(h) $\frac{0}{5}$

(i) $\frac{27}{27}$

2. Carry out the indicated operations and write your answer as a rational number.

(a) $\dfrac{\frac{1}{7}}{\frac{7}{2}}$

(d) $\dfrac{\frac{1}{2}}{\frac{1}{3}+\frac{1}{2}}$

(b) $\dfrac{\frac{1}{2}+\frac{1}{3}}{\frac{1}{3}}$

(e) $\frac{2}{5}\cdot\left(\frac{3}{4}-\frac{1}{2}\right)$

(c) $\dfrac{\frac{1}{2}+\frac{1}{3}}{\frac{3}{2}-\frac{1}{2}}$

(f) $\dfrac{\frac{1}{15}+\frac{1}{5}}{\frac{1}{5}}$

3. Solve each of the following equations.

(a) $\frac{3}{4}\cdot x-\frac{2}{3}=\frac{1}{2}$

(b) $\frac{4}{5}\cdot x=\frac{2}{3}-\frac{3}{7}$

(c) $\frac{5}{2}\cdot x+\frac{0}{1}=\frac{5}{3}$

4. Perform the indicated operations and write your answer as a rational number.

(a) $\dfrac{\frac{3}{7}}{\frac{2}{5}}$

(e) $\dfrac{\frac{2}{3}+\frac{1}{2}}{\frac{11}{6}-2}$

(b) $\dfrac{\frac{6}{11}}{\frac{17}{21}}$

(f) $\dfrac{\frac{2}{3}+\frac{1}{4}}{\frac{11}{12}-\frac{2}{3}}$

(c) $\dfrac{\dfrac{7}{15}}{\dfrac{1}{9}}$ (g) $\dfrac{2-\dfrac{1}{5}}{7+\dfrac{1}{3}-\dfrac{1}{20}}$

(d) $\dfrac{\dfrac{16}{1}}{\dfrac{2}{1}}$

5. Determine the following:

(a) $\frac{2}{3} \div \frac{2}{3}$ (e) $\frac{0}{3} \div \frac{1}{5}$

(b) $\frac{2}{3} \div \frac{3}{2}$ (f) $\frac{0}{3} \div \frac{47}{92}$

(c) $\frac{8}{7} \div \frac{3}{14}$ (g) $\frac{24}{1} \div \frac{3}{1}$

(d) $\frac{5}{9} \div \frac{20}{27}$ (h) $24 \div 3$

6. Another way of dividing two ratios is illustrated by

$$\dfrac{\dfrac{2}{3}}{\dfrac{4}{9}} = \dfrac{\dfrac{2}{3}\cdot\dfrac{9}{4}}{\dfrac{4}{9}\cdot\dfrac{9}{4}} = \dfrac{\dfrac{2}{3}\cdot\dfrac{9}{4}}{1} = \dfrac{2}{3}\cdot\dfrac{9}{4} = \dfrac{3}{2}$$

7. What is wrong with the following?

$$\frac{9\cdot 15}{6} = \frac{\overset{3}{\cancel{9}}\cdot\overset{5}{\cancel{15}}}{\underset{2}{\cancel{6}}} = \frac{3\cdot 5}{2} = \frac{15}{2}$$

8. Mike's age is three-fourths of Katy's. If Mike is nine, how old is Katy?

9. Danny spends one-third of his money on popcorn and then spends three-fourths of what is left on a theater ticket. If he has 60¢ left, how much did he begin with?

10. Six men can complete a job in ten days. If four men work on this job for three days, how long will it take three men to finish the job?

11. What part of 4 is $2\frac{1}{2}$? What part of $\frac{3}{4}$ is $\frac{2}{3}$?

12. In Canada, 98 U.S. cents is equivalent to 100 Canadian cents. Also 5 Canadian quarts is equivalent to 4 U.S. quarts. If an Imperial gallon (5 Canada quarts) of gasoline is sold for 50 Canadian cents, what is the corresponding cost for one U.S. gallon (4 U.S. quarts) in American cents?

6.6 ORDERING THE SET R OF RATIONAL NUMBERS

Recall that the concept of a *positive* integer was introduced in Chapter 5. Here it was discovered that the set of positive integers was the same as the set N of natural numbers. Also, we found that the sum and product of two positive integers was positive. Finally, the trichotomy law was proved. This law said that any integer a had to be either positive or equal to zero or its additive inverse ^{-}a had to be positive.

In this section, we shall talk about *positive* rational numbers. It will turn out that the properties previously proved valid for integers are also true for rational numbers.

DEFINITION 6.6.1

Let $\frac{a}{b}$ be any rational number. If the numerator a is positive, then the rational number $\frac{a}{b}$ is said to be *positive*. That is, the rational number $\frac{a}{b}$ is positive if and only if a is positive.

For example, the rational number $\frac{3}{2}$ is positive since 3 is positive.

We are now led to the following important result concerning positive rational numbers.

Theorem 6.6.1

The set R of rational numbers obeys the following two laws:

(I) If $\frac{a}{b}$ and $\frac{c}{d}$ are two positive rational numbers, their sum $\frac{a}{b}+\frac{c}{d}$ and the product $\frac{a}{b}\cdot\frac{c}{d}$ are both positive rational numbers.

(II) For any rational number $\frac{a}{b}$, we have a trichotomy law, namely,

$$\text{either } \frac{a}{b} \text{ is positive or } \frac{a}{b} \text{ is zero or } -\frac{a}{b} \text{ is positive} \tag{6.6.1}$$

These three possibilities are mutually exclusive.

PROOF: (I) Suppose $\frac{a}{b}$ and $\frac{c}{d}$ are two positive rational numbers. Then the integers a and c are positive. Then, by the definition of addition of rational numbers [cf. (6.2.11)], we have

$$\frac{a}{b} + \frac{c}{d} = \frac{a \cdot d + b \cdot c}{b \cdot d} \tag{6.6.2}$$

But $a \cdot d$ and $b \cdot c$ and $b \cdot d$ are positive integers. Thus, $a \cdot d + b \cdot c$ is positive. Hence, the rational number

$$\frac{a \cdot d + b \cdot c}{b \cdot d} \tag{6.6.3}$$

is positive since its numerator is positive.

Next, we have

$$\frac{a}{b} \cdot \frac{c}{d} = \frac{a \cdot c}{b \cdot d} \tag{6.6.4}$$

Again, since $a \cdot c$ is positive, the rational number $\frac{a \cdot c}{b \cdot d}$ is positive. Thus, the sum and product of two positive rational numbers are positive rational numbers.

(II) Because of the trichotomy law for the set J of integers, any integer a is either positive, zero, or its additive inverse is positive. Thus, a rational number, since it is the ratio of two integers, is either zero (when its numerator is zero) or positive (when its numerator is positive). Since

$$-\frac{a}{b} = \frac{{}^{-}a}{b} \tag{6.6.5}$$

it is seen that the additive inverse $-\frac{a}{b}$ of a rational number $\frac{a}{b}$ is positive when the numerator a is negative. Since no other possibilities for a can occur, the theorem is proved.

The trichotomy law for the set R of rational numbers leads us to the following definition.

DEFINITION 6.6.2

A rational number $\frac{a}{b}$ is called *negative* if and only if it is neither positive nor zero.

Thus, a rational number $\frac{a}{b}$ is negative if and only if $-\frac{a}{b}$ is positive.

For example, the rational number $\frac{^-2}{3}$ is negative since it is neither positive nor zero. Now, $\frac{^-2}{3}$ is obviously not zero. Since $^-2$ is not a positive integer, then $\frac{^-2}{3}$ cannot be a positive rational number. Hence, it must be a negative rational number.

Because of Theorem 6.6.1, we are led to conclude that the product of a positive rational number with a negative rational number is a negative rational number. Also, the product of two negative rational numbers is a positive rational number.

An easy corollary of the above facts is that the product of any rational number with itself is never negative. That is, for any rational number $\frac{a}{b}$, we have that

$$\frac{a}{b} \cdot \frac{a}{b} \tag{6.6.6}$$

is always either positive or zero. Moreover, $\frac{a}{b} \cdot \frac{a}{b}$ is zero if and only if $\frac{a}{b}$ is zero.

Exercise 6.6.1

1. Determine whether the following rational numbers are positive, zero, or negative.

(a) $\frac{^-3}{2}$ (f) $\frac{2}{5}$

(b) $\frac{6}{7}$ (g) $\frac{4}{3}$

(c) 0 (h) 1

(d) $^-4$ (i) 2

(e) $\frac{^-5}{3}$ (j) $^-3$

After the concept of a positive integer was introduced, we then discussed a way of ordering the integers. Now that we have defined a positive rational number, we are ready to introduce an ordering relation for the rational numbers.

Definition 6.6.3

Let $\frac{a}{b}$ and $\frac{c}{d}$ be any two rational numbers. We shall say that $\frac{a}{b}$ *is less than* $\frac{c}{d}$, written as

$$\frac{a}{b} < \frac{c}{d} \tag{6.6.7}$$

if and only if the difference $\frac{c}{d} - \frac{a}{b}$ is a positive rational number.

For example, $\frac{2}{3} < \frac{3}{4}$, since

$$\frac{3}{4} - \frac{2}{3} = \frac{9-8}{12} = \frac{1}{12} \tag{6.6.8}$$

is positive.

Also, $\frac{5}{4} < \frac{9}{7}$ since

$$\frac{9}{7} - \frac{5}{4} = \frac{36-35}{28} = \frac{1}{28} \tag{6.6.9}$$

is positive.

Here, as with the integers, there is the relation *greater than*. A rational number $\frac{c}{d}$ *is greater than* $\frac{a}{b}$, written as

$$\frac{c}{d} > \frac{a}{b} \tag{6.6.10}$$

if and only if the difference $\frac{c}{d} - \frac{a}{b}$ is a positive rational number.

For example, $-\frac{1}{2} > -\frac{2}{3}$ since

$$\left(-\frac{1}{2}\right) - \left(-\frac{2}{3}\right) = \frac{^-1}{2} - \frac{^-2}{3} = \frac{^-1}{2} + \left[-\left(\frac{^-2}{3}\right)\right]$$

$$= \frac{^-1}{2} + \frac{2}{3} = \frac{^-3 + 4}{6} = \frac{1}{6} \tag{6.6.11}$$

which is positive.

Of course, it is immediately clear that

$$\frac{a}{b} < \frac{c}{d} \quad \text{if and only if} \quad \frac{c}{d} > \frac{a}{b} \tag{6.6.12}$$

Definition 6.6.4

Let $\frac{a}{b}$ and $\frac{c}{d}$ be two rational numbers. We shall say that $\frac{a}{b}$ *is less than or equal to* $\frac{c}{d}$, written as

$$\frac{a}{b} \leqq \frac{c}{d} \tag{6.6.13}$$

if and only if either $\frac{a}{b} < \frac{c}{d}$ or $\frac{a}{b} = \frac{c}{d}$.

Of course, to say that $\frac{c}{d}$ *is greater than or equal to* $\frac{a}{b}$, written as

$$\frac{c}{d} \geqq \frac{a}{b} \tag{6.6.14}$$

means that either $\frac{c}{d} > \frac{a}{b}$ or $\frac{c}{d} = \frac{a}{b}$.

As before, it is clear that

$$\frac{a}{b} \leqq \frac{c}{d} \quad \text{if and only if} \quad \frac{c}{d} \geqq \frac{a}{b} \tag{6.6.15}$$

For example, we know that

$$\frac{1}{4} < \frac{1}{3} \tag{6.6.16}$$

since

$$\begin{aligned}\frac{1}{3} - \frac{1}{4} = \frac{1}{3} + \left(-\frac{1}{4}\right) = \frac{1}{3} + \frac{^-1}{4} &= \frac{4 \cdot 1 + 3(^-1)}{3 \cdot 4} \\ &= \frac{4 + (^-3)}{12} = \frac{1}{12}\end{aligned} \tag{6.6.17}$$

and $\frac{1}{12}$ is positive. Thus, we can also say that $\frac{1}{3}$ is greater than $\frac{1}{4}$, or $\frac{1}{3} > \frac{1}{4}$, since $\frac{1}{4} < \frac{1}{3}$.

Because of the way in which we have ordered rational numbers, we can display them on a rational line as follows:

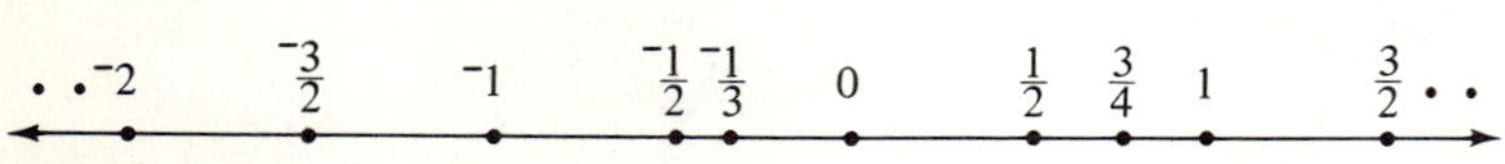

FIG. 6.12

Here, of course, we are using the natural ordering for the set R of rational numbers, so that when a rational number x is placed to the left of a rational number y, then we know $x < y$.

Suppose we have a set of points such that to each point there corresponds exactly one rational number and to each rational number there corresponds exactly one point. Such a set of points is called a *rational line.* In this way, we can represent rational numbers by points on a line. Since we are only considering rational numbers, the set of all the points associated to these rational numbers are said to form a *rational* line.

If we treat the concept of *less than or equal to* as a relation, it will obey four laws, as given in the following theorem.

Theorem 6.6.2

The relation $\leqq$ between rational numbers has the following four properties:

(I) *Reflexive Law:* For any rational number $\frac{a}{b}$, it is true that $\frac{a}{b} \leqq \frac{a}{b}$.

(II) *Anti-Symmetric Law:* Let $\frac{a}{b}, \frac{c}{d}$ be two rational numbers. If

$\frac{a}{b} \leqq \frac{c}{d}$ and $\frac{a}{b} \geqq \frac{c}{d}$, then $\frac{a}{b} = \frac{c}{d}$.

(III) *Transitive Law:* Let $\frac{a}{b}, \frac{c}{d}, \frac{e}{f}$ be three rational numbers. If

$\frac{a}{b} \leqq \frac{c}{d}$ and $\frac{c}{d} \leqq \frac{e}{f}$, then $\frac{a}{b} \leqq \frac{e}{f}$.

(IV) *Trichotomy Law:* For any two rational numbers $\frac{a}{b}$ and $\frac{c}{d}$, we have

$$\text{either} \quad \frac{a}{b} < \frac{c}{d} \quad \text{or} \quad \frac{a}{b} = \frac{c}{d} \quad \text{or} \quad \frac{a}{b} > \frac{c}{d} \tag{6.6.18}$$

For a proof of this result the student is referred to Theorem 5.1.1 in which the word *integer* is replaced by the words *rational number.*

Because every result from Sections 5.2 and 5.3 (which treats the algebra of inequalities and absolute value) is a consequence of Theorem 5.1.1, we are able to say that these results are equally true of the relation $\leqq$ involving *rational numbers.*

Thus, for example, if $\frac{a}{b} \leqq \frac{c}{d}$ and if $\frac{e}{f}$ is *positive,* it follows upon multiplying by $\frac{e}{f}$ that

$$\frac{a}{b}\cdot\frac{e}{f} \leqq \frac{c}{d}\cdot\frac{e}{f} \tag{6.6.19}$$

Had $\frac{e}{f}$ been negative rational number, then the inequality of (6.6.19) would have been reversed.

As another example, if $\frac{a}{b}$ is a known positive rational number and if

$$|x| = \frac{a}{b} \tag{6.6.20}$$

then

$$x = \frac{a}{b} \quad \text{or} \quad x = \frac{^{-}a}{b} \tag{6.6.21}$$

EXAMPLE 6.6.1

Find all x for which

$$3x + 5 \leqq 9 - 4x \tag{6.6.22}$$

Here, we use the same rules as we did in Chapter 5. Thus, adding $^{-}5$ to both sides gives

$$3x \leqq 4 - 4x \tag{6.6.23}$$

Adding $4x$ to both sides, we have

$$7x \leqq 4 \tag{6.6.24}$$

Now $\frac{1}{7}$ is a positive rational number. Multiplying both sides by $\frac{1}{7}$, we have

$$x \leqq \frac{4}{7} \tag{6.6.25}$$

Thus, all rational numbers less than or equal to $\frac{4}{7}$ will be solutions of (6.6.22).

The solution set for Example 6.6.1 is given in Fig. 6.13.

FIG. 6.13

EXAMPLE 6.6.2

Find all x for which

$$\left|\frac{4}{3} + x\right| \leqq \frac{2}{5} \tag{6.6.26}$$

Here, we know that

$$\frac{^-2}{5} \leqq \frac{4}{3} + x \leqq \frac{2}{5} \tag{6.6.27}$$

Add $\frac{^-4}{3}$. Then

$$\frac{^-2}{5} + \frac{^-4}{3} \leqq x \leqq \frac{2}{5} + \frac{^-4}{3} \tag{6.6.28}$$

or

$$\frac{^-26}{15} \leqq x \leqq \frac{^-14}{15} \tag{6.6.29}$$

Thus, all rational numbers between $\frac{^-26}{15}$ and $\frac{^-14}{15}$ are solutions. Figure 6.14 gives the solution set in the rational line.

FIG. 6.14

We can also state the following additional result.

Theorem 6.6.3

Let $a > 0$, $b > 0$ be two positive integers. If

$$b < a \tag{6.6.30}$$

then

$$0 < \frac{1}{a} < \frac{1}{b} \qquad (6.6.31)$$

For example, since $3 < 5$, then by Theorem 6.6.3 we know that $\frac{1}{5} < \frac{1}{3}$.

We now treat some rather complicated inequality problems. The student should study each example carefully.

EXAMPLE 6.6.3

Find all x for which

$$\frac{2}{x} < \frac{1}{3} \qquad (6.6.32)$$

Since no denominator of a rational number can be zero, we know that $x \neq 0$. By the trichotomy law,

$$\text{either} \quad x > 0 \quad \text{or} \quad x < 0 \qquad (6.6.33)$$

Let us consider each of these separately. Suppose $x > 0$. If we multiply by x, we obtain

$$2 \leqq \frac{x}{3} \qquad (6.6.34)$$

since x is positive. If we multiply by 3, then

$$6 \leqq x \qquad (6.6.35)$$

Thus, we have that

$$x > 0 \quad \text{and} \quad x \geqq 6 \qquad (6.6.36)$$

The intersection of these two sets, $x > 0$ and $x \geqq 6$, is

$$x \geqq 6 \qquad (6.6.37)$$

Suppose $x < 0$. If we multiply by x, we obtain

$$2 \geqq \frac{x}{3} \qquad (6.6.38)$$

the inequality being reversed since x is negative. If we multiply by 3, then

$$6 \geqq x \qquad (6.6.39)$$

Thus, we have

$$x < 0 \quad \text{and} \quad x \leqq 6 \qquad (6.6.40)$$

The intersection of these two sets is

$$x < 0 \tag{6.6.41}$$

Since we know that $x > 0$ *or* $x < 0$, it follows that the set union of the two solutions given by (6.6.37) and (6.6.41) is the complete solution.

Thus, the solution S of the problem (6.6.32) is

$$S = \{x \mid x < 0 \text{ or } x \geqq 6\} \tag{6.6.42}$$

That is, any rational number less than zero or greater than or equal to six will make (6.6.32) a true statement. Figure 6.15 shows this solution set on the rational line.

FIG. 6.15

EXAMPLE 6.6.4

Find all x for which

$$\frac{3}{x-2} \geqq 2 \tag{6.6.43}$$

This problem is much like the previous example, and is treated the same way.

It is known that the denominator $x - 2$ cannot be zero. This means that $x \neq 2$. By the trichotomy law, we must have

$$\text{either} \quad x > 2 \quad \text{or} \quad x < 2 \tag{6.6.44}$$

Again let us treat these two possibilities separately. Suppose $x > 2$. This means that the quantity $x - 2 > 0$. Multiply (6.6.43) by $x - 2$. Then, we obtain

$$3 \geqq 2(x-2) \tag{6.6.45}$$

or

$$3 \geqq 2x - 4 \tag{6.6.46}$$

or

$$7 \geqq 2x \tag{6.6.47}$$

or

$$\frac{7}{2} \geqq x \tag{6.6.48}$$

Thus, we know that

$$x > 2 \quad \text{and} \quad x \leqq \frac{7}{2} \tag{6.6.49}$$

Here, the intersection is

$$2 < x \leqq \frac{7}{2} \tag{6.6.50}$$

Next, suppose $x < 2$. This means that $x - 2 < 0$. If we multiply (6.6.43) by $x - 2$, we have

$$3 \leqq 2(x - 2) \tag{6.6.51}$$

the inequality sign being reversed, since $x - 2$ is negative. Now,

$$3 \leqq 2x - 4 \tag{6.6.52}$$

or

$$7 \leqq 2x \tag{6.6.53}$$

or

$$\frac{7}{2} \leqq x \tag{6.6.54}$$

Thus, we have

$$x < 2 \quad \text{and} \quad \frac{7}{2} \leqq x \tag{6.6.55}$$

The intersection of these two sets is the empty set $\varnothing$.

Now, we know that either $x < 2$ *or* $x > 2$. This indicates that we should find the union of the solutions found under these two situations. Thus, the complete solution S is

$$S = \left\{x \mid 2 < x \leqq \frac{7}{2}\right\} \cup \varnothing \tag{6.6.56}$$

or

$$S = \left\{x \mid 2 < x \leqq \frac{7}{2}\right\} \tag{6.6.57}$$

That is, any rational number between 2 and $\frac{7}{2}$, including $\frac{7}{2}$, makes (6.6.43) a true statement. See Fig. 6.16.

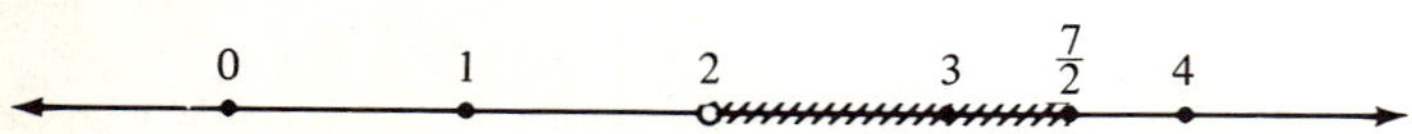

FIG. 6.16

Exercise 6.6.2

1. Find all rational numbers x for which

(a) $|x| = \frac{5}{2}$ (c) $|x| = 0$

(b) $|x| = \frac{1}{2}$ (d) $|x| = -\frac{2}{3}$

Show the solutions on the rational line.

2. Find all rational numbers x for which

(a) $|3 - 2x| \leqq 2$ (c) $|2x + 4| \geqq 3$
(b) $|4x + 1| = 3$ (d) $|x + 4| < 0$

Show the solutions on the rational line.

3. Find all rational numbers x for which

(a) $3x + 5 \leqq 2x - 7$ (e) $17x + 1 = \frac{9}{2}x - \frac{5}{3}$

(b) $^{-}3x - \frac{2}{3} \leqq \frac{3}{2}x + 1$ (f) $\frac{3x}{2} + 1 \geqq 2x - \frac{1}{3}$

(c) $9x - \frac{7}{2} \leqq \frac{4}{3}x + 2$ (g) $4x - 1 \leqq 2x + \frac{3}{4}$

(d) $(^{-}8x - 10) + \frac{2}{3} \leqq 5x + 1$

Show the solutions on the rational line.

4. Find all rational numbers x for which

(a) $\frac{2}{x} \geqq 3$ (c) $2x - 3 \geqq \frac{2}{3}$

(b) $\frac{x}{2} \leqq 4$ (d) $\frac{3}{2} \leqq \frac{4}{x}$

Show the solutions on the rational line.

5. Find all x for which $\frac{3}{x + 4} \leqq 7$. Show the solutions on the rational line.

6. Find all x for which $\frac{9}{x-3} \geqq 2$. Show the solutions on the rational line.

7. Find all x for which $\frac{3}{x} \leqq 2$ and $5 - x \geqq 7$. Show the solutions on the rational line.

8. Solve the following for x.

(a) $\frac{2x-3}{2} \leqq 5$

(b) $\frac{2}{x} \leqq 4$

(c) $\frac{3x+1}{x-2} < 1$

(d) $\frac{x}{x+2} > 4$

(e) $\left|x - \frac{2}{3}\right| = \frac{1}{2}$

(f) $\left|1 - \frac{1}{2}x\right| < 5$

(g) $\left|\frac{3x+2}{4}\right| = 6$

(h) $\left|\frac{2x+1}{3}\right| < 4$

(i) $\left|\frac{3x-1}{5}\right| > 1$

(j) $\left|\frac{5x-4}{3}\right| \leqq 7$

Show the solutions on the rational line.

9. (a) Mike has $\frac{3}{4}$ of a pound of fishing worms. If he gives his friend Dan $\frac{1}{3}$ of these worms, how many worms will he have left?

(b) Katy has eaten $\frac{5}{6}$ of a box of candy. She gives Mary $\frac{1}{2}$ of the remaining candy. How much candy did Mary receive?

10. Find the value of

(a) $-\frac{2}{3} \div \frac{1}{4}$

(b) $-\frac{2}{7} \div \frac{^-6}{5}$

(e) $\frac{1}{2} \cdot \left(\frac{2}{3} \div \frac{1}{4}\right) \cdot \frac{1}{2}$

(f) $\frac{1}{2} + \left(\frac{2}{3} \div \frac{3}{4}\right)$

(c) $\left(\frac{1}{2}\cdot\frac{2}{3}\right) \div \left(\frac{1}{4}\cdot\frac{1}{2}\right)$

(d) $\left(\frac{1}{2}\cdot\frac{2}{3}\right) \div \left(\frac{1}{4}\cdot\frac{1}{2}\right)$

(g) $\left(\frac{1}{2} - \frac{1}{3}\cdot\frac{1}{4}\right) \div 2$

(h) $\left(\frac{3}{2}\cdot\frac{^-4}{7} + \frac{3}{4}\right) \div \frac{5}{4}$

11. Mike has 75¢. If he gives $\frac{1}{3}$ of this to Dan and $\frac{2}{5}$ of it to Kate, how much did each one receive? How much does Mike have left?

12. Even though the set N of natural numbers is a proper subset of the set R of rational numbers, it turns out that the cardinality of N and R is the *same*. That is, N and R have the same cardinal number. As a hint, for showing this the student should notice that we can display the rational numbers in the following manner:

$$
\begin{array}{ccccccccccccccc}
0, & \frac{1}{1}, & \frac{^-1}{1}, & \frac{2}{1}, & \frac{^-2}{1}, & \frac{1}{2}, & \frac{^-1}{2}, & \frac{3}{1}, & \frac{^-3}{1}, & \frac{1}{3}, & \frac{^-1}{3}, & \frac{4}{1}, & \frac{1}{4}, & \frac{^-4}{1}, & \frac{^-1}{4}, \cdots \\
\updownarrow & \updownarrow & \updownarrow & \updownarrow & \updownarrow & \updownarrow & \updownarrow & \updownarrow & \updownarrow & \updownarrow & \updownarrow & \updownarrow & & & \\
1, & 2, & 3, & 4, & 5, & 6, & 7, & 8, & 9, & 10, & 11, & 12, & \cdots & &
\end{array}
$$

13. Find a rational number between $\frac{2}{3}$ and $\frac{3}{4}$. If $\frac{a}{b} < \frac{c}{d}$, can you find a rational number between $\frac{a}{b}$ and $\frac{c}{d}$?

14. Does the rational number $\frac{a+c}{b+d}$ lie between $\frac{a}{b}$ and $\frac{c}{d}$, with $\frac{a}{b} < \frac{c}{d}$? Verify your result.

7 The Real Number System R#

7.1 THE DENSENESS PROPERTY OF RATIONAL NUMBERS

In Chapter 6 we defined the meaning of the rational line:

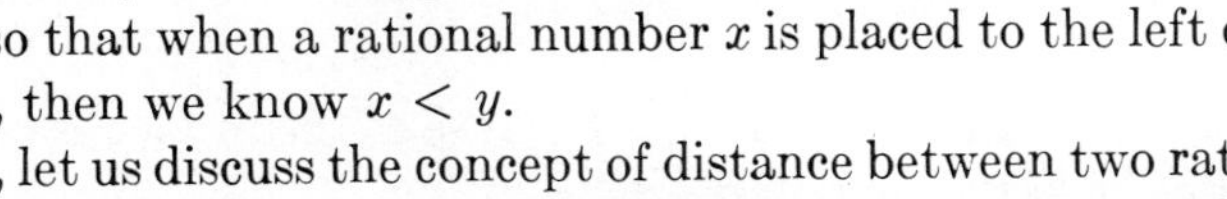

FIG. 7.1

Here, of course, we are using the natural ordering for the set R of rational numbers so that when a rational number x is placed to the left of a rational number y, then we know $x < y$.

Next, let us discuss the concept of distance between two rational points on such a line. It will be seen that this definition coincides with what most of us would agree is the natural way of measuring distance between two points.

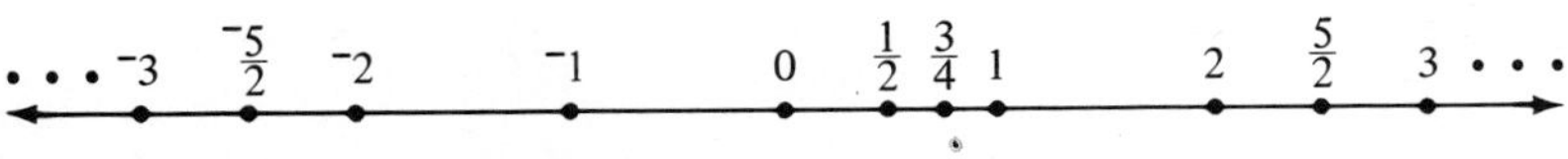

FIG. 7.2

DEFINITION 7.1.1

If r_1 and r_2 are two rational numbers associated with two points P_1 and P_2 on a rational line, the *distance d from P_1 to P_2*, written as $d(P_1, P_2)$, is

$$d = d(P_1, P_2) = |r_2 - r_1| \tag{7.1.1}$$

For example, the distance from $\frac{5}{2}$ to $\frac{3}{4}$ is

$$d = \left|\frac{5}{2} - \frac{3}{4}\right| = \left|\frac{7}{4}\right| = \frac{7}{4} \tag{7.1.2}$$

EXAMPLE 7.1.1

Find the distance from $\frac{3}{2}$ to $-\frac{1}{4}$.

Here, we must compute the absolute value of the *difference* between these two numbers. Thus,

$$\left|\frac{3}{2} - \left(-\frac{1}{4}\right)\right| = \left|\frac{3}{2} + \frac{1}{4}\right| = \left|\frac{7}{4}\right| = \frac{7}{4} \tag{7.1.3}$$

Thus, the distance from $\frac{3}{2}$ to $-\frac{1}{4}$ is $\frac{7}{4}$.

Note: In this section, only the theoretical considerations of distance are discussed. In Chapter 10, the concepts of measurement and various units of measurement will be introduced.

Because of the one-to-one correspondence between rational numbers and points on the rational line, we can use the words *point* and *number* interchangeably. Thus, we can speak of the point $\frac{3}{2}$ or the point $-\frac{3}{2}$ without confusing the issue.

The following result is easy to see because of the properties of absolute value discussed in earlier chapters.

Theorem 7.1.1

The distance between points on a rational line has the following three properties:

(I) *Positive Definite*

If P_1 and P_2 are two points, then

$$d(P_1, P_2) \geqq 0 \quad \text{and} \quad d(P_1, P_2) = 0 \quad \text{if and only if} \quad P_1 = P_2 \tag{7.1.4}$$

(II) *Symmetric*

If P_1 and P_2 are two points, then

$$d(P_1, P_2) = d(P_2, P_1) \tag{7.1.5}$$

(III) *Triangular Inequality*

If P_1, P_2, P_3 are three points, then

$$d(P_1, P_3) \leqq d(P_1, P_2) + d(P_2, P_3) \tag{7.1.6}$$

PROOF: Let r_1, r_2, r_3 be the rational numbers associated to the points P_1, P_2, P_3, respectively. Now,

$$|r_2 - r_1| \geqq 0 \quad \text{and} \quad |r_2 - r_1| = 0 \quad \text{if and only if} \quad r_2 = r_1 \tag{7.1.7}$$

Thus, distance is positive definite. Also,

$$|r_2 - r_1| = |r_1 - r_2| \tag{7.1.8}$$

so that distance is symmetric. Finally,

$$\begin{aligned} |r_3 - r_1| &= |(r_3 - r_2) + (r_2 - r_1)| \\ &\leqq |r_3 - r_2| + |r_2 - r_1| \end{aligned} \tag{7.1.9}$$

so that distance obeys the triangular inequality.

We are now ready to introduce a concept related to that of distance which is of importance for the study of geometry.

DEFINITION 7.1.2

Let P_1 and P_3 be two distinct points. *The point P_2 is said to be between P_1 and P_3 if and only if*

$$d(P_1, P_3) = d(P_1, P_2) + d(P_2, P_3) \tag{7.1.10}$$

For example, the point $\frac{2}{3}$ is between $\frac{1}{2}$ and $\frac{3}{4}$. For, if $P_1 = \frac{1}{2}$, $P_2 = \frac{2}{3}$, $P_3 = \frac{3}{4}$, then

$$\begin{aligned} d(P_1, P_3) &= \left|\frac{3}{4} - \frac{1}{2}\right| = \left|\frac{1}{4}\right| = \frac{1}{4} \\ d(P_1, P_2) &= \left|\frac{2}{3} - \frac{1}{2}\right| = \left|\frac{1}{6}\right| = \frac{1}{6} \\ d(P_2, P_3) &= \left|\frac{3}{4} - \frac{2}{3}\right| = \left|\frac{1}{12}\right| = \frac{1}{12} \end{aligned} \tag{7.1.11}$$

and

$$d(P_1, P_2) + d(P_2, P_3) = \frac{1}{6} + \frac{1}{12} = \frac{3}{12} = \frac{1}{4} = d(P_1, P_3) \tag{7.1.12}$$

This can easily be seen by looking at these points on the rational line.

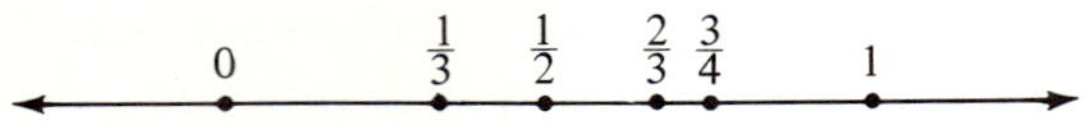

FIG. 7.3

The point $\frac{1}{3}$ is not between $\frac{1}{2}$ and $\frac{3}{4}$. For, if $P_1 = \frac{1}{2}$, $P_2 = \frac{1}{3}$, $P_3 = \frac{3}{4}$, then

$$\begin{aligned} d(P_1, P_3) &= \left|\frac{3}{4} - \frac{1}{2}\right| = \left|\frac{1}{4}\right| = \frac{1}{4} \\ d(P_1, P_2) &= \left|\frac{1}{3} - \frac{1}{2}\right| = \left|-\frac{1}{6}\right| = \frac{1}{6} \\ d(P_2, P_3) &= \left|\frac{3}{4} - \frac{1}{3}\right| = \left|\frac{5}{12}\right| = \frac{5}{12} \end{aligned} \tag{7.1.13}$$

and

$$d(P_1, P_2) + d(P_2, P_3) = \frac{1}{6} + \frac{5}{12} = \frac{7}{12} \neq \frac{1}{4} = d(P_1, P_3) \tag{7.1.14}$$

Thus, $\frac{1}{3}$ is not between $\frac{1}{2}$ and $\frac{3}{4}$. See Fig. 7.3 above.

Next, we introduce a property of the rational numbers which is sometimes difficult to understand. Simply stated, this property says that no matter what two different rational numbers one chooses, there is at least one other rational number between them.

Theorem 7.1.2

Between any two distinct rational numbers, there can be found another rational number. That is, the set R of rational numbers is *dense in itself*.

PROOF: Suppose a and b are two distinct rational numbers. We may assume that

$$a < b \tag{7.1.15}$$

Consider the rational number

$$\frac{a + b}{2} \tag{7.1.16}$$

We shall show that $\frac{a + b}{2}$ is between a and b.

First of all,

$$a < \frac{a + b}{2} \tag{7.1.17}$$

because

$$\frac{a + b}{2} - a = \frac{b - a}{2} > 0 \tag{7.1.18}$$

since $b > a$. Also,

$$\frac{a + b}{2} < b \tag{7.1.19}$$

because

$$b - \frac{a + b}{2} = \frac{b - a}{2} > 0 \tag{7.1.20}$$

since $b > a$. From (7.1.17) and (7.1.19), we have that

$$a < \frac{a + b}{2} < b \tag{7.1.21}$$

Thus, we have found one rational number between the two rational numbers a and b.

DEFINITION 7.1.3

If a and b are two rational numbers with $a < b$, the rational number $\frac{a + b}{2}$ is called the *midpoint of a and b* since it is an equal distance from both a and b.

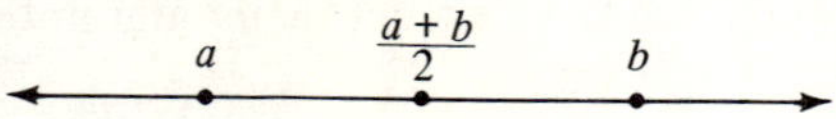

FIG. 7.4

The denseness of the set R of rational numbers is illustrated in Fig. 7.5.

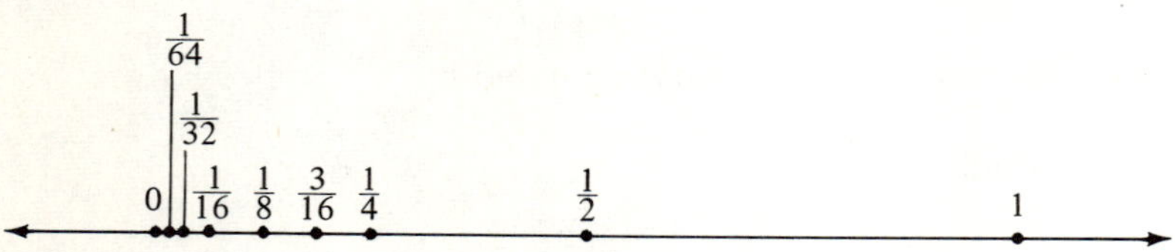

FIG. 7.5

Exercise 7.1

1. Find the distance d between the following points. Sketch on a rational line.

   (a) $P_1 = \frac{2}{3}, P_2 = -\frac{3}{2}$

   (b) $P_1 = -\frac{1}{2}, P_2 = 0$

   (c) $P_1 = \frac{4}{3}, P_2 = -\frac{4}{3}$

   (d) $P_1 = -\frac{1}{2}, P_2 = -\frac{4}{3}$

   (e) $P_1 = \frac{4}{5}, P_2 = \frac{9}{2}$

2. For the following three points, determine whether or not the point P_2 is between P_1 and P_3. Sketch on a rational line.

   (a) $P_1 = -\frac{2}{3}, P_2 = \frac{1}{2}, P_3 = \frac{1}{3}$

   (b) $P_1 = \frac{1}{4}, P_2 = \frac{2}{3}, P_3 = \frac{3}{4}$

   (c) $P_1 = -\frac{3}{2}, P_2 = -\frac{7}{4}, P_3 = -1$

(d) $P_1 = \frac{3}{2}, P_2 = \frac{5}{4}, P_3 = \frac{1}{4}$

(e) $P_1 = -2, P_2 = -\frac{5}{2}, P_3 = -\frac{7}{4}$

3. Find the midpoint of the following rational numbers. Sketch on a rational line.

(a) $\frac{1}{2}, \frac{3}{4}$

(b) $\frac{2}{3}, 1$

(c) $0, \frac{1}{4}$

(d) $-\frac{1}{3}, -\frac{1}{5}$

(e) $\frac{1}{2}, -\frac{1}{2}$

4. (a) Is there a largest rational number less than 3? Why?
   (b) Is there a smallest rational number greater than 2? Why?

5. Prove that the rational number $\frac{a+c}{b+d}$ lies between the rational numbers $\frac{a}{b}$ and $\frac{c}{d}$, with $\frac{a}{b} < \frac{c}{d}$.

★7.2 LEAST UPPER BOUND; GREATEST LOWER BOUND

Let $X \neq \emptyset$ be a subset of the rational number system R and let $\leqq$ denote the natural ordering for R.

Definition 7.2.1

If there is an element u in X so that

$$\text{every } x \text{ in } X \leqq u \tag{7.2.1}$$

then u is called an *absolute maximum to* X. If there is an element m in X so that

$$m \leqq \text{every } x \text{ in } X \tag{7.2.2}$$

then m is termed an *absolute minimum to* X.

Theorem 7.2.1

If a set X has an absolute maximum u (or an absolute minimum m), then it is unique.

PROOF: Suppose the set X has two absolute maxima, say u_1 and u_2, with $u_1 \neq u_2$. Then by definition,

$$u_1 \in X, \qquad u_2 \in X \tag{7.2.3}$$

But by equation (7.2.1) we have

$$u_1 \leqq u_2, \qquad u_2 \leqq u_1 \tag{7.2.4}$$

By the anti-symmetric law,

$$u_1 = u_2 \tag{7.2.5}$$

Hence, a set X can have at most one absolute maximum.

DEFINITION 7.2.2

An element b, either in X or *not* in X, is called an *upper bound to* X if and only if

$$x \leqq b \text{ for every } x \text{ in } X \tag{7.2.6}$$

Similarly, an element a, either in X or *not* in X, is termed a *lower bound* a *to* X if and only if

$$a \leqq x \text{ for every } x \text{ in } X \tag{7.2.7}$$

EXAMPLE 7.2.1

Let the ordering be the natural ordering and let X be the set

$$X = \{2, 3, 5, 8\} \tag{7.2.8}$$

Here the absolute maximum is 8 because

$$8 \in X \quad \text{and} \quad 2 \leqq 8, \qquad 3 \leqq 8, \qquad 5 \leqq 8, \qquad 8 \leqq 8 \tag{7.2.9}$$

The absolute minimum is 2 since

$$2 \in X \quad \text{and} \quad 2 \leqq 2, \qquad 2 \leqq 3, \qquad 2 \leqq 5, \qquad 2 \leqq 8 \tag{7.2.10}$$

Some upper bounds to X are

$$8,\ 9,\ \frac{17}{2},\ \frac{25}{2},\ \cdots \tag{7.2.11}$$

Some lower bounds to X are

$$0,\ 2,\ -\frac{1}{2},\ \frac{5}{4},\ \cdots \tag{7.2.12}$$

DEFINITION 7.2.3

Let $X \neq \varnothing$ be a nonempty subset X of the rational number system R. Suppose X has an upper bound. Then the set B of upper bounds b to X is not empty. If B has an absolute minimum L, then L is called the *least upper bound L of X*, written as

$$L = \text{lub}\,[X] \tag{7.2.13}$$

Suppose X has no upper bound. Then the *least upper bound to* X is defined as *plus infinity*, and we write

$$\text{lub}\,[X] = +\infty \tag{7.2.14}$$

Here $+\infty$ is a *symbol* used to denote the least upper bound of a set X which has no upper bounds.

Similarly, suppose X has a lower bound. Then the set A of lower bounds a to X is not empty. If A has an absolute maximum l, then l is termed a *greatest lower bound l* to X, written as

$$l = \text{glb}\,[X] \tag{7.2.15}$$

If X has no lower bound, then the *greatest lower bound to* X is defined as *minus infinity*, and we write

$$\text{glb}\,[X] = -\infty \tag{7.2.16}$$

Again, $-\infty$ is a symbol denoting the greatest lower bound of a set which has no lower bounds.

EXAMPLE 7.2.2

Consider the set X of rational numbers defined by

$$X = \left\{x \,\middle|\, -\frac{1}{2} < x \leqq 3\right\} \tag{7.2.17}$$

Clearly, 3 is the absolute maximum of X since $3 \in X$, and every element in X is less than or equal to 3.

Now, X has *no* absolute minimum. Clearly, $-\frac{1}{2}$ will not do since $-\frac{1}{2} \notin X$. The denseness property of R tells us there is no rational number "closest" to $-\frac{1}{2}$. For, if there were one, say a, then the midpoint of $-\frac{1}{2}$

and a would be a rational number even closer to $-\frac{1}{2}$. Thus, X has *no* absolute minimum.

Some upper bounds to X are $3, \frac{7}{2}, \frac{31}{10}$, etc. That is, the set B of upper bounds is

$$B = \{x \mid x \geqq 3\} \tag{7.2.18}$$

Since lower bounds to X are $-\frac{1}{2}, -\frac{3}{4}, {}^-1, {}^-7$, etc., the set A of lower bounds to X is

$$A = \left\{x \,\middle|\, x \leqq -\frac{1}{2}\right\} \tag{7.2.19}$$

The absolute minimum of B is 3. Thus,

$$\text{lub } [X] = 3 \tag{7.2.20}$$

The absolute maximum of A is $-\frac{1}{2}$. Thus,

$$\text{glb } [X] = -\frac{1}{2} \tag{7.2.21}$$

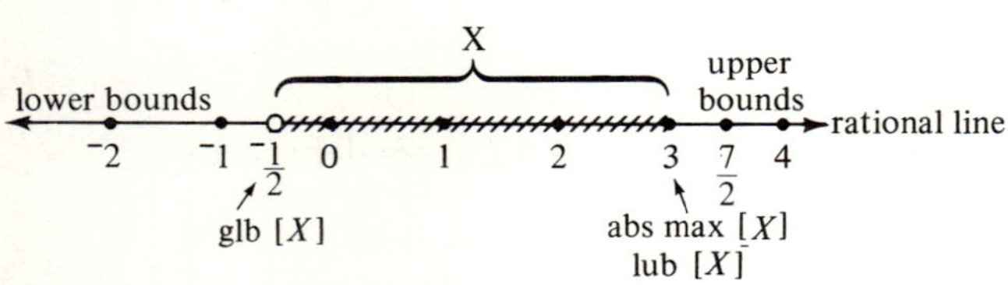

FIG. 7.6

Exercise 7.2

*1. Find the absolute maximum, absolute minimum, set B of upper bounds, set A of lower bounds, least upper bound, and greatest lower bound for the following sets of rational numbers. Sketch each set and label your answers on a number line.

(a) $X = \left\{x \,\middle|\, \frac{2}{3} < x < 3\right\}$ (d) $X = \left\{x \,\middle|\, x > \frac{1}{2}\right\}$

(b) $X = \left\{x \middle| -\frac{2}{5} < x < \frac{2}{3}\right\}$ (e) $X = \left\{x \middle| x < \frac{5}{4}\right\}$

(c) $X = \{x \mid -1 < x < 0\}$ (f) $X = \{x \mid 0 < x\}$

⋆2. Is 3 an upper bound for the set of all negative rational numbers? Why? Is it a least upper bound? Why?

⋆3. Prove that if a set X has a least upper bound it is unique.

⋆4. Give an example of a set X which has a least upper bound in X. Give an example of a set X which has a least upper bound not contained in X.

7.3 A DEFICIENCY OF THE SET R OF RATIONAL NUMBERS

In Chapter 3, the concepts of natural number and whole number were discussed. It soon became evident that the set W of whole numbers had many deficiencies. For example, every element in W does not possess an additive inverse. Because of this, we extended the set W. This new set we called the set J of integers.

Now, the integers, we soon discovered, also possessed deficiencies. Not every integer possessed a multiplicative inverse—for example, there is no integer which when multiplied by 2, gives 1. That is, 2 has no multiplicative inverse. To remedy this, we defined the set R of rational numbers.

Thus, we have so far extended our number systems from the set N of natural numbers, to the set W of whole numbers, to the set J of integers, to the set R of rational numbers. Also, the various properties associated with each of these number systems have been discussed.

We are now ready to make the last extension that this text will deal with. This is the extension of the set R of rational numbers to the *set $R\#$ of real numbers.*

Let us now discuss a deficiency of the set R of rational numbers, which leads us to extend it so as to remove the deficiency. The following result points out to us this deficiency.

Theorem 7.3.1

There is no rational number x whose square is 2. That is, the equation

$$x^2 = 2 \tag{7.3.1}$$

has *no* rational solution.

PROOF: Assume the contrary. Then suppose there is a rational number $\frac{a}{b}$ (remember that $b \geqq 1$ and gcd $\{a, b\} = 1$) so that

$$\frac{a}{b} \cdot \frac{a}{b} = 2 \tag{7.3.2}$$

If we multiply by $b^2 = b \cdot b$, we obtain

$$a^2 = 2 \cdot b^2 \tag{7.3.3}$$

Now (7.3.3) tells us that a^2 is an even integer. Hence, a itself must be even. This means that

$$a = 2k \tag{7.3.4}$$

where k is some integer. Replace the a of (7.3.3) by $2k$. Then

$$2k \cdot 2k = 2b^2 \tag{7.3.5}$$

or

$$2k^2 = b^2 \tag{7.3.6}$$

But this says that b^2 is an even integer. Hence, b is an even integer.

Now, we know that both a and b are even so that both a and b are divisible by 2. But

$$\text{gcd } \{a, b\} = 1 \tag{7.3.7}$$

This is impossible.

Hence, there is no rational number whose square is 2.

Theorem 7.3.1 points up a deficiency of rational numbers we would like to be able to remove. That is, we would like to be able to solve the equation

$$x^2 = 2 \tag{7.3.8}$$

Before doing this, let us try to find rational numbers whose square is close to 2. That is, let us attempt to find an approximate rational solution to the equation

$$x^2 = a, \qquad a > 0 \tag{7.3.9}$$

where a is *any* positive integer.

For example, consider the equation

$$x^2 = 2 \tag{7.3.10}$$

If we try 1 as a solution, $1^2 = 1$ is too small; and if we try 2 as a solution,

$2^2 = 4$ is too large. Thus, our solution x lies between 1 and 2, so that

$$1 < x < 2 \tag{7.3.11}$$

Let us try the midpoint of 1 and 2, namely $\frac{3}{2}$. Then

$$\frac{3}{2} \cdot \frac{3}{2} = \frac{9}{4} > 2 \tag{7.3.12}$$

Thus, $\frac{3}{2}$ is also too large, so the solution x is between 1 and $\frac{3}{2}$,

$$1 < x < \frac{3}{2} \tag{7.3.13}$$

Let us try the midpoint of 1 and $\frac{3}{2}$, namely $\frac{5}{4}$. Then

$$\frac{5}{4} \cdot \frac{5}{4} = \frac{25}{16} < 2 \tag{7.3.14}$$

Thus, $\frac{5}{4}$ is too small, so the solution x is between $\frac{5}{4}$ and $\frac{3}{2}$, that is,

$$\frac{5}{4} < x < \frac{3}{2} \tag{7.3.15}$$

By continuing these steps, we can get a relatively good approximation for the solution x of the equation $x^2 = 2$.

By this process, we obtain *rational approximations* to the solution of $x^2 = 2$.

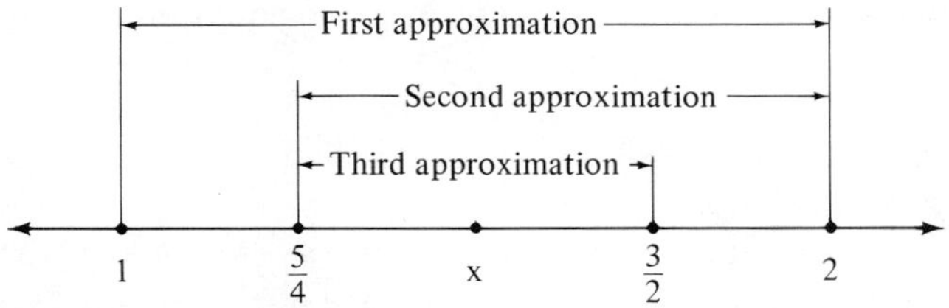

FIG. 7.7

Exercise 7.3

1. Using the equation $x^2 = 3$, $x > 0$, find a fourth approximation for x, starting with $1 < x < 2$. Sketch your results.

2. Using the equation $x^2 = 5$, $x > 0$, find a fifth approximation for x, beginning with $2 < x < 3$. Sketch your results.

7.4 THE REAL LINE

As was seen in Section 7.3, the rational line has "holes" or "gaps" in it, such as the point x where $x^2 = 2$. Although there is a point on the number line corresponding to each rational number, there is not a rational number corresponding to each point on the line. Numbers corresponding to these holes or gaps on the rational line are called *irrational numbers.*

The solution of $x^2 = 2$ is an example of an irrational number.

If we take the union of those points corresponding to rational numbers and those points corresponding to irrational numbers, the resulting line is called a *real line.* The numbers corresponding to these points on a real line are called *real numbers.* The set of real numbers is denoted by $R\#$.

If S denotes the set of irrational numbers, we have

$$S \cap R = \varnothing, \qquad S \cup R = R\# \tag{7.4.1}$$

Now, since the irrational numbers fill in the gaps left by the rational numbers, we speak of the set $R\#$ of real numbers being *complete.* It is this *completeness property* of $R\#$ which removes the aforementioned deficiency of the rational numbers.

Thus, we now have the fundamental relationship that

$$N \subset W \subset J \subset R \subset R\# \tag{7.4.2}$$

Now that we can solve equations such as $x^2 = a$, $a \geqq 0$, the following notation is introduced.

Definition 7.4.1

If $a \geqq 0$ is a real number, the *square root of a,* written as $\sqrt{a}$, is defined to be the nonnegative solution x of the equation

$$x^2 = a, \qquad x \geqq 0 \tag{7.4.3}$$

For example, $\sqrt{9} = 3$ since the positive solution of the equation $x^2 = 9$ is $x = 3$.

The above technique is only one of many ways we could have introduced the concept of a real number. Section 7.5 will deal with a different way of dealing with this concept.

Exercise 7.4[1]

1. A consequence of the discussion presented in Section 7.4 is that an equation of the form

$$x^n = a$$

where $n \geqq 2$ is a positive integer and $a \geqq 0$ is a real number, always has a unique positive solution for x in $R\#$. (The case for $n = 2$ has already been discussed.) This unique, positive solution x is called the *principal* nth *root of* a and is denoted by

$$x = \sqrt[n]{a}$$

If $a < 0$ and n is odd, the equation

$$x^n = a$$

has a unique negative solution for x in $R\#$ called the *principal* nth *root of* a, denoted by $\sqrt[n]{a}$. For example,

$$\sqrt[4]{16} = 2, \qquad \sqrt[3]{-27} = -3$$

Is there a solution for $x^n = a$ if $n \geqq 2$ is an even integer and $a < 0$ is a negative real number? Why? Is this a deficiency of the real number system $R\#$? Give examples of this.

2. Show that $3 + \sqrt{5}$ is irrational. [HINT: See Theorem 7.3.1.]

3. If r is any rational number, is $r + \sqrt{2}$ irrational? What about $r \cdot \sqrt{2}$? This shows there are infinitely many irrational numbers.

★4. If $X \subseteqq R$ (the set of rational numbers), find the absolute maximum, absolute minimum, set B of upper bounds, set A of lower bounds, the least upper bound, and the greatest lower bound. Sketch your results on the rational line.

(a) $X = \{x \mid \sqrt{3} < x \leqq 3\}$ (d) $X = \left\{x \,\middle|\, -1 \leqq x \leqq \frac{2}{3}\right\}$

(b) $X = \{x \mid x < \sqrt{2}\}$ (e) $X = \left\{x \,\middle|\, -\sqrt{2} < x \leqq \frac{1}{2}\right\}$

(c) $X = \{x \mid x \geqq 0\}$

★5. For $X \subseteqq R\#$ (the set of real numbers) answer the *same* questions as given in problem 4. Sketch your results on the real line.

[1] Exercises or problems designated by ★ are to be omitted if previous related optional section was omitted.

*6. In problem 4(b), the set X was bounded from above, but had *no* least upper bound. In problem 5(b), the set X was bounded from above and had a least upper bound. A distinguishing characteristic of real numbers is that any set of real numbers that has an upper bound also has a least upper bound. Is the same statement valid for lower bounds and greatest lower bounds? Give two examples to illustrate your answer.

*7. The "hole" or "gap" discussed in Section 7.4 can be illustrated in the following way. Let R denote the set of rational numbers. Let A and B be two nonempty subsets of R such that

(a) $A \cap B = \varnothing$
(b) $A \cup B = R$
(c) if $a \in A$, $b \in B$, then $a < b$
(d) A has no absolute maximum, B has no absolute minimum

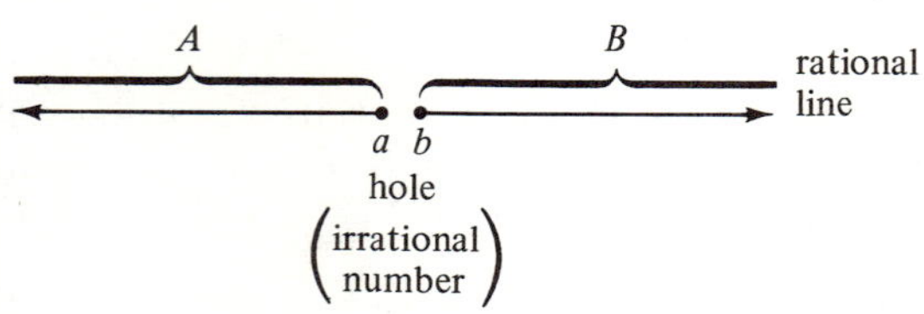

Prove that there cannot be found two subsets A and B of $R\#$, the real number system, with the above four properties. This shows there are *no* "holes" or "gaps" in the real line—that is, it is complete. [HINT: Assume there are two such sets A and B and show that B has no least upper bounds. This contradicts problem 6.]

7.5 DECIMAL REPRESENTATION FOR REAL NUMBERS

It becomes evident very quickly that the representation of a rational number as the ratio of two integers is difficult to work with as far as arithmetical computations are concerned. For example, it is difficult to compare or even add the two rational numbers,

$$\frac{385421}{462183} \quad \text{and} \quad \frac{654021}{843120} \tag{7.5.1}$$

Because of this difficulty, the concept of a *decimal fraction* is introduced.

DEFINITION 7.5.1

A *decimal fraction* is defined as a fraction whose denominator is an integral power of 10. For example:

$$\frac{36931}{100} \tag{7.5.2}$$

is a *decimal fraction* since its denominator 100 is an integral power of 10, namely $10 \cdot 10 = 10^2$.

If we think of 10^3 as being equal to $10 \cdot 10 \cdot 10$, and $10^4 = 10 \cdot 10 \cdot 10 \cdot 10$, etc., then any decimal fraction can take the form

$$\frac{x}{10^n} \tag{7.5.3}$$

where x is an integer and $n \geqq 1$ is a positive integer. Here, n is called the *power* of 10.

Note that $\frac{1}{5}$ is not a decimal fraction but $\frac{20}{100}$ is, even though $\frac{20}{100} = \frac{1}{5}$, that is, they are two forms of the same fraction. Evidently then, the *form* which expresses a fraction is quite important. The student should note that not all fractions can be expressed as decimal fractions; $\frac{1}{3}$ is an example of this.

Instead of writing a decimal fraction with its denominator as a power of ten, we shall introduce the following convention. A point, called the *decimal point*, is inserted between two digits of the numerator so that the number of places to the right of this point corresponds to the power of ten in the denominator.

For example, the decimal fraction $\frac{2312}{1000} = \frac{2312}{10^3}$ can now be written as 2.312; the decimal fraction $\frac{663201}{10^4}$ is 66.3201.

The operations of addition and multiplication of decimal fractions are illustrated by the following examples.

EXAMPLE 7.5.1

Find the sum of

$$4.16,\ 32.97,\ 625.203 \tag{7.5.4}$$

METHOD I: $$4.16 + 32.97 + 625.203 = \frac{416}{100} + \frac{3297}{100} + \frac{625203}{1000}$$

$$= \frac{4160}{1000} + \frac{32970}{1000} + \frac{625203}{1000}$$

$$= \frac{4160 + 32970 + 625203}{1000}$$

$$= \frac{662333}{1000} = 662.333$$

This method reverts back to the representation of a fraction as the ratio of two integers.

METHOD II:

$$\begin{array}{r} 4.16 \\ 32.97 \\ \underline{625.203} \\ 662.333 \end{array}$$

The second method is done by placing every number to be added in such a way that the decimal points lie under one another and then adding in the ordinary way. It is because of Method I that this easier Method II is acceptable and valid.

EXAMPLE 7.5.2

Find the product of

$$2.31 \text{ and } 62.2 \tag{7.5.5}$$

METHOD I: $$(2.31)\cdot(62.2) = \frac{231}{100}\cdot\frac{622}{10} = \frac{143682}{1000}$$

$$= 143.682$$

METHOD II:

$$\begin{array}{r} 62.2 \\ \underline{2.31} \\ 622 \\ 18\,660 \\ \underline{124\,400} \\ 143.682 \end{array}$$

In Method II, we multiply the two decimal fractions disregarding the decimal point. Then, in the answer, we count off the sum of the number of digits present up to the decimal point in each factor multiplied. Then, in 62.2 we have 1 digit and in 2.31 we have 2 digits counted before reaching the decimal point. In the product, then, we count off 3 digits and place the decimal point after the third digit.

Now, it is obvious that not every fraction can be written so that its denominator is a power of 10. In fact, only those rational numbers whose denominators are exactly powers of 2 or 5 can be expressed as decimal fractions. If the denominator contains prime factors other than 2 and 5, the rational number cannot be expressed as a decimal fraction. For example, $\frac{2}{3}$ cannot be expressed as a decimal fraction. We can express $\frac{2}{3}$, *approximately*, by actually dividing 3 into 2. Then,

$$\begin{array}{r} 0.6666 \\ 3\overline{)2.0000} \\ \underline{1\,8} \\ 20 \\ \underline{18} \\ 20 \\ \underline{18} \\ 2 \end{array} \tag{7.5.6}$$

It is clear that the quotient will continue to be 6 and the remainder 2 indefinitely. Then, we can represent $\frac{2}{3}$ by

$$\frac{2}{3} = 0.6666\cdots \tag{7.5.7}$$

This is an example of a *repeating decimal.*

Another example is $\frac{41}{99}$. Here

$$\begin{array}{r} .4141 \\ 99\overline{)41.000} \\ \underline{396} \\ 140 \\ \underline{99} \\ 410 \\ \underline{396} \\ 140 \\ \underline{99} \\ 41 \end{array} \tag{7.5.8}$$

Thus, $\frac{41}{99}$ can be represented by the repeating decimal

$$\frac{41}{99} = .4141\cdots \tag{7.5.9}$$

The decimal representation of a decimal fraction is called a *terminating decimal.*

It turns out that any rational number can be represented by either a terminating decimal or a repeating decimal.

Let us now show how to determine the rational number once its decimal representation is given. The decimal 0.33 was obtained from the rational number $\frac{33}{100}$. $\left(\text{The student should be warned that } 0.33 \neq \frac{1}{3}\right)$. However, $\frac{1}{3} = 0.333\cdots$, and we see that the repeating cycle is 3, which we will denote with a bar above it. We want to show that indeed $0.3\bar{3}\cdots$ is equal to $\frac{1}{3}$. Let

$$\begin{aligned} N &= 0.3\bar{3}\cdots \\ 10N &= 3.\bar{3}\cdots \end{aligned} \tag{7.5.10}$$

Subtract the first from the second; we get

$$\begin{aligned} 10N - N &= 3.\bar{3}\cdots - 0.3\bar{3}\cdots \\ 9N &= 3 \\ N &= \frac{3}{9} = \frac{1}{3} \end{aligned} \tag{7.5.11}$$

EXAMPLE 7.5.3

Find $0.45\overline{723}$ as a ratio of two integers. Let

$$\begin{aligned} N &= 0.45\overline{723} \\ 100N &= 45.\overline{723} \\ 100{,}000N &= 45{,}723.\overline{723} \end{aligned} \tag{7.5.12}$$

Subtracting,

$$99{,}900N = 45{,}678$$

$$N = \frac{45{,}678}{99{,}900} = \frac{7{,}613}{16{,}650} \tag{7.5.13}$$

The general procedure is as follows.

First, if the cycle does not begin with the decimal point, multiply by the necessary power of 10 to bring the decimal point immediately to the left of the cycle. In the above example, this was accomplished when we multiplied N by 100:

$$100N = 45.\overline{723}\cdots \tag{7.5.14}$$

Next, we count the number of digits in the cycle and multiply the above number by a power of 10 corresponding to this number of digits. Thus,

$$10^3 \cdot 100N = 45{,}723.\overline{723}\cdots \tag{7.5.15}$$

Now, the last number is larger than the previous. If we take the difference we will get an integer that is a multiple of N.

EXAMPLE 7.5.4

Find the rational number corresponding to $0.9\overline{9}\cdots$. Let N be that number. Then

$$\begin{aligned} N &= 0.9\overline{9}\cdots \\ 10N &= 9.\overline{9}\cdots \\ 10N - N &= 9.\overline{9}\cdots - 0.9\overline{9}\cdots \\ 9N &= 9 \\ N &= 1 \end{aligned} \tag{7.5.16}$$

Thus, the rational number corresponding to $0.9\overline{9}\cdots$ is 1, and we must write

$$1 = 0.9\overline{9}\cdots \tag{7.5.17}$$

Note that it is incorrect to list only a finite number of 9's.

EXAMPLE 7.5.5

Find the rational number corresponding to $0.250\overline{0}\cdots$.

$$\begin{aligned} N &= 0.250\overline{0}\cdots \\ 100N &= 25 \\ N &= \frac{25}{100} = \frac{1}{4} \end{aligned} \tag{7.5.18}$$

EXAMPLE 7.5.6

Find the rational number corresponding to $0.249\overline{9}\cdots$. Here

$$\begin{aligned} N &= 0.249\overline{9}\cdots \\ 100N &= 24.9\overline{9}\cdots \\ 1000N &= 249.\overline{9}\cdots \\ 900N &= 225 \\ N &= \frac{225}{900} = \frac{1}{4} \end{aligned} \tag{7.5.19}$$

Comparing the last two examples suggests the possibility of expressing a rational number which has a terminating decimal, as a repeating decimal without the necessity of a repeating cycle of zeros. For example,

$$10.5 = 10.4999\bar{9}\cdots \qquad (7.5.20)$$

Now, it is easy to construct a decimal representation which is neither of the above, that is, it *neither repeats nor terminates,* for example, .32465314 . . . . If we consider *all* such decimal representations, the ones that *terminate,* the ones that *repeat,* and those that do *neither,* we have another formulation for the concept of a *real number*. That is, we can define a real number to be any decimal representation.

Next, let us look at a way of approximating, by decimals, irrational numbers.

EXAMPLE 7.5.7

Find the square root of 52134.3.

(1) First of all, starting with the decimal point, group the digits in pairs as follows:

$$5 \quad 21 \quad 34 \quad . \quad 30 \quad 00 \qquad (7.5.21)$$

(2) Starting with the first such pair (here just 5), write the largest integer whose square is less than or equal to it. (Here $2^2 = 4$ and $3^2 = 9$; thus we choose 2.)

$$\begin{array}{l} \quad 2 \\ \sqrt{5 \quad 21 \quad 34 \quad . \quad 30} \end{array} \qquad (7.5.22)$$

(3) Multiply 2 by itself and subtract this from the first pair

$$\begin{array}{l} \quad 2 \\ \sqrt{5 \quad 21 \quad 34 \quad . \quad 30} \\ \quad \underline{4} \\ \quad 1 \end{array} \qquad (7.5.23)$$

(4) Double 2 and write 4 <u>?</u>. Thus,

$$\begin{array}{l} \qquad\quad 2 \\ \qquad\sqrt{5 \quad 21 \quad 34 \quad . \quad 30} \\ \qquad\quad \underline{4} \\ \qquad\quad 1 \\ \underline{4\ ?} \end{array} \qquad (7.5.24)$$

(5) Bring down the next pair (here 21). Thus,

$$\begin{array}{l} \qquad\quad 2 \\ \qquad\sqrt{5 \quad 21 \quad 34 \quad . \quad 30} \\ \qquad\quad \underline{4} \\ \qquad\quad 1 \quad 21 \\ \underline{4\ ?} \end{array} \qquad (7.5.25)$$

(6) Determine how many 4_?_'s there are in 121. Here if we say 3, we must replace the _?_ by a 3. If we do this, we see that 4_3_ times 3 will exceed 121; so we choose 2. Then,

```
            2   2
          √5   21   34  .  30          (7.5.26)
            4
            -
            1  21
  4 2
```

(7) Write 2 above 21. Multiply 4_2_ by 2 and subtract from 121. Then

```
            2   2       .
          √5   21   34  .  30          (7.5.27)
            4
            -
            1  21
               84
               --
  4 2          37
```

(8) Now, repeat the above steps, beginning at step 4. Then

```
            2   2    8  .
          √5   21   34  .  30          (7.5.28)
            4
            -
            1  21
  4 2          84
               --
               37   34
  44 8         35   84
               -------
                1   50
```

and

```
            2   2    8  .   3
          √5   21   34  .  30
            4
            -
            1  21
  4 2          84                      (7.5.29)
               --
               37   34
  44 8         35   84
               -------
                1   50   30
  456 3         1   36   89
               ------------
                    13   41
```

Thus, 228.3 is an approximation to the square root of 52134.3. Of course, we could have continued this process indefinitely, obtaining a decimal representation to as many decimal places as desired.

EXAMPLE 7.5.8

Find $\sqrt{2}$ to 3 decimal places. Here

$$
\begin{array}{r r r r r r}
 & 1 & . & 4 & 1 & 4 \\ \cline{2-6}
 & \sqrt{2} & . & 00 & 00 & 00 \\
 & 1 & & & & \\ \cline{2-2}
 & 1 & & 00 & & \\
2\,\underline{4} & & & 96 & & \\ \cline{2-4}
 & & & 4 & 00 & \\
28\,\underline{1} & & & 2 & 81 & \\ \cline{4-5}
 & & & 1 & 19 & 00 \\
282\,\underline{4} & & & 1 & 12 & 96 \\ \cline{4-6}
 & & & & 6 & 04
\end{array}
\tag{7.5.30}
$$

Thus, 1.414 is a four-digit approximation to $\sqrt{2}$.

The above technique for approximating square roots is called a *square root algorithm.*

Exercise 7.5

1. Which of the following rational numbers can be expressed as decimal fractions? For those that are, write out the decimal representation as a terminating decimal.

(a) $\frac{3}{5}$ (f) $\frac{3}{11}$

(b) $\frac{3}{4}$ (g) $\frac{5}{2}$

(c) $\frac{5}{6}$ (h) $\frac{3}{16}$

(d) $\frac{8}{6}$ (i) $\frac{2}{25}$

(e) $\frac{5}{9}$

2. Find the sum and product of

(a) 4.21, 3.9	(e) 0.0011, 91.8
(b) 632.2, 41	(f) 88.5, 124
(c) 0.0023, 4.23	(g) 98.1, 2.2
(d) 9.86, 0.12	(h) 0.0031, 2.31

3. Write the following rational numbers as either repeating or terminating decimals.

(a) $\frac{9}{2}$	(g) $\frac{2}{13}$
(b) $\frac{2}{9}$	(h) $\frac{5}{3}$
(c) $\frac{21}{17}$	(i) $\frac{6}{7}$
(d) $\frac{1}{8}$	(j) $\frac{15}{22}$
(e) $\frac{2}{7}$	(k) $\frac{21}{16}$
(f) $\frac{5}{6}$	(l) $\frac{8}{9}$

4. Find approximations of the following irrational numbers to 5 decimal places.

(a) $\sqrt{3}$	(e) $\sqrt{56.1}$
(b) $\sqrt{21}$	(f) $\sqrt{100.253}$
(c) $\sqrt{13.214}$	(g) $\sqrt{421.22}$
(d) $\sqrt{22551.1}$	(h) $\sqrt{0.003211}$

5. Find the rational number given the following

(a) 0.25	(f) 3.38
(b) 0.125	(g) $5.99\overline{9}\cdots$
(c) 0.7	(h) $.\overline{4}\cdots$
(d) 0.5	(i) $0.31\overline{7}\cdots$
(e) $0.\overline{509}\cdots$	

7.6 PERCENT

Thus far in our study of fractions, we have discussed two different ways of writing or representing fractions. First, we can write a fraction as the ratio of two integers, such as $\frac{1}{4}$ or $\frac{1}{3}$. A second way of writing a fraction is as a decimal, such as 0.25 or $0.33\overline{3}\cdots$.

In this section, we shall look at a third way of representing fractions.

If a person places \$100 in a bank which advertises 4 percent interest, paid annually, then at the end of one year, the amount of money received for interest is \$4. We arrive at the figure of \$4 by multiplying $100 \cdot \frac{4}{100} = 4$.

Thus, if in a fraction the denominator is 100, and the numerator is P, that is,

$$\frac{P}{100} \tag{7.6.1}$$

then, we can say P *percent* and write $P\%$.

For example, we can write

$$25\% = \frac{25}{100}, \qquad 50\% = \frac{50}{100}, \qquad 10\% = \frac{10}{100} \tag{7.6.2}$$

In this way, we can represent the fraction $\frac{23}{100}$ by

$$23\% \quad \text{or} \quad 0.23 \quad \text{or} \quad \frac{23}{100} \tag{7.6.3}$$

EXAMPLE 7.6.1

Find 28% of 500. Here,

$$28\% \text{ of } 500 = \frac{28}{100} \cdot 500 = 28 \cdot 5 = 140 \tag{7.6.4}$$

EXAMPLE 7.6.2

What percent of 25 is 5? Here we have

$$P\% \text{ of } 25 = 5 \tag{7.6.5}$$

Then we have

$$\frac{P}{100}\cdot 25 = 5 \quad \text{or} \quad P = 20 \tag{7.6.6}$$

Then, 20% of 25 is 5.

EXAMPLE 7.6.3

144 is 16% of what number? Here we have

$$16\% \text{ of } x = 144 \tag{7.6.7}$$

or

$$\frac{16}{100}\cdot x = 144 \tag{7.6.8}$$

or

$$16x = 14400 \tag{7.6.9}$$

or

$$x = 900 \tag{7.6.10}$$

Then, 16% of 900 is 144.

Exercise 7.6

1. Express the following fractions as both decimals and percents.

(a) $\frac{32}{100}$ (f) $\frac{18}{75}$

(b) $\frac{4}{25}$ (g) $\frac{13}{20}$

(c) $\frac{16}{10}$ (h) $\frac{108}{400}$

(d) $\frac{18}{50}$ (i) $\frac{21}{300}$

(e) $\frac{3}{4}$

2. Express the following percents both as fractions in lowest terms and as decimals.

(a) 22% (e) 20%
(b) 14% (f) 422%

(c) 12% (g) 12.5%
(d) 125% (h) 85%

3. Solve the following:

   (a) What is 15% of 90?
   (b) What is 40% of 1600?
   (c) What percent of 248 is 31?
   (d) What percent of 160 is 4?
   (e) Find 65% of 780.
   (f) Find 122% of 12.

4. A man deposits \$25 in a bank account at 4% interest per year. After one year, what is his balance? After two years (if he does not withdraw any funds)?

5. A man purchases a home for \$25,000. He must pay closing costs of 2%. How much will closing costs be? If he sells the home for \$26,000, he must pay the real estate man 6% commission. What is his true profit?
   [HINT: total cost = 25,000 + closing costs
   money received on sale = 26,000 − real estate commission
   profit = sale money − total cost]

8 Numeration Systems

8.1 NUMERALS AND NUMBERS

In our everyday use of numbers we give little thought to the symbols that are being used. Quite often people confuse concepts such as *numeral* and *number*. It is important for us to be able to distinguish between numbers and numerals. Students may argue that the distinction is not important, and in many instances this is true. However, a good deal will be gained by understanding the difference between number and numeral.

We shall call a symbol which names (represents, symbolizes, denotes) a number a *numeral*. A number is the abstract concept which we have been discussing in Chapters 3–7. Numerals may be marks made on paper, chiseled on stone, or scratched on wood, but these are never numbers. For instance, IV and 4 are different numerals which represent the *same number*. A numeral is nothing more than a representation of a number. It is easier to write 6 than it is to write //////.

8.2 NUMERATION SYSTEMS

A numeration system is a means for denoting or naming numbers. It involves two concepts: a set of symbols, and some laws to enable us to denote numbers by combinations of symbols.

The first numeration systems probably used a mark, a notch, or something similar to denote the number one. Successive numbers were usually denoted by repetitions of the same mark. People soon realized the difficulties encountered in using repeated marks to represent a number. To represent large numbers it was necessary to make many marks, which was time consuming, cumbersome, and tiring. Also, the actual number was not readily distinguishable (that is, one could not see immediately how many marks there were). Thus, it became desirable to have different symbols denote numbers of different sizes.

Let's look at a numeration system which is not well known—the ancient Egyptian system of numerals.

TABLE 8.1

ANCIENT EGYPTIAN SYSTEM OF NUMERALS

Ancient Egyptian Numeral	Name of Number	Decimal Numeral
/	One	1
∩	Ten	10
𓍢	One hundred	100
𓆼	One thousand	1000
𓂭	Ten thousand	10000

By looking at Table 8.1, one can see that the Egyptian numerals were based on the number 10 (that is, grouped by factors of 10). To write a numeral such as 2354, one would use two 𓆼's to represent 2000, three 𓍢's to represent 300, five ∩'s to represent 50, and four /'s. Taking all of these and putting them in any order, we have

2354 = 𓆼𓆼 𓍢𓍢𓍢 ∩∩∩∩∩ //// (8.2.1)

To write forty-two, we need four ∩'s and two /'s. Since the position of the numerals does not matter, we could write forty-two any of the ways listed below.

∩∩∩∩// ∩∩// ∩∩ ∩/∩∩/∩

/∩∩∩∩/ ∩∩/∩∩/ (8.2.2)

Today, there are two numeration systems which are familiar to most people of the world. They are the Roman and Hindu-Arabic systems. Let us discuss both of these in some detail.

In the Roman system, the basic set of numerals is

I (one); V (five); X (ten); L (fifty);

C (one hundred); D (five hundred); M (one thousand)

(8.2.3)

The rule for forming other numbers from this basic set is as follows.

When two basic numerals are placed next to each other, and when the left one is smaller than the right one, the rule says to subtract the left one from the right one; if the left one is larger than or equal to the right one, the rule is to add the two.

For example,

IX is the symbol for nine (ten − one) (8.2.4)

XV is the symbol for fifteen (ten + five) (8.2.5)

XIV is the symbol for fourteen [ten + (five − one)] (8.2.6)

Note: When three or more symbols are placed together, always subtract before adding.

In the Hindu–Arabic system, the *digit* is the basic symbol. The digits, as mentioned before, are the ten integers,

$$0, 1, 2, 3, 4, 5, 6, 7, 8, 9 \tag{8.2.7}$$

To represent an integer larger than 9 in this system, we place two or more digits next to each other with the understanding that each digit has a *place value*. This place value is explained by the following examples.

Consider the number forty-two. Since forty-two can be thought of as 4 tens added to 2 ones, we represent forty-two by 42. That is, the digit 4 is given a place value of ten; the digit 2 a place value of one. For the integer three hundred twenty-seven, we see that it consists of 3 hundreds, 2 tens, and 7 ones. Hence, we write it as 327, in which the digit 3 is given a place value of hundreds, 2 a place value of tens, and 7 a place value of ones. If four digits are placed together, such as with 6254, the digit 6 is given a place value of thousands, 2 a place value of hundreds, and so on.

Interested students can refer to books on the history of mathematics for a further study of ancient Egyptian and other systems of numeration.

8.3 EXPONENTS

Before we can continue our study of numeration systems, it is necessary for us to discuss the concept of an *exponent*. As will be seen later in this chapter, the use of exponents will make the study of representation of numbers much easier.

Definition 8.3.1

Let $n \geqq 1$ be a natural number and let x be a real number. The expression *x raised to the power n*, written as

$$x^n \tag{8.3.1}$$

in which x is called the *base* and n is the *exponent,* means

$$(1) \quad x^1 = x$$

$$(2) \quad \text{If } x^{n-1} \text{ is known, then } x^n = x \cdot x^{n-1} \tag{8.3.2}$$

EXAMPLE 8.3.1

Find 2^5. Here,

$$2^5 = 2^{5-1} \cdot 2$$

$$2^5 = 2^4 \cdot 2 \quad \text{but} \quad 2^4 = 2^{4-1} \cdot 2 = 2^3 \cdot 2$$

$$\text{Substituting,} \quad 2^5 = 2^3 \cdot 2 \cdot 2 \quad \text{but} \quad 2^3 = 2^{3-1} \cdot 2 = 2^2 \cdot 2 \tag{8.3.3}$$

$$\text{Therefore,} \quad 2^5 = 2^2 \cdot 2 \cdot 2 \cdot 2 \quad \text{but} \quad 2^2 = 2^{2-1} \cdot 2 = 2^1 \cdot 2 = 2 \cdot 2$$

$$2^5 = \underbrace{2 \cdot 2 \cdot 2 \cdot 2 \cdot 2}_{\text{five factors}} = 32$$

EXAMPLE 8.3.2

Find 10^6. Here,

$$\begin{aligned} 10^6 &= 10^5 \cdot 10 \\ 10^6 &= 10^4 \cdot 10 \cdot 10 \\ 10^6 &= 10^3 \cdot 10 \cdot 10 \cdot 10 \qquad (8.3.4) \\ 10^6 &= 10^2 \cdot 10 \cdot 10 \cdot 10 \cdot 10 \\ 10^6 &= 10 \cdot 10 \cdot 10 \cdot 10 \cdot 10 \cdot 10 = 1{,}000{,}000 \end{aligned}$$

Notice that 2^5 is equal to the product of five factors of 2 and 10^6 is equal to the product of 10 six times. This is true in general for

$$\begin{aligned} b^n &= b^{n-1} \cdot b \qquad \text{but} \quad b^{n-1} = b^{n-2} \cdot b \\ b^n &= b^{n-2} \cdot b \cdot b \quad \text{but} \quad b^{n-2} = b^{n-3} \cdot b \\ b^n &= b^{n-3} \cdot b \cdot b \cdot b \qquad (8.3.5) \\ &\vdots \\ b^n &= \underbrace{b \cdot b \cdot b \cdot b \cdots b}_{n \text{ factors}} \end{aligned}$$

Note: The above definition is a restricted one since both rational and irrational numbers can also be considered for exponents, that is, $b^{1/4}$ or $b^{\sqrt{2}}$ and so on. However, we will not work with such exponents.

Let us now discuss the meaning of raising a real number x to the power zero and to a negative integral power.

DEFINITION 8.3.2

Let $m \geqq 1$ be a positive integer, so that ^-m will be a negative integer. Let x be a real number.

(1) If $x \neq 0$, then x *raised to the power* 0, written as x^0, is defined by

$$x^0 = 1, \qquad x \neq 0 \tag{8.3.6}$$

(2) The expression x *raised to the power* ^-m, written as

$$x^{-m} \tag{8.3.7}$$

means

$$x^{-m} = \frac{1}{x^m} \tag{8.3.8}$$

EXAMPLE 8.3.3
Find 6^{-3}.

$$6^{-3} = \frac{1}{6^3} = \frac{1}{6 \cdot 6 \cdot 6} = \frac{1}{216} \tag{8.3.9}$$

EXAMPLE 8.3.4
Find 3^{-2}.

$$3^{-2} = \frac{1}{3^2} = \frac{1}{3 \cdot 3} = \frac{1}{9} \tag{8.3.10}$$

EXAMPLE 8.3.5

(1) $2^0 = 1$

(2) $(^-5)^0 = 1$ (8.3.11)

EXAMPLE 8.3.6
Evaluate $(^-3)^{-4}$. Here,

$$(^-3)^{-4} = \frac{1}{(^-3)^4} = \frac{1}{(^-3)(^-3)(^-3)(^-3)}$$

$$= \frac{1}{9 \cdot 9} = \frac{1}{81} \tag{8.3.12}$$

The following properties of exponents are useful in dealing with problems involving powers. Let x, y be two real numbers and let m, n represent two integers.

(I) $x^m \cdot x^n = x^{m+n}$

That is, when multiplying two powers with the same base, retain the base and add the exponents.

For example,

$$2^3 \cdot 2^4 = 2^{3+4} = 2^7 \qquad (8.3.13)$$

(II) $$(x^n)^m = x^{n \cdot m}$$

That is, when raising a power to a power, retain the base and multiply the exponents.

For example,

$$(3^2)^{-1} = 3^{2 \cdot (-1)} = 3^{-2} = \frac{1}{3^2} = \frac{1}{9} \qquad (8.3.14)$$

(III) $$x^m \cdot y^m = (x \cdot y)^m$$

That is, to find the product of two bases raised to the same power, multiply the bases and retain the exponent.

For example,

$$3^2 \cdot 4^2 = (3 \cdot 4)^2 = 12^2 = 144 \qquad (8.3.15)$$

(IV) $$\frac{x^m}{x^n} = x^{m-n}$$

For example,

$$\frac{3^5}{3^3} = 3^{5-3} = 3^2 = 9 \qquad (8.3.16)$$

and

$$\frac{4^2}{4^5} = 4^{2-5} = 4^{-3} = \frac{1}{4^3} = \frac{1}{64} \qquad (8.3.17)$$

The following examples will help to illustrate how these four laws can be used.

EXAMPLE 8.3.7

Simplify the expression $\dfrac{5^3 \cdot 5^2 \cdot 5^{-3}}{5^3}$.

$$\begin{aligned}\frac{5^3 \cdot 5^2 \cdot 5^{-3}}{5^3} &= 5^{[3+2+(-3)]} \cdot 5^{-3} \\ &= 5^2 \cdot 5^{-3} \\ &= 5^{[2+(-3)]} = 5^{-1} = \frac{1}{5}\end{aligned} \qquad (8.3.18)$$

EXAMPLE 8.3.8

Simplify the expression $\dfrac{3^7 \cdot 2^3 \cdot 5^2}{3^8 \cdot 2 \cdot 5^2}$.

$$\begin{aligned}
\frac{3^7 \cdot 2^3 \cdot 5^2}{3^8 \cdot 2 \cdot 5^2} &= 3^7 \cdot 2^3 \cdot 5^2 \cdot 3^{-8} \cdot 2^{-1} \cdot 5^{-2} \\
&= 3^7 \cdot 3^{-8} \cdot 2^3 \cdot 2^{-1} \cdot 5^2 \cdot 5^{-2} \\
&= 3^{[7+(-8)]} \cdot 2^{[3+(-1)]} \cdot 5^{[2+(-2)]} \\
&= 3^{-1} \cdot 2^2 \cdot 5^0 = 3^{-1} \cdot 2^2 \cdot 1 \\
&= 3^{-1} \cdot 2^2 = \frac{1}{3} \cdot 4 = \frac{4}{3}
\end{aligned} \tag{8.3.19}$$

Another way is

$$\begin{aligned}
\frac{3^7 \cdot 2^3 \cdot 5^2}{3^8 \cdot 2 \cdot 5^2} &= \frac{3^7}{3^8} \cdot \frac{2^3}{2^1} \cdot \frac{5^2}{5^2} \\
&= 3^{(7-8)} \cdot 2^{(3-1)} \cdot 5^{(2-2)} \\
&= 3^{-1} \cdot 2^2 \cdot 5^0 \\
&= 3^{-1} \cdot 2^2 \cdot 1 \\
&= 3^{-1} \cdot 2^2 = \frac{1}{3} \cdot 4 = \frac{4}{3}
\end{aligned} \tag{8.3.20}$$

EXAMPLE 8.3.9

Simplify $\dfrac{3^2 \cdot 5^{-3} \cdot 2^4}{2^3 \cdot 3^{-2} \cdot 5^4}$.

$$\begin{aligned}
\frac{3^2 \cdot 5^{-3} \cdot 2^4}{2^3 \cdot 3^{-2} \cdot 5^4} &= (3^2 \cdot 5^{-3} \cdot 2^4) \cdot (2^3 \cdot 3^{-2} \cdot 5^4)^{-1} \\
&= (3^2 \cdot 5^{-3} \cdot 2^4) \cdot (2^3)^{-1} \cdot (3^{-2})^{-1} \cdot (5^4)^{-1} \\
&= (3^2 \cdot 5^{-3} \cdot 2^4) \cdot (2^{3(-1)}) \cdot (3^{(-2)(-1)}) \cdot (5^{4(-1)}) \\
&= (3^2 \cdot 5^{-3} \cdot 2^4) \cdot (2^{-3}) \cdot (3^2) \cdot (5^{-4}) \\
&= (3^2 \cdot 3^2) \cdot (5^{-3} \cdot 5^{-4}) \cdot (2^4 \cdot 2^{-3}) \\
&= (3^{[2+2]}) \cdot (5^{[(-3)+(-4)]}) \cdot (2^{[4+(-3)]}) \\
&= 3^4 \cdot 5^{-7} \cdot 2^1 = 81 \cdot \frac{1}{5^7} \cdot 2 \\
&= 81 \cdot \frac{1}{78125} \cdot 2 = \frac{162}{78125}
\end{aligned} \tag{8.3.21}$$

8.4 SCIENTIFIC NOTATION

Scientific notation is often used to express extremely large or small numbers in a compact form. In scientific notation a number is represented as a product of a number greater than or equal to 1 and less than 10, times a power of 10. For example, to express the number 2000 in scientific notation we would write 2×10^3; $0.000025 = 2.5 \times 10^{-5}$, where 0.000025 is the standard notation and 2.5×10^{-5} is the scientific notation.

EXAMPLE 8.4.1

Change each of the following from standard notation to scientific notation.

(a) 0.00781 $\qquad 0.00781 = 7.81 \times 10^{-3}$

(b) 341.6 $\qquad 341.6 = 3.416 \times 10^2 \qquad (8.4.1)$

(c) 1,330,000 $\qquad 1{,}330{,}000 = 1.33 \times 10^6$

EXAMPLE 8.4.2

Change each of the following from scientific notation to standard notation.

(a) 3.51×10^8 $\qquad 3.51 \times 10^8 = 3.51 \times 100{,}000{,}000 = 351{,}000{,}000$

(b) 7.6×10^{-4} $\qquad 7.6 \times 10^{-4} = 7.6 \times 0.0001 = 0.00076 \qquad (8.4.2)$

(c) 8.1×10^0 $\qquad 8.1 \times 10^0 = 8.1 \times 1 = 8.1$

Exercise 8.4

1. Simplify each of the following.

(a) $\dfrac{5^3 \cdot 5^2}{5^2}$ $\qquad$ (d) $2(3 \cdot 2)^2$

(b) $\dfrac{3^2 \cdot 2^3}{2^2}$ $\qquad$ (e) $\dfrac{2^0 \cdot 2^2}{2^2}$

(c) $5^{-3} \cdot 5^3$ $\qquad$ (f) $3^{-3} \cdot 3^2$

2. Change each of the following from standard notation to scientific notation.

(a) 1,730,000 $\qquad$ (c) 0.000078

(b) 282,000,000 $\qquad$ (d) 0.000498

3. Change each of the following from scientific notation to standard notation.

(a) 3.5×10^5 (c) 7.2×10^6
(b) 2.8×10^{-3} (d) 8.1×10^{-6}

4. Simplify each of the following.

(a) $\dfrac{3 \cdot 2^5 \cdot 5^1}{3^{-3} \cdot 2^4}$ (d) $\dfrac{3^5 \cdot 2^4 \cdot 6^{-3}}{6^{-4} \cdot 5^3 \cdot 3^7}$

(b) $\dfrac{3^2 \cdot 2^4}{6^2}$ (e) $\dfrac{7^2 \cdot 3^3 \cdot 5^4}{3^2 \cdot 5^{-2}}$

(c) $\dfrac{4^7 \cdot 8^6 \cdot 7}{8^2 \cdot 4^8 \cdot 8^{-3}}$ (f) $\dfrac{6^2 \cdot 8^4}{4^2 \cdot 9^2}$

5. Change each of the following from standard notation to scientific notation.

(a) 003056 (c) 23,605
(b) 120,111,000,000 (d) 0.0000000825

6. Change each of the following from scientific notation to standard notation.

(a) 4.3×10^{-5} (c) 7.21×10^8
(b) 1.09×10^5 (d) 9.892×10^{-8}

8.5 DECIMAL SYSTEM: GRUNDAL SYSTEMS

The decimal system of numeration is a *place value* system consisting of ten digits 0, 1, 2, 3, 4, 5, 6, 7, 8, and 9. We refer to this as the *base ten* because it contains ten digits. When a number is written in the decimal system, such as 3421, we are actually referring to the number

$$3421 = 3 \cdot 10^3 + 4 \cdot 10^2 + 2 \cdot 10^1 + 1 \cdot 10^0 \tag{8.5.1}$$

(This last form of the number is called the *expanded form* of the number.)

EXAMPLE 8.5.1

Write the expanded form of the following in base 10.

(a) 54710

$$\begin{aligned} 54710 &= 5 \cdot 10{,}000 + 4 \cdot 1000 + 7 \cdot 100 \\ &\quad + 1 \cdot 10 + 0 \cdot 1 \\ &= 5 \cdot 10^4 + 4 \cdot 10^3 + 7 \cdot 10^2 \\ &\quad + 1 \cdot 10^1 + 0 \cdot 10^0 \end{aligned} \tag{8.5.2}$$

(b) 378

$$378 = 3\cdot 100 + 7\cdot 10 + 8\cdot 1$$
$$= 3\cdot 10^2 + 7\cdot 10^1 + 8\cdot 10^0 \qquad (8.5.3)$$

(c) 1332459

$$1332459 = 1\cdot 1{,}000{,}000 + 3\cdot 100{,}000 + 3\cdot 10{,}000 + 2\cdot 1000 + 4\cdot 100 + 5\cdot 10 + 9\cdot 1$$
$$= 1\cdot 10^6 + 3\cdot 10^5 + 3\cdot 10^4 + 2\cdot 10^3 + 4\cdot 10^2 + 5\cdot 10^1 + 9\cdot 10^0 \qquad (8.5.4)$$

If we use our imagination, it is not hard to conceive of other systems, which use place value, but which contain more or less than ten digits. Such systems we call *grundal systems* (cf. German *Grund*, meaning base, foundation).

The *base five* consists of five digits,

$$0, 1, 2, 3, 4 \qquad (8.5.5)$$

from which all other numbers are formed using a place value based on five.

We indicate the base in which a number is written by placing the base in parentheses subscript to the number.

Thus, for example, in base five, the expanded form of $2341_{(\text{five})}$ is

$$2341_{(\text{five})} = 2\cdot 5^3 + 3\cdot 5^2 + 4\cdot 5^1 + 1\cdot 5^0 \qquad (8.5.6)$$

In other words, the base five system is based upon a place value which uses powers of five, whereas the base ten uses powers of ten.

The *base eight* consists of eight digits

$$0, 1, 2, 3, 4, 5, 6, 7 \qquad (8.5.7)$$

from which all other numbers are formed by using a place value based on eight.

For example, in base eight, the expanded form of $372_{(\text{eight})}$ is

$$372_{(\text{eight})} = 3\cdot 8^2 + 7\cdot 8^1 + 2\cdot 8^0 \qquad (8.5.8)$$

Again, we see that in base eight the place value is given by a power of eight.

In the base twelve, there are twelve digits. So that ambiguity does not arise, we must invent symbols to represent the two additional digits that must be used. In base twelve, the twelve digits are

$$0, 1, 2, 3, 4, 5, 6, 7, 8, 9, t, e \qquad (8.5.9)$$

where the digit t is the eleventh digit and e is the twelfth digit.

Here, the place value is based on powers of twelve so that

$$te5_{(\text{twelve})} = t\cdot 12^2 + e\cdot 12^1 + 5\cdot 12^0 \qquad (8.5.10)$$

The actual values of t and e are discussed later on in this chapter. Until then, they should be thought of merely as symbols denoting digits. Just as in base ten the symbol 9 stands for the digit after 8, in base twelve t should be thought of as a symbol for the digit following 9.

Exercise 8.5

1. Write the following base ten numbers in expanded form.

(a) $6023_{(\text{ten})}$	(d) $9060_{(\text{ten})}$
(b) $5421_{(\text{ten})}$	(e) $602_{(\text{ten})}$
(c) $521_{(\text{ten})}$	(f) $320_{(\text{ten})}$

2. Write the following base five numbers in expanded form.

(a) $321_{(\text{five})}$	(d) $4321_{(\text{five})}$
(b) $3123_{(\text{five})}$	(e) $21_{(\text{five})}$
(c) $40_{(\text{five})}$	(f) $34_{(\text{five})}$

3. Write the following base twelve numbers in expanded form.

(a) $t_{(\text{twelve})}$	(d) $11_{(\text{twelve})}$
(b) $12t_{(\text{twelve})}$	(e) $1e_{(\text{twelve})}$
(c) $501_{(\text{twelve})}$	(f) $t05_{(\text{twelve})}$

4. Write the following numbers in expanded form.

(a) $203_{(\text{four})}$	(d) $89_{(\text{twelve})}$
(b) $56_{(\text{eight})}$	(e) $101_{(\text{two})}$
(c) $568_{(\text{nine})}$	(f) $426_{(\text{seven})}$

5. Write the following expanded forms as numbers in the appropriate base.

(a) $3 \cdot 5^2 + 2 \cdot 5^1$	(d) $3 \cdot 4^3 + 2 \cdot 4 + 1 \cdot 4^0$
(b) $6 \cdot 7^2$	(e) $1 \cdot 2^2 + 1 \cdot 2 + 1 \cdot 2^0$
(c) $8 \cdot 12^3 + 5 \cdot 12 + t \cdot 12^0$	(f) $6 \cdot 8^3 + 1 \cdot 8^2 + 5 \cdot 8^0$

8.6 THE CONVERSION OF INTEGERS FROM BASES OTHER THAN TEN TO BASE TEN

Thus far we have concerned ourselves merely with various ways of representing numbers. We discussed the decimal system and systems which use a base different from ten. Now let us develop ways for changing from a base other than ten to base ten. That is, if we are given, for example, the base eight representation of an integer, the *base eight numeral,* we would like to know this integer's representation in base ten. In other words, if we know the representation of an integer in the base eight, what is its representation in the base ten?

It is important to remember that even though the representation of a number will "look" different depending on the base it is written in, each of the different representations are the numerals or symbols for the *same* number. That is, when we write or represent a number in two different bases, the representation will change, but the number itself does not change. For example, IV and 4 are two different representations of the number four.

To change the representation of an integer from a base other than ten to its base ten representation, do the following.

(1) Write the integer in its expanded form.
(2) Perform the operations indicated.
(3) The resulting integer is the base ten representation.

EXAMPLE 8.6.1

Write $3410_{(\text{five})}$ in base ten. Here,

$$\begin{aligned} 3410_{(\text{five})} &= 3 \cdot 5^3 + 4 \cdot 5^2 + 1 \cdot 5^1 + 0 \cdot 5^0 \\ &= 3 \cdot 125 + 4 \cdot 25 + 5 + 0 \\ &= 375 + 100 + 5 = 480 \end{aligned} \tag{8.6.1}$$

Thus the base ten representation of $3410_{(\text{five})}$ is $480_{(\text{ten})}$. Remember $3410_{(\text{five})}$ and $480_{(\text{ten})}$ are two different ways of representing or writing the *same* integer.

EXAMPLE 8.6.2

Write $472_{(\text{eight})}$ in base ten. Here,

$$\begin{aligned} 472_{(\text{eight})} &= 4 \cdot 8^2 + 7 \cdot 8 + 2 \\ &= 4 \cdot 64 + 7 \cdot 8 + 2 = 256 + 56 + 2 \\ &= 314 \end{aligned} \tag{8.6.2}$$

Thus,

$$472_{(\text{eight})} = 314_{(\text{ten})} \tag{8.6.3}$$

In bases with more than ten digits, such as base twelve, it is necessary to be able to represent the new digits by their corresponding base ten representation. In base twelve, t is the digit after 9. In base ten, the integer after 9 is represented by 10. Thus,

$$t_{(\text{twelve})} = 10_{(\text{ten})} \tag{8.6.4}$$

Similarly, since $e_{(\text{twelve})}$ is the digit after $t_{(\text{twelve})}$, then

$$e_{(\text{twelve})} = 11_{(\text{ten})} \tag{8.6.5}$$

EXAMPLE 8.6.3

Write $t5_{(\text{twelve})}$ in base ten. Here,

$$t5_{(\text{twelve})} = t \cdot 12 + 5$$
$$= 10 \cdot 12 + 5 = 120 + 5 = 125 \qquad (8.6.6)$$

Thus,

$$t5_{(\text{twelve})} = 125_{(\text{ten})} \qquad (8.6.7)$$

EXAMPLE 8.6.4

Write $10_{(\text{twelve})}$ in base ten. Here,

$$10_{(\text{twelve})} = 1 \cdot 12 + 0 = 12 \qquad (8.6.8)$$

Thus,

$$10_{(\text{twelve})} = 12_{(\text{ten})} \qquad (8.6.9)$$

EXAMPLE 8.6.5

Write $10_{(\text{five})}$ in base ten. Here,

$$10_{(\text{five})} = 1 \cdot 5 + 0 = 5_{(\text{ten})} \qquad (8.6.10)$$

Exercise 8.6

Write the following integers in the base ten.

1. (a) $372_{(\text{eight})}$ (b) $471_{(\text{nine})}$ (c) $256_{(\text{seven})}$
2. (a) $te1_{(\text{twelve})}$ (b) $67_{(\text{eight})}$ (c) $321_{(\text{five})}$
3. (a) $37_{(\text{twelve})}$ (b) $1e0_{(\text{twelve})}$ (c) $1011_{(\text{two})}$
4. (a) $4201_{(\text{six})}$ (b) $6723_{(\text{eight})}$ (c) $11001_{(\text{two})}$
5. (a) $3t_{(\text{twelve})}$ (b) $tt_{(\text{twelve})}$ (c) $1238_{(\text{nine})}$

8.7 CONVERSION OF INTEGERS FROM BASE TEN TO OTHER BASES

In the last section, we learned a method for converting the representation of an integer from bases other than ten to base ten. In this section, we shall develop a technique for finding the representation of an integer in a base other than ten if we know its representation in base ten. This is accomplished by using an algorithmic process very similar to the Euclidean algorithm discussed in Chapter 5.

Suppose it is desired to write the integer $513_{(\text{ten})}$ in the base five. We begin by dividing 513 by 5. We then obtain a quotient 102 and a remainder 3. That is,

$$513 = 5 \cdot 102 + 3 \qquad (8.7.1)$$

Repeat the process—dividing the quotient 102 by 5. Thus

$$102 = 5\cdot 20 + 2 \tag{8.7.2}$$

Dividing 20 by 5, we obtain

$$20 = 5\cdot 4 + 0 \tag{8.7.3}$$

and

$$4 = 5\cdot 0 + 4 \tag{8.7.4}$$

Once a quotient of 0 is obtained, we use the previous divisions to write $513_{(\text{ten})}$ as powers of 5. Thus,

$$\begin{aligned} 20 &= 4\cdot 5 + 0 \\ 102 &= 20\cdot 5 + 2 \\ 513 &= 5\cdot 102 + 3 = 5\cdot[20\cdot 5 + 2] + 3 \\ &= 20\cdot 5^2 + 2\cdot 5 + 3 \\ &= 4\cdot 5\cdot 5^2 + 2\cdot 5 + 3 \\ &= 4(5^3) + 0(5^2) + 2(5) + 3 \end{aligned} \tag{8.7.5}$$

Thus,

$$513_{(\text{ten})} = 4023_{(\text{five})} \tag{8.7.6}$$

Notice that the digits 4, 0, 2, 3 are the remainders (in reverse order) obtained by dividing 513 by 5.

EXAMPLE 8.7.1

Write $905_{(\text{ten})}$ in base eight. Here we repeatedly divide 905 by 8. Thus,

$$\begin{aligned} 905 &= 8\cdot 113 + 1 \\ 113 &= 8\cdot 14 + 1 \\ 14 &= 8\cdot 1 + 6 \\ 1 &= 8\cdot 0 + 1 \end{aligned} \tag{8.7.7}$$

Since a quotient of zero has been obtained, we see that

$$905_{(\text{ten})} = 1611_{(\text{eight})} \tag{8.7.8}$$

where the digits 1, 6, 1, 1 are the remainders written in reverse order.

We can check this result by using the resubstituting process. Thus,

$$\begin{aligned} 14 &= 8\cdot 1 + 6 \\ 113 &= 8\cdot 14 + 1 = 8\cdot[8 + 6] + 1 = 8^2 + 6\cdot 8 + 1 \\ 905 &= 8\cdot 113 + 1 = 8\cdot[8^2 + 6\cdot 8 + 1] + 1 \\ &= 8^3 + 6\cdot 8^2 + 8 + 1 \end{aligned} \tag{8.7.9}$$

Since $8^3 + 6\cdot 8^2 + 1\cdot 8 + 1$ is the expanded form of $1611_{(\text{eight})}$, we see that

$$905_{(\text{ten})} = 1611_{(\text{eight})} \tag{8.7.10}$$

EXAMPLE 8.7.2

Write $851_{(\text{ten})}$ in base twelve. Here we divide by 12. Thus,

$$\begin{aligned} 851 &= 12\cdot 70 + 11 \\ 70 &= 12\cdot 5 + 10 \\ 5 &= 12\cdot 0 + 5 \end{aligned} \tag{8.7.11}$$

The remainders, since they represent *digits* in base twelve, are really 5, *t*, and *e*. Thus,

$$851_{(\text{ten})} = 5te_{(\text{twelve})} \tag{8.7.12}$$

EXAMPLE 8.7.3

Write $529_{(\text{ten})}$ in base seven. Here we divide by seven. Thus,

$$\begin{aligned} 529 &= 7\cdot 75 + 4 \\ 75 &= 7\cdot 10 + 5 \\ 10 &= 7\cdot 1 + 3 \\ 1 &= 7\cdot 0 + 1 \end{aligned} \tag{8.7.13}$$

Thus,

$$529_{(\text{ten})} = 1354_{(\text{seven})} \tag{8.7.14}$$

We can prove the result by resubstituting. Thus,

$$\begin{aligned} 10 &= 7\cdot 1 + 3 \\ 75 &= 7\cdot 10 + 5 = 7\cdot[7 + 3] + 5 \\ &= 7^2 + 3\cdot 7 + 5 \\ 529 &= 7\cdot 75 + 4 = 7\cdot[7^2 + 3\cdot 7 + 5] + 4 \\ &= 7^3 + 3\cdot 7^2 + 5\cdot 7 + 4. \end{aligned} \tag{8.7.15}$$

Since $7^3 + 3\cdot 7^2 + 5\cdot 7 + 4$ is the expanded form of $1354_{(\text{seven})}$, we see that

$$529_{(\text{ten})} = 1354_{(\text{seven})} \tag{8.7.16}$$

Exercise 8.7

Write the following integers given in base ten in the base indicated.

1. $528_{(\text{ten})}$
   (a) in base five (b) in base eight (c) in base twelve
2. $625_{(\text{ten})}$
   (a) in base two (b) in base twelve (c) in base nine
3. $4203_{(\text{ten})}$
   (a) in base twelve (b) in base eight (c) in base seven

4. $6211_{(ten)}$
   (a) in base five (b) in base twelve (c) in base eight

5. $157_{(ten)}$
   (a) in base two (b) in base six (c) in base five

6. $561_{(ten)}$
   (a) in base two (b) in base twelve (c) in base nine

8.8 OPERATIONS IN BASE FIVE

Up until now, we have concerned ourselves with writing integers in a given base and with changing from one base to another base.

Let us now consider the problem of adding, subtracting, multiplying, and dividing integers which are written in a base other than base ten.

To do this, we must first write down addition and multiplication tables for addition and multiplication of the digits of a given base.

For example, the digits of base five are

$$0,\ 1,\ 2,\ 3,\ 4 \tag{8.8.1}$$

To find $2 + 4$, we proceed as follows:

$$2_{(five)} + 4_{(five)} = 2_{(ten)} + 4_{(ten)} = 6_{(ten)} = 11_{(five)} \tag{8.8.2}$$

Thus, in base five,

$$2 + 4 = 11 \tag{8.8.3}$$

To find $3 \cdot 4$ in base five,

$$3_{(five)} \cdot 4_{(five)} = 3_{(ten)} \cdot 4_{(ten)} = 12_{(ten)} = 22_{(five)} \tag{8.8.4}$$

Thus, in base five

$$3 \cdot 4 = 22 \tag{8.8.5}$$

We can summarize the results of adding and multiplying all possible combinations of the digits in base five, by the tables.

TABLE 8.2
Addition Table, Base Five

+	0	1	2	3	4
0	0	1	2	3	4
1	1	2	3	4	10
2	2	3	4	10	11
3	3	4	10	11	12
4	4	10	11	12	13

TABLE 8.3
Multiplication Table, Base Five

·	0	1	2	3	4
0	0	0	0	0	0
1	0	1	2	3	4
2	0	2	4	11	13
3	0	3	11	14	22
4	0	4	13	22	31

The student is urged not to attempt to memorize these tables; he should derive the needed computation in the way indicated above.

By using the above two tables, we can perform the four operations of addition, multiplication, subtraction, and division in base five.

For example, to compute

$$423_{(\text{five})} + 234_{(\text{five})} \tag{8.8.6}$$

we write the two integers in expanded form and use the distributive and associative laws. Thus,

$$\begin{aligned} 423_{(\text{five})} + 234_{(\text{five})} &= (4\cdot5^2 + 2\cdot5 + 3) + (2\cdot5^2 + 3\cdot5 + 4) \\ &= (4 + 2)\cdot5^2 + (2 + 3)\cdot5 + (3 + 4) \\ &= 11\cdot5^2 + 10\cdot5 + 12 \\ &= (1\cdot5 + 1)\cdot5^2 + (1\cdot5 + 0)\cdot5 + 1\cdot5 + 2 \\ &= 1\cdot5^3 + 1\cdot5^2 + 1\cdot5^2 + 0\cdot5 + 1\cdot5 + 2 \\ &= 1\cdot5^3 + 2\cdot5^2 + 1\cdot5 + 2 \\ &= 1212_{(\text{five})} \end{aligned} \tag{8.8.7}$$

If the above example is studied carefully, it is seen that the above can be more easily accomplished in the following way:

$$\begin{array}{r} {\scriptstyle 1\ 1\ \ } \\ 4\ 2\ 3 \\ +2\ 3\ 4 \\ \hline 1\ 2\ 1\ 2_{(\text{five})} \end{array} \tag{8.8.8}$$

This is accomplished by adding only columns and carrying the excess fives. In (8.8.8), the far right column is added, giving $12_{(\text{five})}$, or 1 five and 2 units. The 1 five is carried to the next column (the fives column) and the 2 is written down. The process is then repeated. We then add the digits of column two, 1 + 2 + 3, giving 11. We write down the 1 and carry the other 1 to the next column. And so on.

Of course, the reason for the above technique is (8.8.7).

EXAMPLE 8.8.1

Evaluate

$$3043_{(\text{five})} + 4122_{(\text{five})} \tag{8.8.9}$$

Here,

$$\begin{array}{r} {\scriptstyle 1\ 1} \\ 3\ 0\ 4\ 3 \\ +4\ 1\ 2\ 2 \\ \hline 1\ 2\ 2\ 2\ 0_{(\text{five})} \end{array} \tag{8.8.10}$$

To check our answer, suppose we do the computation in base ten. Then,

$$\begin{aligned} 3043_{(\text{five})} + 4122_{(\text{five})} &= 398_{(\text{ten})} + 537_{(\text{ten})} \\ &= 935_{(\text{ten})} = 12{,}220_{(\text{five})} \end{aligned} \tag{8.8.11}$$

Of course, the above way is much longer since one must convert to base ten, perform the computation, and then reconvert to base five. However, it does provide a good check.

To multiply, in the base five, we again write the integers involved in expanded form, and use the laws of exponents.

For example, to find the product

$$23_{(\text{five})} \cdot 42_{(\text{five})} \tag{8.8.12}$$

we write

$$\begin{aligned} 23_{(\text{five})} \cdot 42_{(\text{five})} &= (2\cdot5 + 3)\cdot(4\cdot5 + 2) \\ &= 2\cdot5(4\cdot5 + 2) + 3(4\cdot5 + 2) \\ &= 2\cdot4\cdot5^2 + 2\cdot2\cdot5 + 3\cdot4\cdot5 + 3\cdot2 \\ &= (1\cdot5 + 3)\cdot5^2 + 4\cdot5 + (2\cdot5 + 2)\cdot5 + 1\cdot5 + 1 \\ &= 1\cdot5^3 + 3\cdot5^2 + 4\cdot5 + 2\cdot5^2 + 2\cdot5 + 1\cdot5 + 1 \\ &= 1\cdot5^3 + (3 + 2)5^2 + (4 + 2 + 1)5 + 1 \\ &= 1\cdot5^3 + 1\cdot5\cdot5^2 + (1\cdot5 + 2)\cdot5 + 1 \\ &= 1\cdot5^3 + 1\cdot5^3 + 1\cdot5^2 + 2\cdot5 + 1 \\ &= 2\cdot5^3 + 1\cdot5^2 + 2\cdot5 + 1 = 2121_{(\text{five})} \end{aligned} \tag{8.8.13}$$

The above technique leads to the following method:

$$\begin{array}{r} {}^{1} \\ 4\ 2 \\ \times 2\ 3 \\ \hline 2\ 3\ 1 \\ 1\ 3\ 4 \\ \hline 2\ 1\ 2\ 1_{(\text{five})} \end{array} \tag{8.8.14}$$

In the above method, we multiply the 3 of 23 by the 2 of 42, obtaining 11. We write the 1 and carry 1. Then multiply 3 by the 4 of 42, obtaining 22 and add the 1 carried, giving 23. Repeat this process using the 2 of 23, but write the product over one place to the left since the 2 represents 2 fives and so the product must lie in the fives column. Add the two results. Since $23_{(\text{five})} = 2 \cdot 5 + 3$, this gives $2121_{(\text{five})}$.

We can check this answer by transferring the computation to base ten. Thus,

$$\begin{aligned} 42_{(\text{five})} \cdot 23_{(\text{five})} &= 22_{(\text{ten})} \cdot 13_{(\text{ten})} = 286_{(\text{ten})} \\ &= 2121_{(\text{five})} \end{aligned} \tag{8.8.15}$$

EXAMPLE 8.8.2

Compute

$$43_{(\text{five})} \cdot 32_{(\text{five})} \tag{8.8.16}$$

Here

$$\begin{array}{r} 4\ 3 \\ \times 3\ 2 \\ \hline 1\ 4\ 1 \\ 2\ 3\ 4 \\ \hline 3\ 0\ 3\ 1_{(\text{five})} \end{array} \tag{8.8.17}$$

Check:

$$43_{(\text{five})} = 23_{(\text{ten})} \quad \text{and} \quad 32_{(\text{five})} = 17_{(\text{ten})} \tag{8.8.18}$$

Then

$$23_{(\text{ten})} \cdot 17_{(\text{ten})} = 391_{(\text{ten})} \tag{8.8.19}$$

Also,

$$391_{(\text{ten})} = 3031_{(\text{five})} \tag{8.8.20}$$

Exercise 8.8.1

Compute the following and check your result by transferring to base ten.

1. (a) $4311_{(\text{five})} + 2043_{(\text{five})}$ (d) $303_{(\text{five})} + 442_{(\text{five})}$
   (b) $4322_{(\text{five})} + 404_{(\text{five})}$ (e) $113_{(\text{five})} + 332_{(\text{five})}$
   (c) $32_{(\text{five})} + 1024_{(\text{five})}$ (f) $444_{(\text{five})} + 1001_{(\text{five})}$

2. (a) $43_{(\text{five})} \cdot 23_{(\text{five})}$ (d) $303_{(\text{five})} \cdot 24_{(\text{five})}$
(b) $231_{(\text{five})} \cdot 21_{(\text{five})}$ (e) $4001_{(\text{five})} \cdot 32_{(\text{five})}$
(c) $44_{(\text{five})} \cdot 212_{(\text{five})}$ (f) $301_{(\text{five})} \cdot 34_{(\text{five})}$

3. (a) $2110_{(\text{five})} + 404_{(\text{five})} + 1004_{(\text{five})}$
(b) $21_{(\text{five})} \cdot 22_{(\text{five})} + 104_{(\text{five})}$
(c) $234_{(\text{five})} + 41_{(\text{five})} \cdot 20_{(\text{five})}$
(d) $(101_{(\text{five})} + 404_{(\text{five})}) \cdot 22_{(\text{five})}$
(e) $101_{(\text{five})} + 41_{(\text{five})} \cdot 12_{(\text{five})}$
(f) $13_{(\text{five})} \cdot (23_{(\text{five})} + 44_{(\text{five})})$

Let us take up the problem of subtracting two integers written in the base five.

For example, to compute the difference

$$4040_{(\text{five})} - 331_{(\text{five})} \tag{8.8.21}$$

we proceed as follows:

$$\begin{aligned}
4040_{(\text{five})} - 331_{(\text{five})} \\
&= [4 \cdot 5^3 + 0 \cdot 5^2 + 4 \cdot 5 + 0] - [3 \cdot 5^2 + 3 \cdot 5 + 1] \\
&= 4 \cdot 5^3 + 0 \cdot 5^2 + 4 \cdot 5 + 0 - 3 \cdot 5^2 - 3 \cdot 5 - 1 \\
&= 4 \cdot 5^3 + [0 \cdot 5^2 - 3 \cdot 5^2] + [4 \cdot 5 - 3 \cdot 5] + [0 - 1] \\
&= 4 \cdot 5^3 + [0 \cdot 5^2 - 3 \cdot 5^2] + [(4 - 3) \cdot 5] + [0 - 1] \\
&= 4 \cdot 5^3 + [0 \cdot 5^2 - 3 \cdot 5^2] + 1 \cdot 5 + [0 - 1] \\
&= 4 \cdot 5^3 + [0 \cdot 5^2 - 3 \cdot 5^2] + 5 + [0 - 1] \\
&= 4 \cdot 5^3 + [0 \cdot 5^2 - 3 \cdot 5^2] + [5 + 0 - 1] \\
&= 4 \cdot 5^3 + [0 \cdot 5^2 - 3 \cdot 5^2] + 4 \\
&= (3 + 1)5^3 + [0 \cdot 5^2 - 3 \cdot 5^2] + 4 \\
&= [3 \cdot 5^3 + 1 \cdot 5^3] + [0 \cdot 5^2 - 3 \cdot 5^2] + 4 \\
&= 3 \cdot 5^3 + [1 \cdot 5^3 + (0 \cdot 5^2 - 3 \cdot 5^2)] + 4 \\
&= 3 \cdot 5^3 + [5 \cdot 5^2 + 0 \cdot 5^2 - 3 \cdot 5^2] + 4 \\
&= 3 \cdot 5^3 + [(5 + 0 - 3) \cdot 5^2] + 4 \\
&= 3 \cdot 5^3 + [2 \cdot 5^2] + 4 \\
&= 3204_{(\text{five})}
\end{aligned} \tag{8.8.22}$$

The steps outlined in (8.8.22) explain why the borrowing technique works.

For example, to compute

$$4040_{(\text{five})} - 331_{(\text{five})} \tag{8.8.23}$$

we proceed as follows:

$$\begin{array}{cccc} {\scriptstyle 3} & {\scriptstyle 10} & {\scriptstyle 3} & {\scriptstyle 10} \\ 4 & 0 & 4 & 0 \\ -3 & 3 & 1 & \\ \hline 3 & 2 & 0 & 4_{(\text{five})} \end{array} \tag{8.8.24}$$

This was accomplished by starting in the far right column. Since $0 < 1$, we replace 0 by 10 by borrowing $1 \cdot 5$ from 4. Then we can subtract 1 from 10 leaving 4. Repeat this process in column two. Now, in column two, $3 - 3$ is 0. Proceed to column three. Since $3 > 0$, replace the 0 by 10 by borrowing 1 from 4. This is possible since the place value of column four is one more than that of column three. Then, 3 from 10 gives 2. Since there is a 0 understood in column four of 331, we just write the 3.

Check: To check our result, merely add 331 and 3204. The answer should be 4040.

EXAMPLE 8.8.3

Compute

$$3401_{(\text{five})} - 221_{(\text{five})} \tag{8.8.25}$$

Here

$$\begin{array}{cccc} 3 & 4 & 0 & 1 \\ & -2 & 1 & 1 \\ \hline 3 & 1 & 4 & 0_{(\text{five})} \end{array} \tag{8.8.26}$$

To divide in base five, consider the following example:

$$13433 \div 23 \tag{8.8.27}$$

Here,

$$\begin{array}{r} 300 + 20 + 1 = 321 \\ 23\overline{)13433} \\ \underline{12400} \\ 1033 \\ \underline{1010} \\ 23 \\ \underline{23} \end{array} \tag{8.8.28}$$

That is

$$13433_{(\text{five})} = 23_{(\text{five})} \cdot 321_{(\text{five})} \tag{8.8.29}$$

This should help explain why the following technique works:

$$\begin{array}{r} 3 \\ 2\,3\overline{)1\,3\,4\,3\,3} \\ \underline{1\,2\,4} \\ 1\,0\,3 \end{array} \tag{8.8.30}$$

Here we ask ourselves what is the largest digit which when multiplied by 23 gives a number less than or equal to 13. Since the answer is 0, we ask the same question about 134. Here, by trial and error, we find the digit is 3. Then, 3 times 23 gives 124. Subtract 124 from 134, giving 10. Bring down the next digit 3 and repeat the above process on 103. Then,

$$\begin{array}{r} 3\,2\,1 \\ 2\,3\overline{)1\,3\,4\,3\,3} \\ \underline{1\,2\,4} \\ 1\,0\,3 \\ \underline{1\,0\,1} \\ 2\,3 \\ \underline{2\,3} \end{array} \tag{8.8.31}$$

EXAMPLE 8.8.4

Compute

$$23332_{(\text{five})} \div 32_{(\text{five})} \tag{8.8.32}$$

Here

$$\begin{array}{r} 4\,0\,1 \\ 3\,2\overline{)2\,3\,3\,3\,2} \\ \underline{2\,3\,3} \\ 3 \\ \underline{0} \\ 3\,2 \\ \underline{3\,2} \end{array} \tag{8.8.33}$$

Note, in the above result, the necessity of using a 0 to denote the fact that the largest digit which can be multiplied by 32 and give a result less than or equal to 3 is 0.

Exercise 8.8.2

Compute the following in base five and check your result by transferring to the base ten and performing the computation in base ten.

1. (a) $301_{(\text{five})} - 20_{(\text{five})}$ (d) $1044_{(\text{five})} - 313_{(\text{five})}$
   (b) $411_{(\text{five})} - 22_{(\text{five})}$ (e) $3013_{(\text{five})} - 214_{(\text{five})}$
   (c) $3221_{(\text{five})} - 402_{(\text{five})}$ (f) $103_{(\text{five})} - 34_{(\text{five})}$

2. (a) $1231_{(\text{five})} \div 4_{(\text{five})}$
   (b) $1414_{(\text{five})} \div 3_{(\text{five})}$
   (c) $42133_{(\text{five})} \div 12_{(\text{five})}$
   (d) $1432_{(\text{five})} \div 4_{(\text{five})}$
   (e) $11203_{(\text{five})} \div 243_{(\text{five})}$
   (f) $11444_{(\text{five})} \div 123_{(\text{five})}$

3. Is there anything wrong with

$$\begin{array}{r} 411_{(\text{five})} \\ \underline{-22_{(\text{five})}} \\ (-1) \\ +(-10) \\ \underline{+(400)} \\ 334_{(\text{five})} \end{array}$$

8.9 OPERATIONS IN OTHER BASES

To add, subtract, multiply, or divide in other bases, the same rules and procedures used in the base ten and in the base five (as outlined in Section 8.8) can be used. It is only necessary to be able to add and multiply the digits of a base in order to perform operations in the base.

For example, in the base two, the digits are 0 and 1. The rules for addition and multiplication are given by Table 8.4.

TABLE 8.4

Addition

+	0	1
0	0	1
1	1	10

Multiplication

·	0	1
0	0	0
1	0	1

EXAMPLE 8.9.1
Compute

$$1011_{(\text{two})} + 101_{(\text{two})} \tag{8.9.1}$$

Here

$$\begin{array}{r} {\scriptstyle 1\ 1\ 1} \\ 1\ 0\ 1\ 1 \\ \underline{+1\ 0\ 1} \\ 1\ 0\ 0\ 0\ 0_{(\text{two})} \end{array} \tag{8.9.2}$$

EXAMPLE 8.9.2
Compute

$$101_{(\text{two})} - 11_{(\text{two})} \tag{8.9.3}$$

Here

$$\begin{array}{r} 1\;0\;1 \\ -1\;1 \\ \hline 1\;0_{(\text{two})} \end{array} \tag{8.9.4}$$

EXAMPLE 8.9.3

Compute

$$101_{(\text{two})} \cdot 10_{(\text{two})} \tag{8.9.5}$$

Here

$$\begin{array}{r} 1\;0\;1 \\ \times 1\;0 \\ \hline 0\;0\;0 \\ 1\;0\;1\;\;\; \\ \hline 1\;0\;1\;0_{(\text{two})} \end{array} \tag{8.9.6}$$

EXAMPLE 8.9.4

Compute

$$1010_{(\text{two})} \div 10_{(\text{two})} \tag{8.9.7}$$

$$\begin{array}{r} 1\;0\;1 \\ 1\;0\overline{)1\;0\;1\;0} \\ \underline{1\;0}\;\;\;\;\; \\ 0\;1\;\;\; \\ \underline{0}\;\;\; \\ 1\;0 \\ \underline{1\;0} \end{array} \tag{8.9.8}$$

Notice that in each of the above examples, the techniques for computation do *not* vary. The only change is in the tables for adding and multiplying.

Exercise 8.9.1

Perform the indicated operations in base two and check your result by transferring the computation to base ten.

1. $1101_{(\text{two})} + 111_{(\text{two})} + 1011_{(\text{two})}$
2. $1101_{(\text{two})} - 111_{(\text{two})}$
3. $1101_{(\text{two})} \cdot 111_{(\text{two})}$
4. $1011_{(\text{two})} \cdot 101_{(\text{two})}$
5. $11011_{(\text{two})} \div 11_{(\text{two})}$
6. $11011_{(\text{two})} - 1101_{(\text{two})}$
7. $11010_{(\text{two})} \div 10_{(\text{two})}$

Next, let us consider the base twelve. The digits of the base twelve are

$$0, 1, 2, 3, 4, 5, 6, 7, 8, 9, t, e \qquad (8.9.9)$$

In order to write the addition rules for base twelve, we need to determine the sum of all possible combinations of these digits. A few of these are worked out below. The rest should be verified by the student.

Consider $4_{(\text{twelve})}$ and $9_{(\text{twelve})}$. Then

$$\begin{aligned} 4_{(\text{twelve})} + 9_{(\text{twelve})} &= 4_{(\text{ten})} + 9_{(\text{ten})} \\ &= 13_{(\text{ten})} = 11_{(\text{twelve})} \end{aligned} \qquad (8.9.10)$$

Also

$$\begin{aligned} 4_{(\text{twelve})} + e_{(\text{twelve})} &= 4_{(\text{ten})} + 11_{(\text{ten})} \\ &= 15_{(\text{ten})} = 13_{(\text{twelve})} \end{aligned} \qquad (8.9.11)$$

Another sum is

$$\begin{aligned} t_{(\text{twelve})} + e_{(\text{twelve})} &= 10_{(\text{ten})} + 11_{(\text{ten})} \\ &= 21_{(\text{ten})} = 19_{(\text{twelve})} \end{aligned} \qquad (8.9.12)$$

Finally,

$$\begin{aligned} e_{(\text{twelve})} + e_{(\text{twelve})} &= 11_{(\text{ten})} + 11_{(\text{ten})} \\ &= 22_{(\text{ten})} = 1t_{(\text{twelve})}. \end{aligned} \qquad (8.9.13)$$

TABLE 8.5
Addition Table, Base Twelve

+	0	1	2	3	4	5	6	7	8	9	*t*	*e*
0	0	1	2	3	4	5	6	7	8	9	*t*	*e*
1	1	2	3	4	5	6	7	8	9	*t*	*e*	10
2	2	3	4	5	6	7	8	9	*t*	*e*	10	11
3	3	4	5	6	7	8	9	*t*	*e*	10	11	12
4	4	5	6	7	8	9	*t*	*e*	10	11	12	13
5	5	6	7	8	9	*t*	*e*	10	11	12	13	14
6	6	7	8	9	*t*	*e*	10	11	12	13	14	15
7	7	8	9	*t*	*e*	10	11	12	13	14	15	16
8	8	9	*t*	*e*	10	11	12	13	14	15	16	17
9	9	*t*	*e*	10	11	12	13	14	15	16	17	18
t	*t*	*e*	10	11	12	13	14	15	16	17	18	19
e	*e*	10	11	12	13	14	15	16	17	18	19	1*t*

EXAMPLE 8.9.5
Evaluate

$$123_{\text{(twelve)}} + 81_{\text{(ten)}} \tag{8.9.14}$$

Here

$$\begin{array}{r} 1\ 2\ 3 \\ +8\ 1 \\ \hline 1\ t\ 4_{\text{(twelve)}} \end{array} \tag{8.9.15}$$

EXAMPLE 8.9.6
Evaluate

$$123_{\text{(twelve)}} - 81_{\text{(twelve)}} \tag{8.9.16}$$

Here

$$\begin{array}{r} ^{12} \\ \not{1}\ \not{2}\ 3 \\ -8\ 1 \\ \hline 6\ 2_{\text{(twelve)}} \end{array} \tag{8.9.17}$$

Exercise 8.9.2

Perform the indicated operations in base twelve and check your answers by transferring the computation to the base ten.

1. $679 + 46$
2. $1t1 + 1t$
3. $t12 + 10t$
4. $67t - 46$
5. $1t1 - 1t$
6. $204 + 695$
7. $9e7 + t25$
8. $60t - 14$
9. $312 - 5t$
10. $18e8t - 5te4$

Let us obtain some of the entries of the multiplication table for the base twelve. First

$$\begin{aligned} 5_{\text{(twelve)}} \cdot 6_{\text{(twelve)}} &= 5_{\text{(ten)}} \cdot 6_{\text{(ten)}} \\ &= 30_{\text{(ten)}} = 26_{\text{(twelve)}} \end{aligned} \tag{8.9.18}$$

$$\begin{aligned} 7_{\text{(twelve)}} \cdot 9_{\text{(twelve)}} &= 7_{\text{(ten)}} \cdot 9_{\text{(ten)}} \\ &= 63_{\text{(ten)}} = 53_{\text{(twelve)}} \end{aligned} \tag{8.9.19}$$

$$\begin{aligned} t_{\text{(twelve)}} \cdot 5_{\text{(twelve)}} &= 10_{\text{(ten)}} \cdot 5_{\text{(ten)}} \\ &= 50_{\text{(ten)}} = 42_{\text{(twelve)}} \end{aligned} \tag{8.9.20}$$

$$\begin{aligned} t_{\text{(twelve)}} \cdot t_{\text{(twelve)}} &= 10_{\text{(ten)}} \cdot 10_{\text{(ten)}} \\ &= 100_{\text{(ten)}} = 84_{\text{(twelve)}} \end{aligned} \tag{8.9.21}$$

Finally,

$$t_{(\text{twelve})} \cdot e_{(\text{twelve})} = 10_{(\text{ten})} \cdot 11_{(\text{ten})}$$

$$= 110_{(\text{ten})} = 92_{(\text{twelve})} \tag{8.9.22}$$

TABLE 8.6
Multiplication Table, Base Twelve

$\cdot$	0	1	2	3	4	5	6	7	8	9	t	e
0	0	0	0	0	0	0	0	0	0	0	0	0
1	0	1	2	3	4	5	6	7	8	9	t	e
2	0	2	4	6	8	t	10	12	14	16	18	$1t$
3	0	3	6	9	10	13	16	19	20	23	26	29
4	0	4	8	10	14	18	20	24	28	30	34	38
5	0	5	t	13	18	21	26	$2e$	34	39	42	47
6	0	6	10	16	20	26	30	36	40	46	50	56
7	0	7	12	19	24	$2e$	36	41	48	53	$5t$	65
8	0	8	14	20	28	34	40	48	54	60	68	74
9	0	9	16	23	30	39	46	53	60	69	76	83
t	0	t	18	26	34	42	50	$5t$	68	76	84	92
e	0	e	$1t$	29	38	47	56	65	74	83	92	$t1$

EXAMPLE 8.9.7
Evaluate

$$123_{(\text{twelve})} \cdot 81_{(\text{twelve})} \tag{8.9.23}$$

Here

$$\begin{array}{r} 1\ 2\ 3 \\ \times 8\ 1 \\ \hline 1\ 2\ 3 \\ 9\ 6\ 0\ \ \ \\ \hline 9\ 7\ 2\ 3_{(\text{twelve})} \end{array} \tag{8.9.24}$$

EXAMPLE 8.9.8
Evaluate

$$12312_{(\text{twelve})} \div 81_{(\text{twelve})} \tag{8.9.25}$$

Here

$$\begin{array}{r} 1\ 9\ 2 \\ 8\ 1\overline{)1\ 2\ 3\ 1\ 2} \\ 8\ 1 \\ \hline 6\ 2\ 1 \\ 6\ 0\ 9 \\ \hline 1\ 4\ 2 \\ 1\ 4\ 2 \\ \hline \end{array} \qquad (8.9.26)$$

Exercise 8.9.3

Perform the indicated operations and check your answers by transferring to base ten.

1. (a) $679_{(\text{twelve})} \cdot 81_{(\text{twelve})}$ (c) $1t1_{(\text{twelve})} \cdot 1t_{(\text{twelve})}$
   (b) $679_{(\text{twelve})} \div t_{(\text{twelve})}$ (d) $293_{(\text{twelve})} \div 7_{(\text{twelve})}$
2. (a) $214e_{(\text{twelve})} \cdot 54_{(\text{twelve})}$ (c) $te41_{(\text{twelve})} \cdot 325_{(\text{twelve})}$
   (b) $e468_{(\text{twelve})} \div 4e_{(\text{twelve})}$ (d) $5732t_{(\text{twelve})} \div 60t_{(\text{twelve})}$
3. Write out the addition table for base 8.
4. Evaluate

   (a) $706_{(\text{eight})} + 66_{(\text{eight})}$ (c) $452_{(\text{eight})} + 65_{(\text{eight})}$
   (b) $205_{(\text{eight})} - 77_{(\text{eight})}$ (d) $6271_{(\text{eight})} - 176_{(\text{eight})}$
5. Write out the multiplication table for base eight.
6. Evaluate

   (a) $612_{(\text{eight})} \cdot 21_{(\text{eight})}$ (c) $64_{(\text{eight})} \cdot 24_{(\text{eight})}$
   (b) $54_{(\text{eight})} \cdot 214_{(\text{eight})}$ (d) $2534_{(\text{eight})} \div 61_{(\text{eight})}$

9 Relations

9.1 INTRODUCTION

The concept of a relation has already been introduced intuitively. For example, in Chapter 2 we spoke of four relations between sets—*equality*, *equivalence*, *subset*, and *proper subset*. In Chapter 3 the notions of *less than* and *less than or equal to* were introduced as relations involving whole numbers. The purpose of this chapter is to discuss *relations* in more depth and to give several applications of this concept.

Before proceeding, however, the student is urged to review Section 2.7. In that section, the product set $A \times B$ of two sets A and B was defined and various properties of the product set $A \times B$ were given.

9.2 A GEOMETRIC INTERPRETATION OF PRODUCT SET

So far in our study of sets of numbers, we have discussed whole numbers, integers, rational numbers, and real numbers. Corresponding to each of these sets, we have introduced a number line. Thus, we have talked about a whole number line, an integer line, a rational line, and a real line. Each of these lines consists of points such that there is a one-to-one correspondence between the points on the line and the members of the corresponding set. The number corresponding to a point on a line is called the *coordinate* of that point.

Consider a horizontal line (it may be the whole number line, the integer line, the rational line, or the real line). On this line, label the numbers associated to its points. Figure 9.1 gives this situation for the integer line.

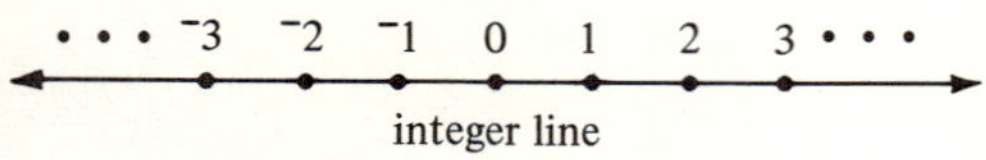

FIG. 9.1

Next, take a vertical line and label the numbers associated to its points. Suppose we agree that the vertical line and the horizontal line meet at the point corresponding to zero on both lines. It is further assumed that the *same* scale is used for both lines. Although the assumption is not necessary for the discussion at hand, by using the *same* scale on both lines the location of points becomes more realistic. For example, in assigning numbers to streets in a city, blocks (going either east–west or north–south) are usually the same part of a mile, that is, 8 blocks to a mile.

Suppose we also agree that the points on the horizontal line to the right of the zero are associated to positive numbers and those to the left of zero to negative numbers. Similarly, suppose we agree that points on the vertical line above zero are associated to positive numbers and those below zero to negative numbers. This is depicted in Fig. 9.2.

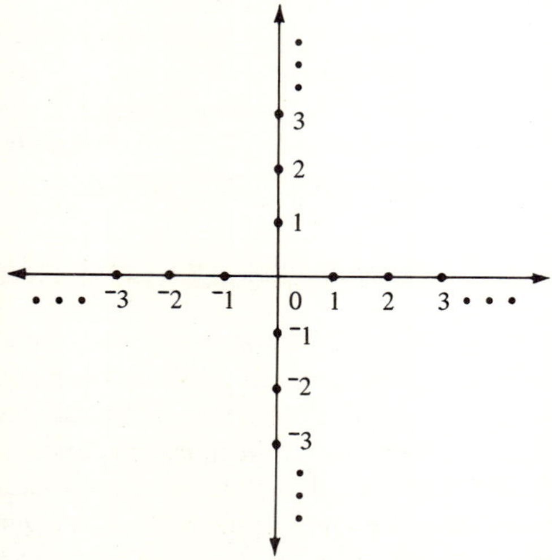

FIG. 9.2

With this convention established, we can now depict the ordered pairs of a product set.

Suppose

$$A = \{1, 2, 3\}, \qquad B = \{3, 4\} \tag{9.2.1}$$

The product set $A \times B$ is

$$A \times B = \{(1, 3), (1, 4), (2, 3), (2, 4), (3, 3), (3, 4)\} \tag{9.2.2}$$

By considering the first entry of the ordered pairs to be on a horizontal line and the second entry to be on a vertical line, we can find the ordered pair. Figure 9.3 shows this.

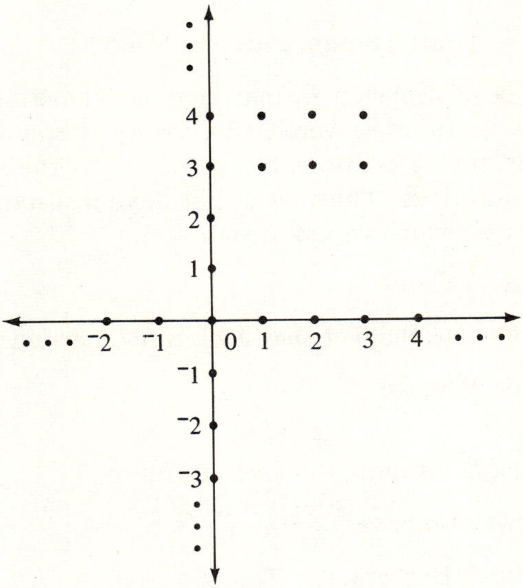

FIG. 9.3

Figure 9.3 is called a *graph* of the product set $A \times B$.

Exercise 9.2

Find the product set $A \times B$ if

1. $A = \{1, 3, 5\}$, $B = \{2, 3\}$.
2. $A = \{^-2, 1, 0\}$, $B = \{1, 2\}$.
3. $A = \{3\}$, $B = \{0, 1\}$.

4. $A = \{1, 2\}$, $B = \{^-3, ^-2\}$.

5. $A = \{^-1, ^-2\}$, $B = \{^-3, ^-4\}$.

6. $A = \{^-1, 0, 1\}$, $B = \{^-1, 0, 1\}$.

Graph each of the product sets.

9.3 RELATIONS

Consider two sets A and B in which the elements of set A are males, say Jack, Tom, and John, and those of set B are females, say Carol, Susan, Barbara, and Mary. Then,

$$A = \{\text{Jack, Tom, John}\}$$

$$B = \{\text{Carol, Susan, Barbara, Mary}\} \qquad (9.3.1)$$

Suppose that a relationship R, marriage, is defined between the elements of these two sets. In other words, suppose we obtain a set of ordered pairs in such a way that the first entry is related to the second entry by the relation R of marriage. Thus, we might have a situation in which Jack is married to Carol, which we can denote by

$$\text{Jack } R \text{ Carol} \qquad (9.3.2)$$

Then we can denote the fact that Jack is *not* married to Susan by

$$\text{Jack } \not{R} \text{ Susan} \qquad (9.3.3)$$

Similarly,

$$\text{Jack } \not{R} \text{ Barbara}, \qquad \text{Jack } \not{R} \text{ Mary} \qquad (9.3.4)$$

Suppose we also have

$$\text{John } R \text{ Barbara}, \qquad \text{Tom } R \text{ Susan} \qquad (9.3.5)$$

Thus, we have the following set C of ordered pairs due to the relation R of marriage.

$$C = \{(\text{Jack, Carol}), (\text{John, Barbara}), (\text{Tom, Susan})\} \qquad (9.3.6)$$

Notice that the set C is a subset of the product set $A \times B$.

Consider the set of all men M, the set of all women W, and the relationship marriage R. Evidently, this relationship can be used as a criterion for deciding whether a pair of human beings (h_1, h_2) in $M \times W$ belong together under the relationship R or not. If h_1 is married to h_2, we write $h_1 \, R \, h_2$ or $(h_1, h_2) \in R$ to mean the pair (h_1, h_2) is in the relationship of marriage.

We notice the relation R involves here the notion of two sets, together with a *rule* for deciding whether or not an element from one of the sets and an element from the other belong together in the relationship.

Going back to our example, we see that any relation can be determined by the set of ordered pairs that are in it. Looked at in this way, R is a *set*; it is, moreover, a subset of a *product set*. Evidently the set of all possible couples of men and women, no matter what their relationship, will belong to the product set $M \times W$. If R denotes the set of those pairs that are married, then

$$R \subseteq M \times W \tag{9.3.7}$$

We can now give the precise meaning of a relation R.

DEFINITION 9.3.1

Let A and B be arbitrary nonempty sets. A *relation R from A to B* is a subset of the product set $A \times B$.

That is, a relation R from A to B consists of ordered pairs in $A \times B$ so that

$$R \subseteq A \times B \tag{9.3.8}$$

If (a, b) is an ordered pair in the product set $A \times B$ and if (a, b) is a member of the relation R from A to B, then we will say that *a is related to b by R* and write

$$a\,R\,b \tag{9.3.9}$$

If a is related to b by R, then b is called *an image of a* and a is termed *a pre-image of b*.

It is important to keep in mind that a relation R from A to B is a subset of $A \times B$. This means that the relation R is a set which may contain *no* ordered pairs of $A \times B$, some ordered pairs of $A \times B$, or all ordered pairs of $A \times B$. That is, *any* subset of $A \times B$ is a relation from A to B.

EXAMPLE 9.3.1

Consider the sets

$$A = \{1, 2, 3\}, \qquad B = \{3, 4\} \tag{9.3.10}$$

The product set $A \times B$ is

$$A \times B = \{(1,3), (1,4), (2,3), (2,4), (3,3), (3,4)\} \tag{9.3.11}$$

One possible relation R from A to B is

$$R = \{(1, 3), (2, 3), (3, 4)\} \tag{9.3.12}$$

This is so since $R \subseteq A \times B$. Another relation S from A to B is

$$S = \{(1, 3), (2, 4), (3, 3)\} \tag{9.3.13}$$

since $S \subseteq A \times B$.

In the above example, for the relation R given in (9.3.12), 3 is an image of 1, 3 is an image of 2, and 4 is an image of 3. Also, 1 is a pre-image of 3, 2 is a pre-image of 3, and 3 is a pre-image of 4.

We can demonstrate the distinction between a product set and a relation by using the graphing techniques given previously. If we use a dot to denote a member of the product set $A \times B$ and a circled dot to denote a member of a relation from A to B, the graphs of the relations R and S in Example 9.3.1 are given by the figures below.

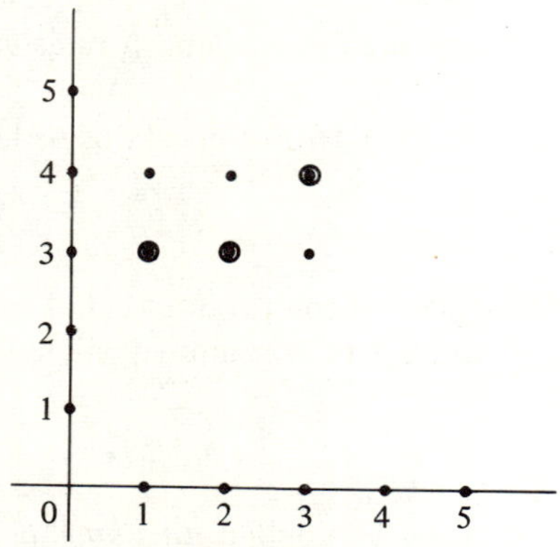

Relation R From A To B

FIG. 9.4

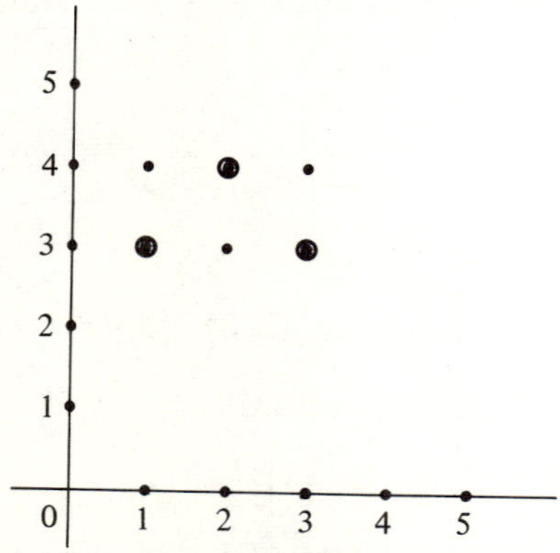

Relation S From A To B

FIG. 9.5

Definition 9.3.2

If R is a relation from A to B, the *domain of* R and the *range of* R, denoted by X_R and Y_R, respectively, are defined as

$$X_R = \{x \mid x \in A \text{ and } (x, y) \in R\}$$

$$Y_R = \{y \mid y \in B \text{ and } (x, y) \in R\} \tag{9.3.14}$$

That is, the domain X_R of R is the set of all pre-images while the range Y_R is the set of all images in a relation R. In other words, the domain X_R consists of the first entries of the ordered pairs of R while the range Y_R is the set of all second entries of ordered pairs in R.

EXAMPLE 9.3.2

Let $A = \{x, z, 2\}$ and $B = \{1, 2, 3, 4\}$. Then the set

$$R = \{(x, 1), (x, 4), (z, 2), (z, 4)\} \tag{9.3.15}$$

is a relation from A to B since it is a subset of $A \times B$. Clearly,

$$X_R = \{x, z\}, \qquad Y_R = \{1, 2, 4\} \tag{9.3.16}$$

Note that $X_R \subseteq A$ and $Y_R \subseteq B$.

Definition 9.3.3

If R is a relation from A to B, the *inverse of* R is a relation from B to A, which is denoted by R^{-1}, and which is defined as

$$R^{-1} = \{(y, x) \mid x\, R\, y\} \tag{9.3.17}$$

EXAMPLE 9.3.3

Consider the sets

$$A = \{1, 2, 3\}, \qquad B = \{2, 4\} \tag{9.3.18}$$

A relation R from A to B is

$$R = \{(1, 4), (1, 2), (3, 2), (3, 4)\} \tag{9.3.19}$$

Here

$$\text{domain } R = X_R = \{1, 3\}, \qquad \text{range } R = Y_R = \{2, 4\} \tag{9.3.20}$$

The inverse of R is the relation

$$R^{-1} = \{(4, 1), (2, 1), (2, 3), (4, 3)\} \tag{9.3.21}$$

Notice that R^{-1} is a relation from B to A since it is a subset of $B \times A$. Furthermore,

$$\text{domain } R^{-1} = X_{R^{-1}} = \{2, 4\}, \qquad \text{range } R^{-1} = Y_{R^{-1}} = \{1, 2\} \tag{9.3.22}$$

Notice that the domain and range of a relation and its inverse are interchanged.

As an example of a relation R and its inverse R^{-1}, consider the relation R, *is a husband of*. If $(a, b) \in R$, then we can say

$$a \text{ is a husband of } b \tag{9.3.23}$$

The inverse relation is $b\, R^{-1}\, a$ and we can say

$$b \text{ is the wife of } a \tag{9.3.24}$$

Thus, the inverse of the relation *is a husband of* would be the relation *is a wife of*.

Exercise 9.3

1. Let $A = \{1, 2, 3\}$ and $B = \{x, y, z, w\}$. Which of the following describe relations from A to B?

   (a) $\{(1, y), (2, x), (3, y)\}$
   (b) $\{(1, x), (2, x), (3, y), (3, z), (1, w)\}$
   (c) $\{(1, 2)\}$
   (d) $\{(1, 3), (x, 3)\}$
   (e) $\varnothing$
   (f) $\{(1, 2), (2, w)\}$

2. Let $A = \{1, 3, 4, 6\}$, $B = \{1, 2, 3\}$. Find the domain and range of the following relations from A to B and graph each one.

   (a) $\{(1, 1), (3, 2), (4, 1)\}$
   (b) $\{(4, 2), (4, 1), (6, 1), (6, 3)\}$
   (c) $\{(1, 2), (3, 3), (3, 1)\}$
   (d) $\{(6, 1), (6, 2), (4, 3)\}$

3. Find the inverse relation R^{-1} for the relations R of problem 2. Find the domain $X_{R^{-1}}$ and range $Y_{R^{-1}}$ of the inverse relation R^{-1}.

4. Find all possible relations from A to B when

   $$A = \{1, 2\}, \qquad B = \{5, 7\}$$

5. Let $A = \{1, 2, 3\}$ and $B = \{1, 2, 3\}$. Write down the ordered pairs of the relation R from A to B defined by

   (a) $R = \{(x, y) \mid y = x + 1\}$
   (b) $R = \{(x, y) \mid y > x\}$
   (c) $R = \{(x, y) \mid y < x\}$

6. A relation R from the set W of whole numbers to W is defined by

   $$R = \{(x, y) \mid y - x \text{ is divisible by two}\}$$

   Which of the following are true?

   (a) $(2, 12) \in R$
   (b) $(0, 5) \in R$

(c) $1\,R\,6$
(d) $6\,R\,8$
(e) $9\,R\,9$

9.4 FUNCTIONS

Among all the relations R associated to a given product set $A \times B$, most are characterized or defined by conditions which are so relaxed that a general theory about them is difficult to develop. In other words, it is usually difficult to assign a rule or law for all relations R, that is, all the subsets of given product set $A \times B$. A *function* is a further restriction of the very general notion of relation.

DEFINITION 9.4.1

Let A and B be two arbitrary nonempty sets. Let f denote a relation from A to B with domain X_f and range Y_f. Let (a, b) and (a, c) be any two members of f. If $(a, b) \in f$ and $(a, c) \in f$ always implies $b = c$, then f is called a *function* f.

That is, a function f is a relation that associates to every element of its domain exactly one element in its range.

Thus, a relation R is a function provided that each first entry of the ordered pairs of R has exactly one image.

EXAMPLE 9.4.1

Consider the sets

$$A = \{1, 2, 3\}, \qquad B = \{3, 4\} \tag{9.4.1}$$

Two possible relations R and S from A to B are

$$R = \{(1, 3), (1, 4), (2, 3)\}$$
$$S = \{(2, 4), (3, 3)\} \tag{9.4.2}$$

The domain of R is

$$X_R = \{1, 2\} \tag{9.4.3}$$

Its range is

$$Y_R = \{3, 4\} \tag{9.4.4}$$

Take notice of the members of R. It is seen that the element 1 in X_R corresponds to both 3 and 4 in Y_R. That is, $(1, 3) \in R$ and $(1, 4) \in R$. Clearly, $3 \neq 4$. This means R is *not* a function.

The domain X_S and range Y_S of the relation S are

$$X_S = \{2, 3\}, \qquad Y_S = \{3, 4\} \tag{9.4.5}$$

Looking at the members of S, it is seen that each member of X_S corresponds to exactly one member of Y_S. Thus, S is an example of a relation that is also a function.

The graph of a relation gives a very useful way of checking whether the relation is a function or not. Figure 9.6 shows the graphs of the relations R and S of Example 9.4.1.

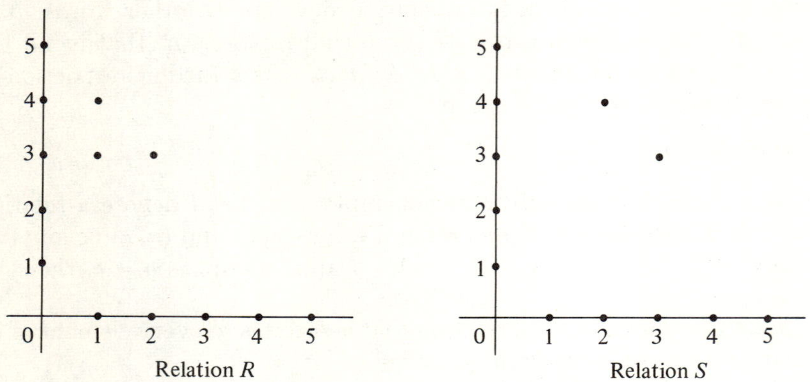

FIG. 9.6

The relation R is not a function since 1 has two images, namely 3 and 4, as can be seen since there are *two* points of the relation R above 1. In general, a relation is *not* a function if above or below any point on the horizontal line there are two or more points. The relation S is a function since corresponding to each point on the horizontal line there is exactly one point above it.

DEFINITION 9.4.2

Let A and B be arbitrary nonempty sets. A *correspondence R in A into B* is a rule or law that assigns to each element a in A at least one element b in B. This correspondence R is denoted by

$$a \xrightarrow{R} b, \qquad a \in A, \qquad b \in B \tag{9.4.6}$$

If a in A corresponds to b in B by R, then b is called an *image of a* and a is termed a *pre-image of b*. If

$$A = \{1, 2, 3\}, \qquad B = \{3, 4\} \tag{9.4.7}$$

one possible correspondence R in A into B is

$$1 \xrightarrow{R} 3, \qquad 2 \xrightarrow{R} 4, \qquad 3 \xrightarrow{R} 3, \qquad 3 \xrightarrow{R} 4 \tag{9.4.8}$$

Figure 9.7 gives a pictorial representation of this mapping.

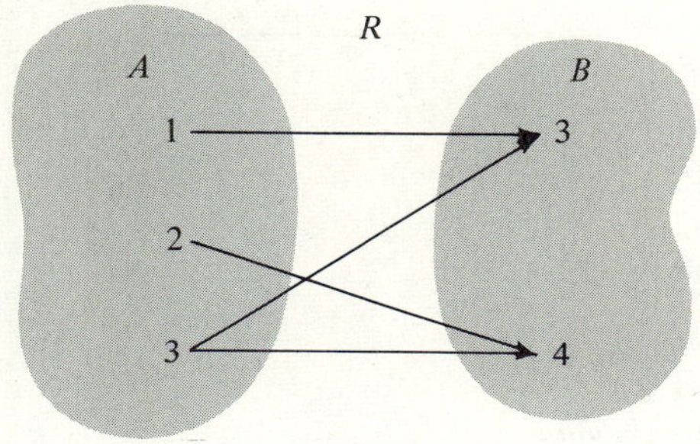

FIG. 9.7

DEFINITION 9.4.3

Let R be a correspondence in A into B. Denote R by

$$a \xrightarrow{R} b, \qquad a \in A, \qquad b \in B \tag{9.4.9}$$

If each element a of A has exactly one image b in B, then the correspondence R is said to be *many-to-one*.

If each element b of B has exactly one pre-image a in A, then the correspondence R is *one-to-many*.

Finally, if every a in A has exactly one image b in B and each element b in B has exactly one pre-image a in A, then the correspondence R is *one-to-one*.

EXAMPLE 9.4.2

Consider the two sets

$$A = \{1, 2, 3, 5\}, \qquad B = \{3, 4, 5\} \tag{9.4.10}$$

A correspondence R in A into B is

$$1 \xrightarrow{R} 3, \qquad 2 \xrightarrow{R} 3, \qquad 3 \xrightarrow{R} 4, \qquad 5 \xrightarrow{R} 5 \tag{9.4.11}$$

Note that the elements 1 and 2 in A have the same element 3 as an image in B. Thus, this correspondence R is many-to-one. Figure 9.8 depicts this characteristic of R.

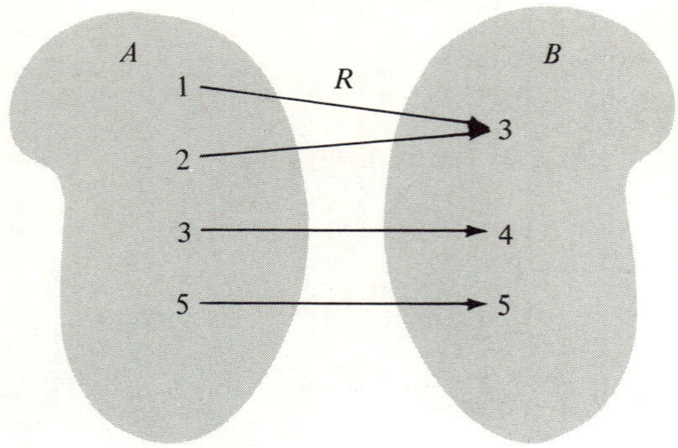

FIG. 9.8

EXAMPLE 9.4.3

Consider the two sets

$$A = \{1, 3, 5\}, \qquad B = \{2, 4, 6, 8\} \tag{9.4.12}$$

A correspondence R in A into B is

$$1 \xrightarrow{R} 2, \qquad 1 \xrightarrow{R} 4, \qquad 3 \xrightarrow{R} 6, \qquad 5 \xrightarrow{R} 8 \tag{9.4.13}$$

Note that the element 1 in A has two images in B, namely, 2 and 4. Thus, the correspondence R defined by (9.4.13) is one-to-many. See Fig. 9.9.

EXAMPLE 9.4.4

Consider the two sets

$$A = \{1, 3, 5\}, \qquad B = \{2, 4, 6\} \tag{9.4.14}$$

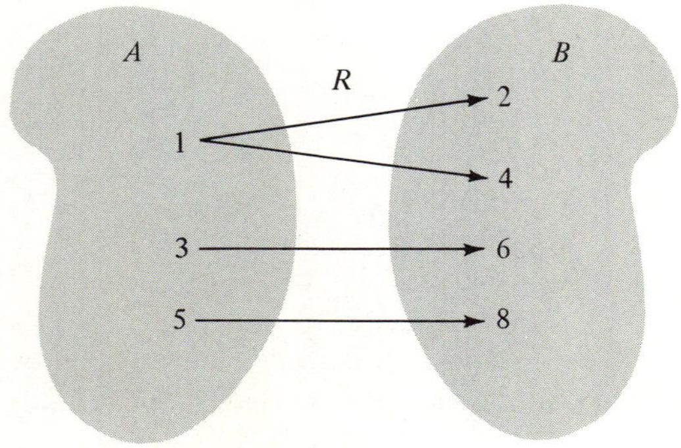

FIG. 9.9

A correspondence R in A into B is

$$1 \xrightarrow{R} 4, \qquad 3 \xrightarrow{R} 2, \qquad 5 \xrightarrow{R} 6 \tag{9.4.15}$$

Note that to each element in A there is exactly one element in B and to each one in B there is one and only one element in A. That is, no element in A has more than one image in B and no element in B has more than one pre-image in A. This means that the correspondence R is one-to-one. See Fig. 9.10.

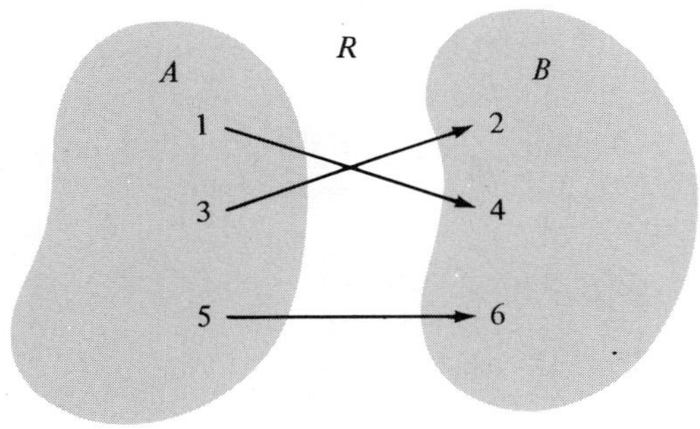

FIG. 9.10

Of course, some correspondences between sets are neither many-to-one nor one-to-many, nor one-to-one. Clearly, a one-to-one correspondence is both one-to-many and many-to-one.

EXAMPLE 9.4.5
Consider the two sets

$$A = \{1, 3, 5\}, \qquad B = \{2, 4, 6\} \tag{9.4.16}$$

A correspondence R in A into B is

$$1 \xrightarrow{R} 2, \qquad 1 \xrightarrow{R} 4, \qquad 3 \xrightarrow{R} 4, \qquad 5 \xrightarrow{R} 6 \tag{9.4.17}$$

Note that the element 1 in A has two images, 2 and 4 in B. Thus, R cannot be many-to-one, nor can it be one-to-one. It is also not one-to-many. Figure 9.11 shows this situation.

From the definition of a correspondence R in A into B and the previous examples, it is concluded that a correspondence R in A into B can be thought of as a relation R whose domain is A and whose range is B. Thus, for $a \in A$, $b \in B$,

$$a \xrightarrow{R} b \text{ is equivalent to } (a, b) \in R \tag{9.4.18}$$

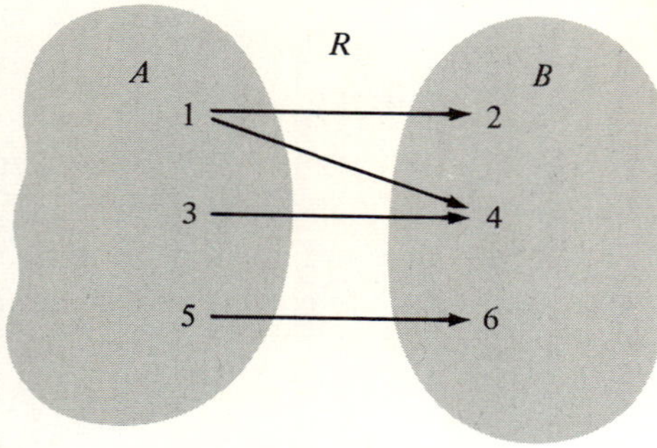

FIG. 9.11

That is, we can think of the first entry a of an ordered pair (a, b) of a relation R as being the pre-image of the second entry b of the ordered pair (a, b) of R.

Understanding this, a function f is merely a correspondence with the property that to each pre-image a there is exactly one image b. This means that a function f is a many-to-one correspondence.

Exercise 9.4

1. Consider the following relations:

$$\text{(i)} \quad R = \{(4, 1), (4, 2), (4, 3), (4, 5)\}$$

$$\text{(ii)} \quad S = \{(1, 1), (1, 2), (2, 2), (1, 4)\}$$

$$\text{(iii)} \quad T = \{(2, 2), (3, 4), (4, 3), (1, 1)\}$$

(a) Which of the above are functions?
(b) Write down the inverses of the above relations.
(c) Which of the inverse relations are functions?
(d) Graph each of the relations R, S, T.

2. Consider the correspondences:

$$\text{(i)} \quad 1 \xrightarrow{R} 3, \quad 2 \xrightarrow{R} 3, \quad 3 \xrightarrow{R} 4, \quad 3 \xrightarrow{R} 5$$

$$\text{(ii)} \quad 4 \xrightarrow{S} 1, \quad 3 \xrightarrow{S} 2, \quad 4 \xrightarrow{S} 2$$

$$\text{(iii)} \quad 6 \xrightarrow{T} 2, \quad 5 \xrightarrow{T} 3, \quad 4 \xrightarrow{T} 4$$

(a) Which of the correspondences are many-to-one?
(b) Which of the correspondences are one-to-one?

(c) Which of the correspondences are one-to-many?
(d) Which of the correspondences are neither?

3. Let f be a function. Consider the inverse f^{-1}. Under what conditions on f will f^{-1} also be a function? [HINT: See Fig. 9.10.]

4. For $A = \{1, 2, 3\}$, $B = \{4, 5, 6\}$, write down all possible one-to-one correspondences in A into B.

5. For $A = \{1, 2\}$, $B = \{3, 4\}$, find two many-to-one correspondences in A into B. Also, find two one-to-many correspondences in A into B.

9.5 EQUIVALENCE RELATIONS

At this point, let us return to the general concept of a relation. In this section, we shall further classify relations into various categories.

From this point forward, we shall concern ourselves only with those relations R defined from a nonempty set A to A itself. That is, if A is an arbitrary nonempty set, we will restrict our study to relations R from A to A. Thus,

$$R \subseteq A \times A \tag{9.5.1}$$

DEFINITION 9.5.1

A relation R from A to A is called *reflexive* if and only if for every x in A it follows that $(x, x) \in R$.

It is important to note the role of the set A in this definition. For a relation R from A to A to be reflexive, *every* member of set A must have itself for an image by R.

EXAMPLE 9.5.1

Let

$$A = \{1, 2, 3\} \tag{9.5.2}$$

A relation R from A to A is given as

$$R = \{(1, 1), (1, 3), (2, 3), (2, 2)\} \tag{9.5.3}$$

This relation R is *not* reflexive since $3 \in A$, but $(3, 3) \notin R$.

EXAMPLE 9.5.2

Let

$$A = \{1, 2\} \tag{9.5.4}$$

A relation R from A to A is given by

$$R = \{(1, 1), (1, 2), (2, 2)\} \tag{9.5.5}$$

This relation R is reflexive since $1 \in A$ and $(1, 1) \in R$, and $2 \in A$ and $(2, 2) \in R$. Thus, for every $x \in A$, we have $(x, x) \in R$. Hence, R is reflexive.

The relation of equality of sets is a reflexive relation since every set is equal to itself. The relation of *less than* for whole numbers is not reflexive since no whole number is less than itself.

Definition 9.5.2

A relation R from A to A is called *symmetric* if and only if whenever $(x, y) \in R$, then $(y, x) \in R$.

Notice that the set A plays no role in determining whether a relation R from A to A is symmetric or not. However, for a relation R from A to A to be symmetric, for *every* ordered pair (x, y) in R, the ordered pair (y, x) must also be in R.

EXAMPLE 9.5.3

If $A = \{1, 2, 3\}$, a relation R from A to A is given by

$$R = \{(1, 1), (1, 3), (2, 2), (3, 1)\} \tag{9.5.6}$$

To determine whether R is symmetric, we must test each ordered pair of R. Obviously, if $(1, 1) \in R$, so is $(1, 1) \in R$. For $(1, 3) \in R$, we see that $(3, 1) \in R$. For $(2, 2) \in R$ we have $(2, 2) \in R$ and finally, for $(3, 1) \in R$, we have $(1, 3) \in R$. Thus, for any ordered pair $(x, y) \in R$, we have $(y, x) \in R$. Hence R is symmetric.

EXAMPLE 9.5.4

For $A = \{1, 2, 3\}$, a relation R from A to A is

$$R = \{(1, 2), (1, 1), (2, 1), (2, 3)\} \tag{9.5.7}$$

To show R is not symmetric, we need only find one ordered pair (x, y) in R for which $(y, x) \notin R$. Here we notice that $(2, 3) \in R$, but $(3, 2) \notin R$. Then R is *not* symmetric.

The relation *is married to* for people is an example of a symmetric relation. However, the relation *knows the name of* for people is not symmetric, since if a knows the name of b, it does not follow that b knows the name of a.

Definition 9.5.3

A relation R from A to A is called *transitive* if and only if whenever (x, y) and (y, z) are in R, then so is (x, z) in R.

Note that the set A plays no role in determining whether a given relation R from A to A is transitive or not.

EXAMPLE 9.5.5

If $A = \{1, 2, 3\}$, consider the relation R from A to A given by

$$R = \{(1, 2), (2, 3), (1, 3), (2, 2)\} \tag{9.5.8}$$

To show that R is transitive, we must look for all the members of R for which the second entry of one is the same as the first entry of the other. Thus,

$$(1, 2) \in R, \qquad (2, 3) \in R \tag{9.5.9}$$

Since $(1, 3) \in R$, we continue to test R. Next,

$$(1, 2) \in R, \qquad (2, 2) \in R \tag{9.5.10}$$

Since $(1, 2) \in R$, we keep testing. However, now there are no members of R left with second entries equal to first entries. Thus, we see that for all $(x, y) \in R$ and $(y, z) \in R$, the ordered pair (x, z) is in R. Hence R is transitive.

EXAMPLE 9.5.6

If $A = \{1, 2, 3\}$, a relation R from A to A is

$$R = \{(1, 2), (2, 3), (1, 3), (2, 1)\} \tag{9.5.11}$$

To show that R is not transitive, we need only find two members of R, (x, y) and (y, z), for which $(x, z) \notin R$. Here

$$(1, 2) \in R, \qquad (2, 1) \in R \tag{9.5.12}$$

but

$$(1, 1) \notin R \tag{9.5.13}$$

Thus R is not transitive.

An example of a nontransitive relation relative to football teams is the relation of "beats." For, if the Bears beat the Colts and the Colts beat the Lions, it does not follow that the Bears will beat the Lions. For sets, *is a subset of* is a transitive relation. This fact was proved in Chapter 2.

The reader will note that the relation *is a subset of* for sets introduced in Chapter 2 is an example of a relation that is both reflexive and transitive. The relation of equivalence of two sets is a relation that is reflexive, symmetric, and transitive. Also, the division relation of natural numbers (see Section 3.5) is reflexive and transitive.

The three classes of relations are independent of each other. That is, a relation may be reflexive without being transitive or symmetric; it may be symmetric without being reflexive or transitive; etc.

Exercise 9.5.1

1. The following relations are defined from A to A where

$$A = \{1, 2, 3, 4\}$$

Determine whether they are reflexive, symmetric, or transitive.

(a) $R = \{(1, 2), (2, 3), (2, 2), (2, 1), (1, 3)\}$
(b) $R = \{(1, 1), (2, 2), (3, 3), (4, 4)\}$
(c) $R = \{(1, 1), (3, 1), (4, 1)\}$
(d) $R = \{(2, 2), (1, 1), (1, 2), (2, 1)\}$
(e) $R = \{(1, 3), (3, 1)\}$
(f) $R = \{(1, 2), (2, 1), (1, 1)\}$
(g) $R = \{(1, 1), (2, 2), (3, 3), (3, 4), (4, 3)\}$
(h) $R = \{(4, 4), (4, 3), (3, 1), (1, 3), (4, 1)\}$

2. Give an example of a relation R from A to A that is

(a) reflexive, but not symmetric or transitive
(b) symmetric, but not reflexive or transitive
(c) transitive, but not reflexive or symmetric
(d) reflexive and symmetric, but not transitive
(e) reflexive and transitive, but not symmetric
(f) symmetric and transitive, but not reflexive
(g) not reflexive, not symmetric, not transitive
(h) reflexive, symmetric, and transitive

You may use a different set A for each problem.

3. If J is the set of integers, define a relation R from J to J by

$$R = \{(x, y) \mid y - x \text{ is divisible by } 2\}$$

Is R reflexive? Is R symmetric? Is R transitive?

4. Define a relation R from J to J by

$$R = \{(x, y) \mid y - x = 2\}$$

Is R reflexive? Symmetric? Transitive?

5. Define a relation R from J to J by

$$R = \{(x, y) \mid x \leqq y\}$$

Is R reflexive? Symmetric? Transitive?

6. Let R be a relation from A to A. Suppose $R = R^{-1}$.
   (a) Is R reflexive? Why?
   (b) Is R symmetric? Why?
7. Let R be a relation from A to A.
   (a) If R is reflexive, is R^{-1} reflexive?
   (b) If R is symmetric, is R^{-1} symmetric?
   (c) If R is transitive, is R^{-1} transitive?
8. Let R and S be two relations from A to A. If R and S are both reflexive,
   (a) is $R \cup S$ reflexive?
   (b) is $R \cap S$ reflexive?
   (c) is $\bar{R}$ reflexive? (The universal set is $A \times A$ here)
9. Let R and S be two relations from A to A. If R and S are both symmetric,
   (a) is $R \cup S$ symmetric?
   (b) is $R \cap S$ symmetric?
   (c) is $\bar{R}$ symmetric?
10. Let R and S be two relations from A to A. If R and S are both transitive,
    (a) is $R \cup S$ transitive?
    (b) is $R \cap S$ transitive?
    (c) is $\bar{R}$ transitive?

Of great interest in mathematics are those relations which are simultaneously reflexive, symmetric, and transitive.

DEFINITION 9.5.4

A relation R from A to A which is reflexive, symmetric, and transitive is called an *equivalence relation*.

EXAMPLE 9.5.7

Let J denote the set of integers. Define a relation R from J to J by

$$R = \{(x, y) \mid y - x \text{ is divisible by } 2\} \tag{9.5.14}$$

Now for any integer x, we know that $x - x = 0$ and we know 0 is divisible by 2. Thus, for any integer x,

$$(x, x) \in R \tag{9.5.15}$$

Hence, R is reflexive.

Next, if $(x, y) \in R$, then $y - x$ is divisible by 2. But if $y - x$ is divisible by 2 so is $x - y$. Thus, whenever $(x, y) \in R$ so is $(y, x) \in R$. That is, R is symmetric.

Finally, suppose $(x, y) \in R$ and $(y, z) \in R$. Then $y - x$ and $z - y$ are both divisible by 2. Hence, their sum must be divisible by 2. But their sum is

$$(y - x) + (z - y) = z - x \tag{9.5.16}$$

This means that $(x, z) \in R$. Hence, R is transitive also.

Since R is reflexive, symmetric, and transitive, it is an equivalence relation.

Let us look at the relation R of Example 9.5.7. Some of the ordered pairs of R are

$$R = \{(0, 0), (2, 0), (4, 2), (3, 1), (^-1, 1), (^-2, 0), \cdots\} \tag{9.5.17}$$

An interesting property of this equivalence relation R is that every integer is related by R to either 0 or 1. Notice also that no integer is related by R to both 0 and 1. Thus, in effect, this equivalence relation R for J to J has divided or partitioned the integers into two classes, those related to 0 and those related to 1. These two classes are usually called *equivalence classes.*

This example leads us to state the following property about equivalence relations.

An equivalence relation R from A to A *decomposes* or *partitions* A into nonempty subsets, called equivalence classes, so that

(1) every element x of A is in at least one equivalence class
(2) every element x of A is in at most one equivalence class
(3) two elements x and y in the *same* equivalence class are related by R.

As an example, consider the relation "is the same height as" between people. Clearly, this relation is an equivalence relation. For, since every person is the same height as himself, this relation is reflexive. Next, if A is the same height as B, then B must be the same height as A. Hence R is symmetric. Finally, if A is the same height as B and B is the same height as C, then A has to be the same height as C. Thus R is transitive.

Now the question is how does this equivalence class R partition the set of people? Clearly, the equivalence classes consist of all people who are the same height. Thus, one equivalence class would consist of people 5′ 10″; another would consist of people 6′ 2″; etc.

Of course,

(1) every person would be in at least one such class
(2) no person could be in more than one such class
(3) two people in the same class would be the same height.

Exercise 9.5.2

1. Let J denote the set of integers. Define a relation R from J to J by

$$R = \{(x, y) \mid y - x \text{ is divisible by } 3\}$$

(a) Show that R is an equivalence relation.
(b) What are the equivalence classes determined by R?

2. Follow the directions of problem 1 for the relation R from J to J defined by

$$R = \{(x, y) \mid y - x \text{ is divisible by } 5\}$$

9.6 THE CONGRUENCE RELATION

In this section, we shall define and give various properties of a *congruence relation*. It will be seen that a congruence relation is an example of an equivalence relation and when applied to the integers has many useful applications.

DEFINITION 9.6.1

Let a and b be any two integers and let $m \geqq 2$ be a positive integer. We say that *a is congruent to b modulo m*, written as

$$a \equiv b \bmod m \qquad (9.6.1)$$

if and only if the difference $a - b$ is divisible by m. For example,

$$5 \equiv 1 \bmod 2 \qquad (9.6.2)$$

since $5 - 1 = 4$ is divisible by 2. Also,

$$17 \not\equiv 2 \bmod 4 \qquad (9.6.3)$$

since $17 - 2 = 15$ is not divisible by 4. Of course,

$$^-3 \equiv 1 \bmod 2 \qquad (9.6.4)$$

since $^-3 - 1 = {}^-4$ is divisible by 2.

What Definition 9.6.1 does is define a relationship between two integers a and b. That is to say, if $a - b$ is divisible by m, then a and b are related; we symbolize this relationship by saying a is congruent to b modulo m, and we write

$$a \equiv b \bmod m \qquad (9.6.5)$$

The symbol $\equiv$ is called a *congruence relation*. The positive integer $m \geqq 2$ is fixed.

EXAMPLE 9.6.1

Some integers that are related modulo 3 are

$$6 \equiv 0 \bmod 3, \qquad 5 \equiv 2 \bmod 3$$

$$7 \equiv 1 \bmod 3, \qquad {}^{-}4 \equiv 2 \bmod 3 \tag{9.6.6}$$

Notice that every integer is congruent to either 0, 1, or 2 modulo 3. This is because when any integer is divided by 3 the remainder is either 0 or 1 or 2. We can now divide the set J of integers in three classes, say $\bar{0}$, $\bar{1}$, $\bar{2}$, with the property that any integer in the class $\bar{0}$ is congruent to 0 modulo 3, any integer in $\bar{1}$ is congruent to 1 modulo 3, and any integer in $\bar{2}$ is congruent to 2 modulo 3.

$\bar{0}$	$\bar{1}$	$\bar{2}$
0, 3, 6 ⁻3, ⁻6, ⁻9, etc.	1, 4, 7 ⁻2, ⁻5, ⁻8, etc.	2, 5, 8 ⁻1, ⁻4, ⁻7 etc.

FIG. 9.12

Exercise 9.6.1

1. Which of the following are true statements?

   (a) $15 \equiv 2 \bmod 9$
   (b) $73 \equiv 3 \bmod 5$
   (c) ${}^{-}15 \equiv 3 \bmod 4$
   (d) ${}^{-}5 \equiv {}^{-}2 \bmod 3$
   (e) $56 \equiv 4 \bmod 13$
   (f) ${}^{-}26 \equiv 2 \bmod 6$

2. For the integers modulo 5, every integer is congruent to either 0, 1, 2, 3, or 4. Write down some members of the classes $\bar{0}$, $\bar{1}$, $\bar{2}$, $\bar{3}$, and $\bar{4}$.

3. Follow the directions of problem 2 for

   (a) the integers modulo 4
   (b) the integers modulo 7
   (c) the integers modulo 12

4. If $a \equiv b \bmod m$, is $b \equiv a \bmod m$? Give several examples to illustrate your answer.

Next, let us show that the relation of congruence $\equiv$ between integers is an equivalence relation.

Theorem 9.6.1

The relation of congruence $\equiv$ between integers modulo m is an equivalence relation. That is, the relation $\equiv$ is

(I) *Reflexive:* $a \equiv a \bmod m$ for any integer a.
(II) *Symmetric:* if $a \equiv b \bmod m$, then $b \equiv a \bmod m$ for all integers a and b.
(III) *Transitive:* if $a \equiv b \bmod m$ and if $b \equiv c \bmod m$, then $a \equiv c \bmod m$ for all integers a, b, c.

PROOF: Recall that two integers a and b are congruent modulo m if and only if $a - b$ is divisible by m. This means there is an integer k such that

$$a - b = k \cdot m \tag{9.6.7}$$

Now,

$$a - a = 0 \tag{9.6.8}$$

Since 0 is divisible by m, the congruence relation $\equiv$ is reflexive.

Next, suppose

$$a \equiv b \bmod m \tag{9.6.9}$$

Then, since $a - b$ is divisible by m, there is an integer k such that

$$a - b = k \cdot m \tag{9.6.10}$$

But if this is so, then by multiplying by $(^-1)$, we see

$$b - a = (^-k) \cdot m \tag{9.6.11}$$

That is, $b - a$ is divisible by m also. Thus,

$$b \equiv a \bmod m \tag{9.6.12}$$

The congruence relation $\equiv$ is symmetric.

Finally, suppose

$$a \equiv b \bmod m, \qquad b \equiv c \bmod m \tag{9.6.13}$$

This means integers k and n exist so that

$$a - b = k \cdot m, \qquad b - c = n \cdot m \tag{9.6.14}$$

Adding the equations in (9.6.14) gives

$$(a - b) + (b - c) = k \cdot m + n \cdot m \tag{9.6.15}$$

Or, upon simplifying,

$$a - c = (k + n)m \tag{9.6.16}$$

This says $a - c$ is divisible by m. Hence

$$a \equiv c \bmod m \tag{9.6.17}$$

Thus the relation $\equiv$ is transitive.

Because the congruence relation $\equiv$ between integers modulo m is an equivalence relation, it induces a partition of the set J of integers into equivalence classes. Two integers will belong to the *same* equivalence class provided they are congruent modulo m.

For example, for the integers modulo 5, there are five equivalence classes $\bar{0}, \bar{1}, \bar{2}, \bar{3}, \bar{4}$. The integers 5 and 10 both belong to $\bar{0}$ since

$$5 \equiv 0 \bmod 5, \qquad 10 \equiv 0 \bmod 5 \tag{9.6.18}$$

Let us now discuss some other properties of congruence relations.

Theorem 9.6.2

The relation of congruence $\equiv$ between integers modulo m has the following properties.

(I) If $a \equiv b \bmod m$ and $c \equiv d \bmod m$, then $a + c \equiv b + d \bmod m$.
(II) If $a \equiv b \bmod m$ and $c \equiv d \bmod m$, then $a \cdot c \equiv b \cdot d \bmod m$.

PROOF: Suppose

$$a \equiv b \bmod m, \qquad c \equiv d \bmod m \tag{9.6.19}$$

This means there are integers k and n so that

$$a - b = k \cdot m, \qquad c - d = n \cdot m \tag{9.6.20}$$

or

$$a = b + k \cdot m, \qquad c = d + n \cdot m \tag{9.6.21}$$

Add the two equations above. Then,

$$\begin{aligned} a + c &= (b + k \cdot m) + (d + n \cdot m) \\ &= (b + d) + (k + n) \cdot m \end{aligned} \tag{9.6.22}$$

or

$$(a + c) - (b + d) = (k + n) \cdot m \tag{9.6.23}$$

This means $(a + c) - (b + d)$ is divisible by m so that

$$(a + c) \equiv b + d \bmod m \tag{9.6.24}$$

Part II of the proof is left as an exercise.

Theorem 9.6.2 tells us that the relation of congruence $\equiv$ is preserved under addition and multiplication.

For example,

$$17 \equiv 2 \bmod 5, \qquad 9 \equiv 4 \bmod 5 \tag{9.6.25}$$

Now,

$$17 + 9 = 26, \qquad 2 + 4 = 6 \tag{9.6.26}$$

and

$$26 \equiv 6 \bmod 5 \tag{9.6.27}$$

EXAMPLE 9.6.2

Consider the integers modulo 5. The equivalence classes are $\bar{0}, \bar{1}, \bar{2}, \bar{3}, \bar{4}$. Now suppose we wish to talk about adding and multiplying these equivalence classes; that is, let us find out what might be the answer to

$$\bar{2} + \bar{4} \tag{9.6.28}$$

for integers modulo 5. Since we know that

$$2 \in \bar{2}, \qquad 4 \in \bar{4} \tag{9.6.29}$$

and since

$$2 + 4 = 6 \quad \text{and} \quad 6 \in \bar{1} \tag{9.6.30}$$

we will conclude that

$$\bar{2} + \bar{4} = \bar{1} \tag{9.6.31}$$

Similarly, to find

$$\bar{2} \cdot \bar{4} \tag{9.6.32}$$

we note that

$$2 \in \bar{2}, \qquad 4 \in \bar{4} \tag{9.6.33}$$

and

$$2 \cdot 4 = 8, \qquad 8 \in \bar{3} \tag{9.6.34}$$

so that

$$\bar{2} \cdot \bar{4} = \bar{3} \tag{9.6.35}$$

The above example leads us to make the following definition.

DEFINITION 9.6.2

Let $\bar{a}$ and $\bar{b}$ be two equivalence classes of integers modulo m. Their *sum* $\bar{a} + \bar{b}$ is defined as the equivalence class $\overline{a + b}$ and their *product* $\bar{a} \cdot \bar{b}$ is defined as the equivalence class $\overline{a \cdot b}$.

The following tables give the rules for addition and multiplication of integers modulo 5.

+	$\bar{0}$	$\bar{1}$	$\bar{2}$	$\bar{3}$	$\bar{4}$
$\bar{0}$	$\bar{0}$	$\bar{1}$	$\bar{2}$	$\bar{3}$	$\bar{4}$
$\bar{1}$	$\bar{1}$	$\bar{2}$	$\bar{3}$	$\bar{4}$	$\bar{0}$
$\bar{2}$	$\bar{2}$	$\bar{3}$	$\bar{4}$	$\bar{0}$	$\bar{1}$
$\bar{3}$	$\bar{3}$	$\bar{4}$	$\bar{0}$	$\bar{1}$	$\bar{2}$
$\bar{4}$	$\bar{4}$	$\bar{0}$	$\bar{1}$	$\bar{2}$	$\bar{3}$

•	$\bar{0}$	$\bar{1}$	$\bar{2}$	$\bar{3}$	$\bar{4}$
$\bar{0}$	$\bar{0}$	$\bar{0}$	$\bar{0}$	$\bar{0}$	$\bar{0}$
$\bar{1}$	$\bar{0}$	$\bar{1}$	$\bar{2}$	$\bar{3}$	$\bar{4}$
$\bar{2}$	$\bar{0}$	$\bar{2}$	$\bar{4}$	$\bar{1}$	$\bar{3}$
$\bar{3}$	$\bar{0}$	$\bar{3}$	$\bar{1}$	$\bar{4}$	$\bar{2}$
$\bar{4}$	$\bar{0}$	$\bar{4}$	$\bar{3}$	$\bar{2}$	$\bar{1}$

FIG. 9.13

The arithmetic of integers modulo m is called *modular arithmetic* or *clock arithmetic* because if we write the elements in a circular pattern, addition and multiplication are performed as a hand would move about a clock.

For example, for the integers modulo 5, we have

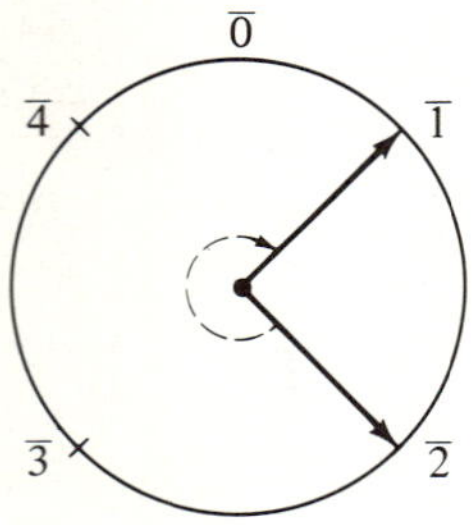

FIG. 9.14

To add $\bar{2}$ and $\bar{4}$, start at $\bar{2}$ and move 4 spaces. The hand ends at $\bar{1}$. Then,

$$\bar{2} + \bar{4} = \bar{1} \bmod 5 \tag{9.6.36}$$

To multiply $\bar{2}$ by $\bar{4}$, start at $\bar{0}$ and move $2 \cdot 4 = 8$ places. See Fig. 9.15. The hand ends up at $\bar{3}$. Thus, $\bar{2} \cdot \bar{4} = \bar{3} \bmod 5$.

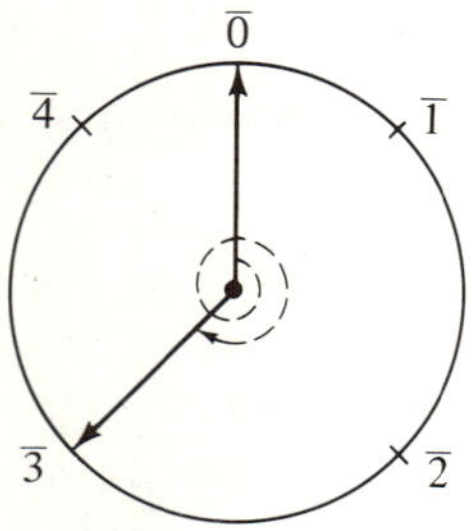

FIG. 9.15

Exercise 9.6.2

1. Write down addition and multiplication tables for

   (a) the integers modulo 2
   (b) the integers modulo 8
   (c) the integers modulo 12

2. Is addition of integers modulo 12 commutative? How can you tell?

3. Is multiplication of integers modulo 8 commutative? How can you tell?

4. Using an example, demonstrate that both addition and multiplication of integers modulo 5 are associative.

5. For the integers modulo 12, find the additive inverses of

(a) 5	(d) 0
(b) 7	(e) 3
(c) 1	(f) 6

6. Follow the above instructions for the integers modulo 9.

7. Using an example, demonstrate that multiplication is distributive over addition for integers modulo 8.

8. Is the product law valid for integers modulo 8? What about for integers modulo 5? Give reasons.

9. Prove part II of Theorem 9.6.2. [HINT: Look at the product of the equations in (9.6.21).]

10. Find the equivalence class x for which

(a) $x \equiv 7 \bmod 3$	(e) $x + 11 \equiv 11 \bmod 11$
(b) $x + 4 \equiv 6 \bmod 5$	(f) $x + 9 \equiv 10 \bmod 2$
(c) $x \equiv 8 \bmod 5$	(g) $x - 9 \equiv 9 \bmod 2$
(d) $x - 3 \equiv 6 \bmod 5$	

11. Find three systems of integers modulo m in which the product law is *not* valid. That is, find systems in which

$$\bar{a} \cdot \bar{b} = \bar{0} \quad \text{but} \quad \bar{a} \neq \bar{0}, \qquad \bar{b} \neq \bar{0}.$$

What can you conjecture about m?

12. A famous theorem due to Fermat (1601–1655) states that if $a \not\equiv 0 \bmod p$, p a prime integer, then

$$a^p \equiv a \bmod p$$

Use this result to do the following problem.

Find the remainder when 2^{25} is divided by 5. [HINT: $2^{25} = (2^5)^5$.]

13. The days of the week can be interpreted as a modulo 7 system in which Sunday is 0, Monday is 1, Tuesday is 2, · · ·, Saturday

is 6. In 1963, July 4 fell on a Thursday and it is the 185th day of the year. On what day did Christmas (the 359th day), New Year's, Halloween (the 304th day), and Washington's birthday (the 53rd day) fall?

An application of congruence relations is a technique for checking computational work. The technique given here is called *casting out nines.*

We will illustrate this method by the following examples.

EXAMPLE 9.6.3

The sum of the integers 5061, 2033, 6112 is 13,206. Use the method of casting out nines to check this computation.

Here we look at each integer to be added and find the sum of their digits.

Thus,

$$\begin{array}{ll} 5061 & 5 + 0 + 6 + 1 = 12 \\ 2033 & 2 + 0 + 3 + 3 = 8 \\ 6112 & 6 + 1 + 1 + 2 = 10 \end{array} \tag{9.6.37}$$

Now, add these numbers and determine the equivalence class their sum belongs to, modulo 9. Thus,

$$12 + 8 + 10 = 30 \equiv 3 \bmod 9 \tag{9.6.38}$$

Finally, repeat the process for the sum of the digits of the answer. Thus,

$$13{,}206 \qquad 1 + 3 + 2 + 0 + 6 = 12 \equiv 3 \bmod 9 \tag{9.6.39}$$

Since both are congruent to 3 modulo 9, we conclude that the answer 13,206 is *probably* correct. Had they not both been congruent to the same number modulo 9, we would conclude the answer 13,206 was *wrong.*

The reason we can only conclude *probable correctness* is illustrated in the following example.

EXAMPLE 9.6.4

The product of 23 and 32 is 745. Use casting out nines to check this computation.

Here the sums of the digits of 23 and 32 are

$$\begin{array}{ll} 23 & 2 + 3 = 5 \\ 32 & 3 + 2 = 5 \end{array} \tag{9.6.40}$$

Since we are multiplying 23 by 32, we will multiply 5 by 5 and see what it is congruent to modulo 9. Thus,

$$5 \cdot 5 = 25 \equiv 7 \bmod 9 \tag{9.6.41}$$

Now, the sum of the digits of the answer is

$$745 \qquad 7 + 4 + 5 = 16 \equiv 7 \bmod 9 \tag{9.6.42}$$

Since these are both congruent to 7 mod 9, we conclude the answer is probably correct. Note that the correct answer is 736, *not* 745.

EXAMPLE 9.6.5

The difference 7432 − 246 is 7196. Use the method of casting out nines to check the answer.

Here the sums of the digits of 7432 and 246 are

$$7432 \qquad 7 + 4 + 3 + 2 = 16$$

$$246 \qquad 2 + 4 + 6 = 12 \tag{9.6.43}$$

Since we are subtracting, we subtract 12 from 16 and see what it is congruent to modulo 9. Thus

$$16 - 12 = 4 \equiv 4 \bmod 9 \tag{9.6.44}$$

For the answer, we have

$$7196 \qquad 7 + 1 + 9 + 6 = 23 \equiv 5 \bmod 9 \tag{9.6.45}$$

Since we obtain two different results, we conclude the answer 7196 is definitely wrong.

The technique of casting out nines can be used for checking computations involving the four rational operations of addition, multiplication, subtraction, and division.

Finally, if numbers are written in a base other than ten, an analogous "casting out" process can be used. For example, in the base five, we cast out fours; in the base eight, sevens are cast out; in the base twelve, we would cast out *e*'s; in general, in the base g, we cast out $(g - 1)$'s.

Exercise 9.6.3

1. Are the answers to the following computations *absolutely correct, probably correct, absolutely wrong, probably wrong?* Use the method of casting out nines to determine your answer.

(a)
$$\begin{array}{r} 2\,055 \\ 3\,022 \\ +6\,120 \\ \hline 10{,}197 \end{array}$$

(b)
$$\begin{array}{r} 4103 \\ 1902 \\ +\ 921 \\ \hline 7026 \end{array}$$

(f) $\dfrac{2366}{91} = 26$

(g) $32{,}611 \cdot 7146 = 258{,}373{,}806$

(h) $\dfrac{12{,}495}{21} = 559$

(i) $456 - 132 = 234$

(c)
$$\begin{array}{r} 1871 \\ 1463 \\ +1891 \\ \hline 5225 \end{array}$$

(d) $456 \cdot 132 = 60{,}912$

(e) $28{,}964 \cdot 1431 = 41{,}444{,}784$

(j) $4451 - 205 = 4246$

(k) $7840 - 244 = 7596$

(l) $\dfrac{444{,}717}{34{,}209} = 31$

(m) $\dfrac{4641}{17} = 271$

2. Are the following computations in the base indicated *absolutely correct, probably correct, absolutely wrong,* or *probably wrong?* Use method of casting out to determine your answer.

(a)
$$\begin{array}{r} 503_{(\text{six})} \\ +201_{(\text{six})} \\ \hline 1204_{(\text{six})} \end{array}$$

(b)
$$\begin{array}{r} 307_{(\text{eight})} \\ +504_{(\text{eight})} \\ \hline 1011_{(\text{eight})} \end{array}$$

(c)
$$\begin{array}{r} 62e_{(\text{twelve})} \\ +5t1_{(\text{twelve})} \\ \hline 1210_{(\text{twelve})} \end{array}$$

(d)
$$\begin{array}{r} 523_{(\text{eight})} \\ -201_{(\text{eight})} \\ \hline 232_{(\text{eight})} \end{array}$$

(e)
$$\begin{array}{r} 203_{(\text{five})} \\ \times\ 21_{(\text{five})} \\ \hline 4331_{(\text{five})} \end{array}$$

(f)
$$\begin{array}{r} 402_{(\text{five})} \\ \times\ 13_{(\text{five})} \\ \hline 10{,}131_{(\text{five})} \end{array}$$

★9.7 A DEFINITION OF INTEGERS BY MEANS OF EQUIVALENCE CLASSES

In Chapter 4, a deficiency of the set W of whole numbers, namely $\{0, 1, 2, \cdots\}$ was recognized and, because of that deficiency, the whole number system was enlarged. The new system, which did not have the deficiencies present in the set W, we called the set J of integers.

However, we never really defined in a rigorous way what an integer is. Rather, we merely stated various properties of integers; that is to say, the integers were described, not defined.

The purpose of this section is to give a rigorous definition of an integer.

Definition 9.7.1

Consider the set N of natural numbers and the product set $N \times N$. Any two elements (a, b) and (c, d) of $N \times N$ are said to be *equivalent,* written as

$$(a, b) \sim (c, d) \tag{9.7.1}$$

if and only if

$$a + d = b + c \tag{9.7.2}$$

For example,

$$(3, 2) \sim (6, 5) \tag{9.7.3}$$

since

$$3 + 5 = 2 + 6 \tag{9.7.4}$$

Also,

$$(4, 1) \nsim (3, 2) \tag{9.7.5}$$

since

$$4 + 2 \neq 1 + 3 \tag{9.7.6}$$

The above definition of equivalence between elements of $N \times N$ is a relation between ordered pairs of natural numbers. The following theorem tells us this relation $\sim$ is also an equivalence relation.

Theorem 9.7.1

The relation of equivalence $\sim$ between ordered pairs (a, b) of $N \times N$ is an equivalence relation. That is, the relation of equivalence $\sim$ is

(I) *Reflexive:* $(a, b) \sim (a, b)$ for $a, b \in N$.
(II) *Symmetric:* If $(a, b) \sim (c, d)$, then $(c, d) \sim (a, b)$ for $a, b, c, d \in N$.
(III) *Transitive:* If $(a, b) \sim (c, d)$ and $(c, d) \sim (e, f)$, then $(a, b) \sim (e, f)$ for $a, b, c, d, e, f \in N$.

PROOF: First of all,

$$(a, b) \sim (a, b) \tag{9.7.7}$$

since

$$a + b = b + a \tag{9.7.8}$$

Next, suppose

$$(a, b) \sim (c, d) \tag{9.7.9}$$

This means that

$$a + d = b + c \tag{9.7.10}$$

Using the commutative property of natural numbers (cf. Chapter 3), we see that

$$c + b = d + a \tag{9.7.11}$$

This means that

$$(c, d) \sim (a, b) \tag{9.7.12}$$

Thus, the relation $\sim$ is symmetric.

Finally, suppose

$$(a, b) \sim (c, d), \qquad (c, d) \sim (e, f) \tag{9.7.13}$$

Then

$$a + d = b + c, \qquad c + f = d + e \tag{9.7.14}$$

Add these two equations. Then

$$(a + d) + (c + f) = (b + c) + (d + e) \tag{9.7.15}$$

Using the associative and commutative laws,

$$(a + f) + (c + d) = (b + e) + (c + d) \tag{9.7.16}$$

Apply the cancellation law for addition. Then

$$(a + f) = (b + e) \tag{9.7.17}$$

This means that

$$(a, b) \sim (e, f) \tag{9.7.18}$$

Hence, the relation $\sim$ of equivalence is transitive.

Since the relation of equivalence $\equiv$ between elements of $N \times N$ is an equivalence relation, then it partitions $N \times N$ into equivalence classes. Any two members of a given equivalence class have the property that they are equivalent.

For example, the ordered pairs

$$(3, 2), (2, 1), (4, 3), \cdots, (n + 1, n), \cdots \tag{9.7.19}$$

are in the same equivalence class.

Exercise 9.7.1

★1. Determine which of the following are true statements.

(a) $(5, 3) \sim (7, 5)$
(b) $(15, 2) \sim (5, 4)$
(c) $(7, 7) \sim (4, 4)$
(d) $(8, 15) \sim (1, 8)$
(e) $(9, 10) \sim (1, 1)$
(f) $(24, 36) \sim (13, 1)$
(g) $(24, 36) \sim (1, 13)$

★2. Write down three ordered pairs (a, b) equivalent to each of the following ordered pairs:

(a) $(5, 2)$
(b) $(1, 1)$
(c) $(1, 2)$
(d) $(3, 1)$

Let us now return to the partition of $N \times N$ into equivalence classes. Consider the equivalence class

$$\{(2, 1), (3, 2), (4, 3), \cdots, (n + 1, n), \cdots\} \tag{9.7.20}$$

Notice that each member of this equivalence class has the property that its first entry exceeds its second entry by one. This equivalence class is named the integer *one* and is given the symbol 1.

Consider the equivalence class

$$\{(3, 1), (4, 2), \cdots, (n + 2, n), \cdots\} \tag{9.7.21}$$

Here, each first entry exceeds the second entry by two. The name given this equivalence class is the integer two, denoted by 2.

For the equivalence class

$$\{(1, 1), (2, 2), \cdots, (n, n), \cdots\} \tag{9.7.22}$$

the first entry exceeds the second entry by zero. Thus, the name given this equivalence class is the integer zero, denoted by 0.

For the equivalence class

$$\{(1, 2), (2, 3), (3, 4), \cdots, (n, n + 1) \cdots\} \tag{9.7.23}$$

the second entry exceeds the first entry by one. Hence, the symbol given this equivalence class is the integer $^-1$.

By looking at all the equivalence classes obtained by partitioning $N \times N$, we can define the set J of integers, namely,

$$J = \{\cdots, {}^-2, {}^-1, 0, 1, 2, \cdots\} \tag{9.7.24}$$

Exercise 9.7.2

*1. Write down the integer defined by the following equivalence classes:

(a) $\{(5, 1), (6, 2), (7, 3), \cdots\}$
(b) $\{(1, 3), (2, 4), (3, 5) \cdots\}$
(c) $\{(19, 1), (20, 2), \cdots\}$
(d) $\{(1, 19), (2, 20), \cdots\}$

*2. Write down the equivalence class which defines the following integers:

(a) $^-3$
(b) 5
(c) 0
(d) $^-1$
(e) $^-6$
(f) 15
(g) $^-15$

Now, we are ready to define what is meant by adding and multiplying two integers. In the following definition, we use (a, b) as a representative member of the equivalence class defining the integer.

Definition 9.7.2

Let (a, b) and (c, d) represent two integers, where $a, b, c, d \in N$. The *sum of* (a, b) *and* (c, d), written as

$$(a, b) + (c, d) \tag{9.7.25}$$

is defined to be

$$(a, b) + (c, d) \sim (a + d, b + c) \tag{9.7.26}$$

The *product of* (a, b) *and* (c, d), written as

$$(a, b) \cdot (c, d) \tag{9.7.27}$$

is defined to be

$$(a, b) \cdot (c, d) \sim (a \cdot c + b \cdot d, b \cdot c + a \cdot d) \tag{9.7.28}$$

example 9.7.1

Find the sum and product of 7 and $^-2$ using the above definition.

Here, a member of the equivalence class which names 7 is (8, 1) and a member of the equivalence class which names $^-2$ is (1, 3). (We could have used (10, 3) to represent 7 just as well). Now,

$$(8, 1) + (1, 3) \sim (8 + 1, 1 + 3) \tag{9.7.29}$$

or

$$(8, 1) + (1, 3) \sim (9, 4) \tag{9.7.30}$$

But (9, 4) is a representative of the equivalence class defining 5. Thus,

$$7 + (^-2) = 5 \tag{9.7.31}$$

Finally, to find $7 \cdot (^-2)$,

$$(8, 1) \cdot (1, 3) \sim (8 \cdot 1 + 1 \cdot 3, 1 \cdot 1 + 8 \cdot 3) \tag{9.7.32}$$

or

$$(8, 1) \cdot (1, 3) \sim (11, 25) \tag{9.7.33}$$

Here (11, 25) is a representative from the equivalence class defining $^-14$. Thus,

$$7 \cdot (^-2) = {}^-14 \tag{9.7.34}$$

Exercise 9.7.3

*1. Using Definition 9.7.2, find the sum and product of

(a) $5, {}^-3$
(b) $3, 2$
(c) ${}^-4, {}^-5$
(d) $6, {}^-11$
(e) $4, {}^-1$
(f) $0, 2$
(g) ${}^-2, 2$

*2. Show that the additive identity in J is 0. [HINT: A representative of 0 is $(1, 1)$.]

*3. Show that if (a, b) represents any integer, then $(a, b) + (b, a) \sim (1, 1)$.

*4. Show that addition and multiplication of integers is both commutative and associative.

*5. Show that 1 is the multiplicative identity of J.

*9.8 A DEFINITION OF RATIONAL NUMBERS BY MEANS OF EQUIVALENCE CLASSES

In Chapter 6, a deficiency of the set J of integers, namely,

$$J = \{0, 1, {}^-1, 2, {}^-2, 3, {}^-3, \cdots\} \tag{9.8.1}$$

was noticed. Recall that in J, some elements, 2 for example, do not possess multiplicative inverses. That is, there is no member of J which when multiplied by 2 gives 1. It was this drawback to J which caused us to enlarge J. The new set, of which J was a proper subset and which did not have the aforementioned deficiency of J, was called the set R of rational numbers.

Here, just as in Chapter 4, we never really defined *what* a rational number is, but rather merely gave properties or laws that rational numbers obey. The purpose of this section is to give a rigorous definition for rational numbers.

DEFINITION 9.8.1

Consider the set J of integers and the product set $J \times J$. Any two elements (a, b) and (c, d) of $J \times J$ are said to be *equivalent*, written as

$$(a, b) \sim (c, d) \tag{9.8.2}$$

if and only if

$$a \cdot d = b \cdot c \tag{9.8.3}$$

For example,

$$(2, 3) \sim (4, 6) \tag{9.8.4}$$

since

$$2 \cdot 6 = 3 \cdot 4 \tag{9.8.5}$$

Also,

$$(5, 2) \not\sim (6, 3) \tag{9.8.6}$$

since

$$5 \cdot 3 \neq 2 \cdot 6 \tag{9.8.7}$$

The above definition of equivalence between elements of $J \times J$ is a relation between ordered pairs of integers. It turns out this relation is also an equivalence relation.

Theorem 9.8.1

The relation of equivalence $\sim$ between ordered pairs (a, b) of $J \times J$ is an equivalence relation. That is, the relation of equivalence $\sim$ is

(I) *Reflexive:* $(a, b) \sim (a, b)$ for $a, b \in J$
(II) *Symmetric:* if $(a, b) \sim (c, d)$, then $(c, d) \sim (a, b)$ for $a, b, c, d \in J$
(III) *Transitive:* if $(a, b) \sim (c, d)$ and if $(c, d) \sim (e, f)$, then $(a, b) \sim (e, f)$ for $a, b, c, d, e, f \in J$.

PROOF: First of all

$$(a, b) \sim (a, b) \tag{9.8.8}$$

since

$$a \cdot b = b \cdot a \tag{9.8.9}$$

The relation $\sim$ is thus reflexive.

Next, suppose

$$(a, b) \sim (c, d) \tag{9.8.10}$$

This means that

$$a \cdot d = b \cdot c \tag{9.8.11}$$

Thus, since multiplication of integers is commutative,

$$c \cdot b = d \cdot a \tag{9.8.12}$$

This tells us that

$$(c, d) \sim (a, b) \tag{9.8.13}$$

The relation $\sim$ is symmetric.

Finally, suppose

$$(a, b) \sim (c, d) \qquad (c, d) \sim (e, f) \tag{9.8.14}$$

Then,

$$a \cdot d = b \cdot c \qquad c \cdot f = d \cdot e \tag{9.8.15}$$

Multiply these two equations. Then,

$$(a \cdot d) \cdot (c \cdot f) = (b \cdot c) \cdot (d \cdot e) \tag{9.8.16}$$

Since multiplication of integers is commutative and associative, we have

$$(a \cdot f) \cdot (d \cdot c) = (b \cdot e) \cdot (d \cdot c) \tag{9.8.17}$$

If either c or d is not zero, then by the cancellation law of multiplication,

$$a \cdot f = b \cdot e \tag{9.8.18}$$

This says that

$$(a, b) \sim (e, f) \tag{9.8.19}$$

(If $c = 0$ or $d = 0$ or both c and d are zero, the result is also true. This part is left as an exercise.) Thus, the relation of equivalence $\sim$ is transitive.

Since the relation of equivalence $\sim$ between elements of $J \times J$ is an equivalence relation, then it partitions $J \times J$ into equivalence classes, any two members of a given equivalence class having the property that they are equivalent.

For example, the ordered pairs

$$(2, 1), (4, 2), (^-4, ^-2), (^-2, ^-1), \cdots \tag{9.8.20}$$

are each equivalent to each other. Hence, they belong to the *same* equivalence class.

Exercise 9.8.1

*1. Determine which of the following are true statements.

(a) $(5, 3) \sim (^-5, ^-3)$	(e) $(1, 1) \sim (^-4, ^-4)$
(b) $(4, 1) \sim (16, 4)$	(f) $(^-3, 2) \sim (3, ^-2)$
(c) $(^-3, 2) \sim (^-2, 3)$	(g) $(1, 3) \sim (2, 6)$
(d) $(1, 1) \sim (^-4, 4)$	(h) $(1, 3) \sim (6, 2)$

*2. Write down three ordered pairs (a, b) equivalent to each of the following ordered pairs.

(a) $(5, 2)$	(e) $(^-1, 1)$
(b) $(1, 1)$	(f) $(^-1, ^-2)$

(c) (1, 2)   (g) (1, ⁻2)
(d) (2, 1)

Let us now return to the partition of $J \times J$ into equivalence classes and let us consider all those equivalence classes except the one in which the second member of the ordered pairs is zero. That is, from now on we will only consider pairs (a, b) in which $b \neq 0$.

Consider the equivalence class

$$\{(1, 2), (2, 4), (^-2, ^-4), (^-1, ^-2), \cdots\} \tag{9.8.21}$$

Notice that each member of this equivalence class has the property that its second entry is twice the first entry. This equivalence class is named the *rational number* $\frac{1}{2}$.

Consider the equivalence class

$$\{(3, 1), (6, 2), (^-3, ^-1), \cdots\} \tag{9.8.22}$$

Here the first entry is three times each second entry. This equivalence class is called the *rational number* $\frac{3}{1} = 3$.

The equivalence class

$$\{(1, 3), (2, 6), (^-1, ^-3), \cdots\} \tag{9.8.23}$$

is called the *rational number* $\frac{1}{3}$.

The equivalence class

$$\{(0, 1), (0, ^-1), (0, 2), \cdots\} \tag{9.8.24}$$

is called the *rational number* 0.

Thus, by looking at all the equivalence classes of $J \times J$ (except those which have zero as a second entry), we can define the set R of rational numbers, namely,

$$R = \left\{0, 1, ^-1, 2, ^-2, \frac{1}{2}, \frac{^-1}{2}, \cdots\right\} \tag{9.8.25}$$

Notice that a member (a, b) of an equivalence class of $J \times J$ is nothing more than another name for the ratio of a to b, $\frac{a}{b}$.

Exercise 9.8.2

★1. Write down the rational number defined by the following equivalence classes.

(a) $\{(5, 1), (10, 2), (^-5, ^-1), \cdots\}$
(b) $\{(1, 4), (^-1, ^-4), \cdots\}$
(c) $\{(1, 5), (2, 10), (^-3, ^-15), \cdots\}$
(d) $\{(6, 4), (3, 2), \cdots\}$
(e) $\{(4, 5), (8, 10), \cdots\}$

★2. Write down a few members of the equivalence class which defines the following rational numbers.

(a) $^-4$ (d) 0

(b) $\frac{3}{4}$ (e) 5

(c) $\frac{9}{4}$ (f) $\frac{2}{9}$

At this stage, we can define the meaning of addition and multiplication of rational numbers.

In all subsequent work, the ordered pair (a, b) where $a, b \neq 0$ are two integers, will denote the *best* representative member of the equivalence class defining the rational number. The *best* representative member of an equivalence class is that ordered pair (a, b) in the equivalence class for which

$$b > 0, \qquad \gcd\{a, b\} = 1 \tag{9.8.26}$$

For example, the best representative of the equivalence class

$$\{(2, 3), (^-2, ^-3), (4, 6), \cdots\} \tag{9.8.27}$$

is

$$(2, 3) \tag{9.8.28}$$

since $3 > 0$ and $\gcd\{2, 3\} = 1$.

DEFINITION 9.8.2

Let (a, b) and (c, d) be the best representatives of two rational numbers. The *sum of* (a, b) *and* (c, d), written as

$$(a, b) + (c, d) \tag{9.8.29}$$

is defined to be the best representative of the equivalence class having

$$(a \cdot d + b \cdot c, b \cdot d) \tag{9.8.30}$$

as a member. The *product of* (a, b) *and* (c, d) written as

$$(a, b) \cdot (c, d) \tag{9.8.31}$$

is the best representative of the equivalence class of which

$$(a \cdot c, b \cdot d) \tag{9.8.32}$$

is a member.

EXAMPLE 9.8.1

Find the sum and product of

$$\frac{2}{3} = \{(2, 3), (^-2, ^-3), (4, 6), \cdots\}$$

$$\frac{1}{5} = \{(1, 5), (^-1, ^-5), (2, 10), \cdots\} \tag{9.8.33}$$

Here,

$$\begin{aligned}(2, 3) + (1, 5) &= (2 \cdot 5 + 3 \cdot 1, 3 \cdot 5)\\ &= (10 + 3, 15) = (13, 15)\end{aligned} \tag{9.8.34}$$

Since $15 > 0$ and gcd $\{13, 15\} = 1$, then $(13, 15)$ is the best representative. Thus,

$$\frac{2}{3} + \frac{1}{5} = \frac{13}{15} \tag{9.8.35}$$

Also,

$$(2, 3) \cdot (1, 5) = (2 \cdot 1, 3 \cdot 5) = (2, 15) \tag{9.8.36}$$

Again, since $15 > 0$, and gcd $\{2, 15\} = 1$, we have

$$\frac{2}{3} \cdot \frac{1}{5} = \frac{2}{15} \tag{9.8.37}$$

Exercise 9.8.3

*1. Using the definition of sum and product of rational numbers, find the sum and product of

(a) $5, \frac{2}{3}$

(b) $\frac{^-2}{5}, \frac{1}{8}$

(e) $\frac{^-3}{2}, \frac{^-8}{9}$

(f) $0, \frac{^-1}{3}$

(c) $\frac{4}{3}, \frac{2}{9}$

(d) $\frac{8}{5}, \frac{10}{3}$

(g) $1, \frac{1}{2}$

(h) $\frac{-1}{3}, \frac{-2}{3}$

*2. Show that the multiplicative identity of R is 1. [HINT: Show that if (a, b) is the best representative of any rational number, the product of (a, b) multiplied by $(1, 1)$ is (a, b).]

*3. Show that if (a, b) is the best representative of a rational number, $a > 0$, then (b, a) is its multiplicative inverse. [HINT: Show that their product is 1.]

10 Geometry

10.1 INTRODUCTION

The format of this chapter will be different from the one used in the previous chapters. The reason for this change stems from the different ways in which geometry can be introduced.

One approach is what is known as the *axiomatic method.* In this approach, many concepts and properties are introduced by statements accepted as true without proof, that is, by *axioms.* Then definitions are given and theorems are proved using these assumptions and some basic notions of logic. An example of this approach was given in Section 1.7.

The other approach is what is known as the *intuitive method* or *informal method.* In this approach, assumptions are made but emphasis is placed on an introduction to geometric concepts by appealing to intuition and a knowledge of the physical world. It is this approach that we have decided to use.

Because of the way we have elected to discuss geometry, many formal statements of theorem and proof do not appear. However, important statements are italicized for emphasis and some proofs are given.

A short historical development of geometry will be useful before we begin. Around the year 300 B.C., a Greek named Euclid published a text on geometry called the *Elements.* For over two thousand years, this text was considered a perfect model of logical reasoning. Euclid's development began with basic assumptions from which all theorems were proved.

By the end of the 19th century, mathematicians, probing deeply into the basic ideas of Euclid, had discovered subtle gaps in his reasoning. These gaps arose because tacit assumptions were made by Euclid based on visual evidence and spatial intuition. This probing led to a more rigorous formulation of Euclid's geometry and to a discovery of new geometries, such as projective and Riemannian geometries.

Since the classical Greeks' knowledge of algebra was limited (for example, they dealt only with rational numbers and did not consider an irrational number to be a number), their approach to geometry was nonalgebraic. It was not until the 17th century, when Descartes (1596–1650) and Fermat (1601–1665) realized that algebra could be used to supply methodology to geometry, that what is now known as "coordinate geometry" was introduced.

In Chapter 7 we discussed the real line and associated real numbers to each point on the line. These real numbers were termed the *coordinates* of the points. In Chapter 9, the concept of coordinatizing points on a line was generalized to *two* lines—or *two* dimensions—by using an ordered *pair* of real numbers to represent points.

At this stage, let us return to the concept of a point and distance from one point to another. Recall that if P and Q are any two points, the distance $d = d(P, Q)$ from P to Q is a number having the following three properties:

(I) *Positive Definite:* $d(P, Q) \geqq 0$ and $d(P, Q) = 0$ iff $P = Q$.
(II) *Symmetric:* $d(P, Q) = d(Q, P)$.
(III) *Triangular Inequality:* $d(P, Q) + d(Q, R) \geqq d(P, R)$.

Relative to this last property, if P, Q, R are distinct points on a line for which

$$d(P, Q) + d(Q, R) = d(P, R) \tag{10.1.1}$$

then Q is *between* P and R.

The rest of this chapter deals with defining various geometric figures using the two terms *point* and *betweenness* as abstractions.

10.2 LINES AND ANGLES

In Section 10.1, the notions of *point* and *betweenness* were reviewed and discussed. In this section, we shall use these two concepts in order to define some important geometric ideas.

DEFINITION 10.2.1

Let P and Q be two distinct points. The *line determined by P and Q* is the following set:

(I) The points P and Q
(II) All points X between P and Q.

(III) All points X so that Q is between P and X.
(IV) All points X so that P is between X and Q.

We shall denote the line determined by P and Q by the symbolism

$$\overleftrightarrow{PQ} \tag{10.2.1}$$

Thus, a line consists of the points X in Fig. 10.1.

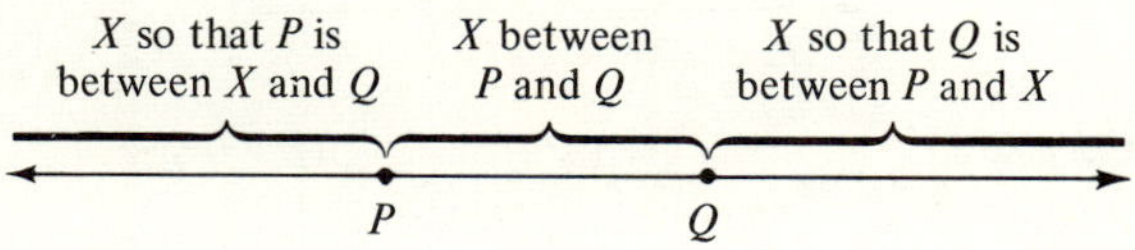

FIG. 10.1

By Def. 10.2.1, the line determined by P and Q depends on the universal set.

EXAMPLE 10.2.1

If the universal set is the set J of integers and if we choose any two distinct integers, say $^-5$ and 2, the line determined by $^-5$ and 2 is the integer line of Chapter 5. It consists of

(I) The integers $^-5$ and 2
(II) The integers X between $^-5$ and 2, namely $^-4$, $^-3$, $^-2$, $^-1$, 0, 1
(III) The integers X so that 2 is between $^-5$ and X, namely, 3, 4, 5, 6, $\cdots$
(IV) The integers X so that $^-5$ is between X and 2, namely, $^-6$, $^-7$, $^-8$, $^-9$, $\cdots$

See Fig. 10.2.

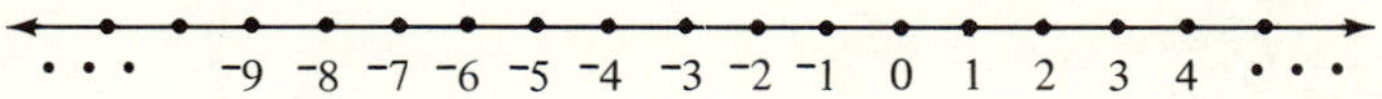

FIG. 10.2

Had we chosen the integers $^-3$ and 4, we would have obtained the same integer line.

If in Ex. 10.2.1, the universal set had been the set R of rational numbers, we would have obtained the rational line of Chapter 6. The real line introduced in Chapter 7 has $R\#$ as universal set.

DEFINITION 10.2.2

Let P and Q be two distinct points. *A closed line segment determined by P and Q* denoted by the symbol

$$\overset{\bullet\!\!-\!\!\bullet}{PQ} \tag{10.2.2}$$

is the set consisting of:

(1) the points P and Q
(2) the points X between P and Q

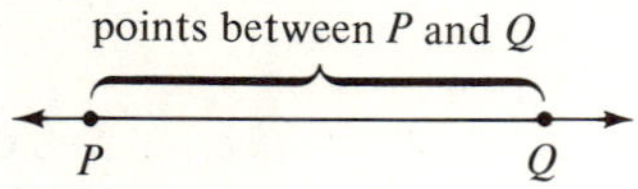

FIG. 10.3

An *open line segment determined by P and Q* denoted by the symbol

$$\overset{\circ\quad\circ}{\overline{PQ}} \qquad (10.2.3)$$

is the set consisting of the points X between P and Q.

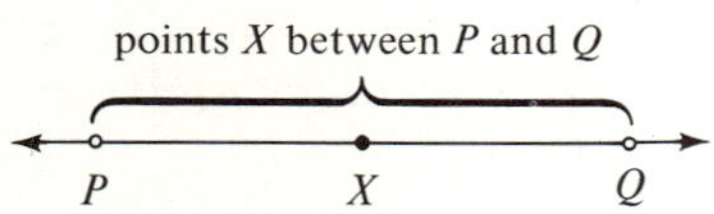

FIG. 10.4

Note that an open line segment $\overset{\circ\quad\circ}{\overline{PQ}}$ does not contain the points P and Q.

EXAMPLE 10.2.2

Consider the real line and the two points $\sqrt{2}$ and 6 on it. The closed line segment determined by $\sqrt{2}$ and 6 consists of all real numbers x for which

$$\sqrt{2} \leqq x \leqq 6 \qquad (10.2.4)$$

The open line segment determined by $\sqrt{2}$ and 6 consists of all real numbers x for which

$$\sqrt{2} < x < 6 \qquad (10.2.5)$$

For either a closed line segment $\overset{\bullet\quad\bullet}{\overline{PQ}}$ or an open line segment $\overset{\circ\quad\circ}{\overline{PQ}}$, the points P and Q are called the *end points.*

DEFINITION 10.2.3

Let P and Q be two distinct points. A *ray determined by P and Q,* denoted by

$$\overrightarrow{PQ} \qquad (10.2.6)$$

is the set consisting of

(I) the points P and Q
(II) the points X between P and Q
(III) the points X so that Q is between P and X

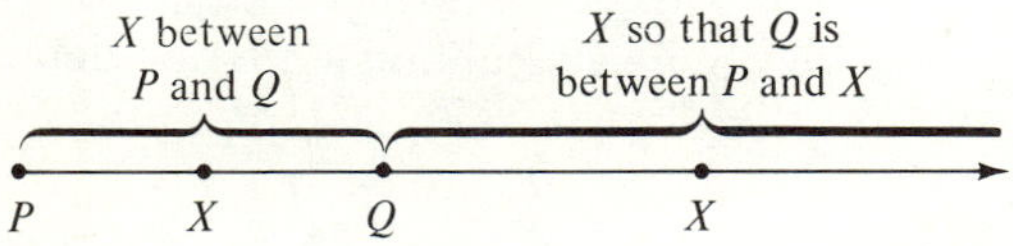

FIG. 10.5

For a ray $\overrightarrow{PQ}$, the point P is called the *vertex of the ray.*

Let us now mention a rule for "comparing" two line segments. For this purpose, we shall let $\overline{PQ}$ denote either a closed or an open line segment.

Definition 10.2.4

Let $\overline{PQ}$ and $\overline{RS}$ denote two closed or open line segments. We say that $\overline{PQ}$ and $\overline{RS}$ are *congruent,* written as

$$\overline{PQ} \cong \overline{RS} \tag{10.2.7}$$

if and only if

$$d(P, Q) = d(R, S) \tag{10.2.8}$$

Next, we would like to discuss the nature of an *angle.* In our study of angles, we will see that there is a difference between *angle* and the *measurement of an angle.* Quite often these two concepts are misunderstood and confused.

By an *angle* is meant the set of points formed by two rays with the *same* vertex. The rays are called the *sides of the angle* and the vertex is the *vertex of the angle.*

Some examples of angles are shown in Fig. 10.6.

Our next problem is to find a way to define the sum of two angles. It should be clear that once we assign a measure to an angle, that is, a number

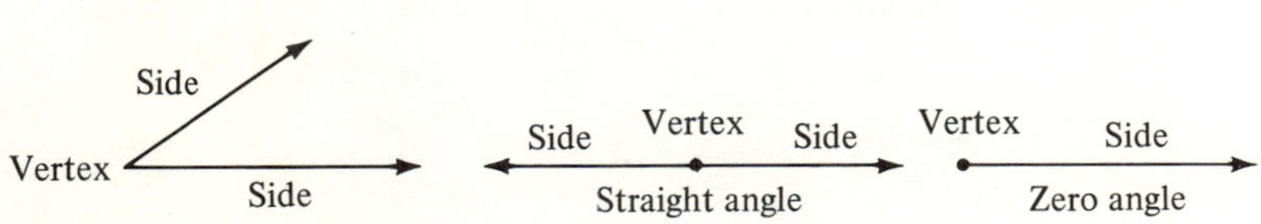

FIG. 10.6

to an angle, then the sum of two such measurements will give us the measurement of the sum of these angles. However, this does not help us determine what is meant by adding two angles.

Suppose A and B are two angles. Then each angle has two sides. See Fig. 10.7. Now to add two angles, we would ordinarily look at the angle formed by coinciding one side of angle A with one side of angle B. Of course, depending upon how we choose the coinciding sides, we get different angles. See Fig. 10.8. This obviously cannot serve as a definition for adding two angles.

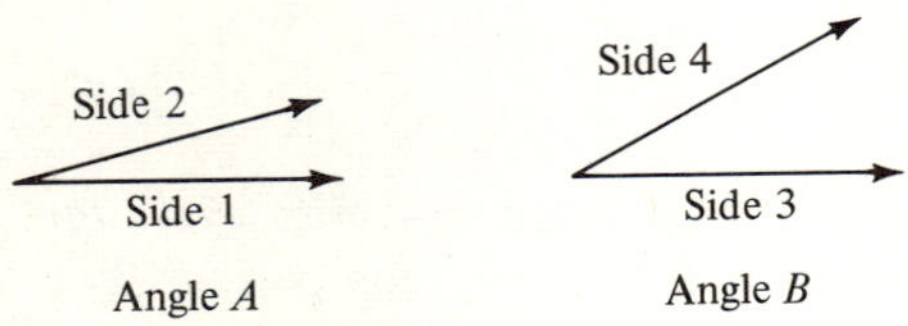

FIG. 10.7

Possibility one:

Possibility two:

Possibility three:

Possibility four:

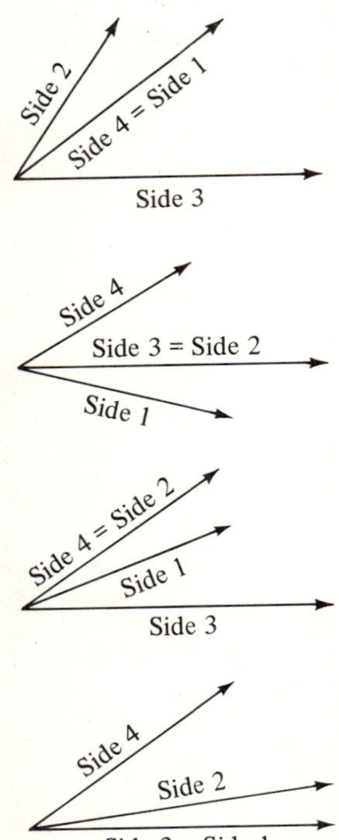

FIG. 10.8

To get around this ambiguity, we introduce the notion of a *directed angle*. A *directed angle* is an angle in which we label one side of the angle the *initial* side and the other side the *terminal* side.

If the shortest rotation necessary for the initial side to coincide with the terminal side is in the counterclockwise direction, the angle is said to be *positively directed*. If the shortest rotation is in the clockwise direction, the angle is *negatively directed*.

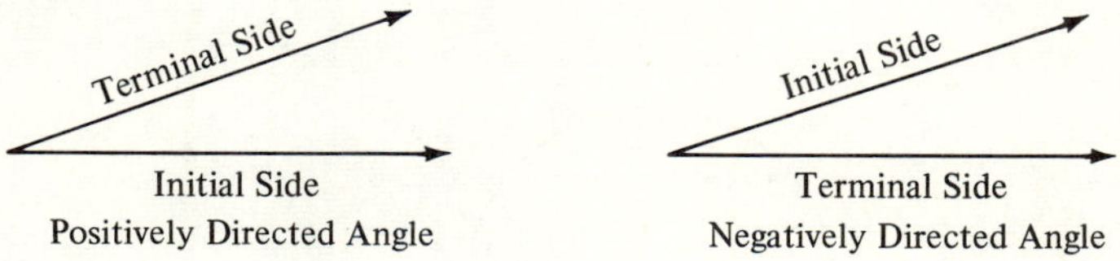

FIG. 10.9

With this convention established, we can proceed to give an unambiguous definition for the sum of two angles. Let A and B be directed angles. The sum of angle A and angle B is that angle C whose initial side coincides with the initial side of A, whose terminal side coincides with the terminal side of B, and the terminal side of A coincides with the initial side of B. See Fig. 10.10. The above definition applies whether the angles A and B are positively directed or negatively directed.

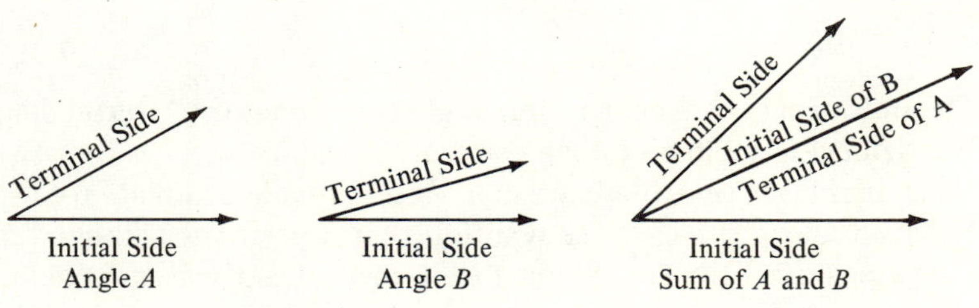

FIG. 10.10

To compare two arbitrary angles A and B, the concept of *congruency* is introduced again. We will agree that two angles A and B are *congruent*, written as

$$\text{angle } A \cong \text{angle } B \tag{10.2.9}$$

if it is possible to transform or move one of the angles into a position of coincidence with the other, without deforming either angle.

Such a transformation is termed a *rigid motion*. In general, any two sets of points are called *congruent* if there is a rigid motion that carries or transforms one of them into the other.

Definition 10.2.5

One complete rotation is the amount of rotation necessary for a ray to coincide with itself after it has been rotated exactly once in the counterclockwise direction.

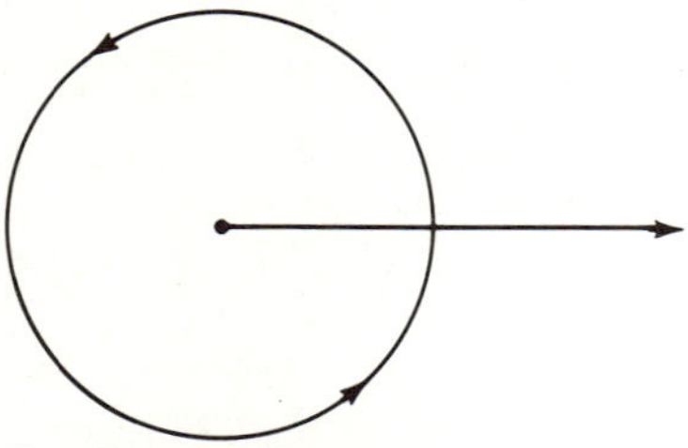

One Complete Rotation

FIG. 10.11

Consider the angle formed from one complete rotation. Divide it into 360 congruent angles. By using one such angle, we are able to make the following definition pertaining to the measurement of an angle.

Definition 10.2.6

The unit of measurement for an angle is a *degree*. *One degree*, denoted by 1°, is defined as 1/360th of one complete rotation.

Thus, it is clear that there are 360 degrees in one complete rotation.

We shall adopt the usual convention that a positively directed angle will have a positive measure assigned to it and a negatively directed angle

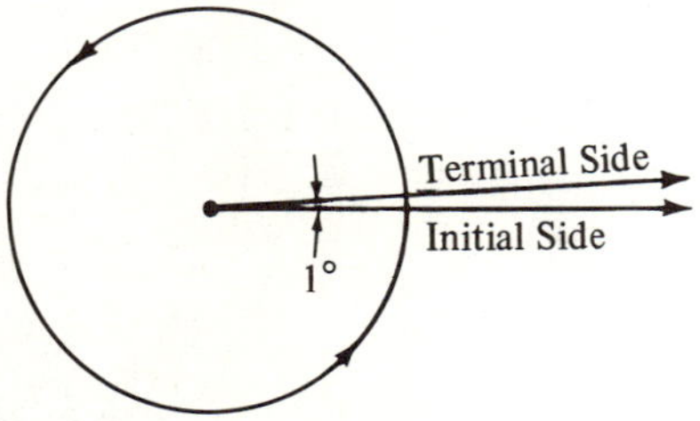

FIG. 10.12

will have a negative measure assigned to it. Because of this convention, it is possible to have two congruent angles whose measures are different. Figure 10.13 gives an example of this situation.

FIG. 10.13

From now on, by *angle* we will mean *positively directed angle* whose measure does not exceed 180°.

Definition 10.2.7

A *right angle* is an angle whose measure is 90°. A *straight angle* is an angle whose measure is 180°. If for example, A is an angle with measure of 90°, we will denote its measure by $m(A) = 90°$.

Thus a right angle is $\frac{1}{4}$ of one complete rotation and a straight angle is $\frac{1}{2}$ of one complete rotation.

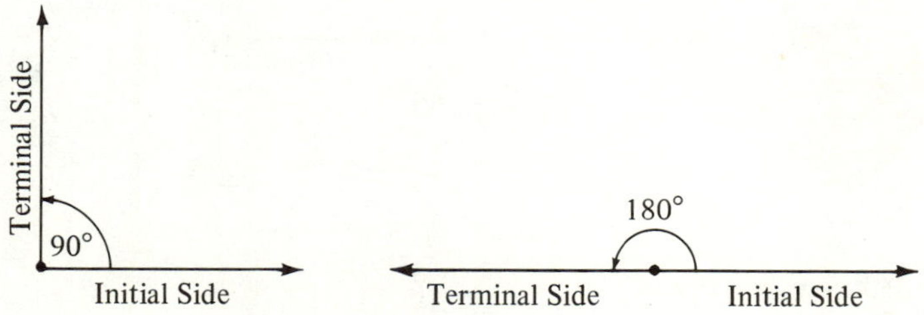

FIG. 10.14

Definition 10.2.8

An angle A is said to be *acute* provided

$$0° < m(A) < 90° \tag{10.2.10}$$

An angle A is *obtuse* if

$$90° < m(A) < 180° \tag{10.2.11}$$

Two angles A and B are said to be *complementary* provided their sum is a right angle, that is, if

$$m(A) + m(B) = 90° \tag{10.2.12}$$

Two angles A and B are said to be *supplementary* provided their sum is a straight angle, that is, if

$$m(A) + m(B) = 180° \tag{10.2.13}$$

For example, the angle 40° is acute. The angle 120° is obtuse.

The angles 40° and 50° are complementary and the angles 60° and 120° are supplementary.

A notation that will be used in the subsequent sections of this chapter is given below.

Suppose r_1 denotes the initial side and r_2 denotes the terminal side of an angle A. Let O denote the vertex of angle A. If P is any point on r_1 different from O and if Q is any point on r_2 different from O, then we will denote angle A by

$$POQ \quad \text{or} \quad QOP \tag{10.2.14}$$

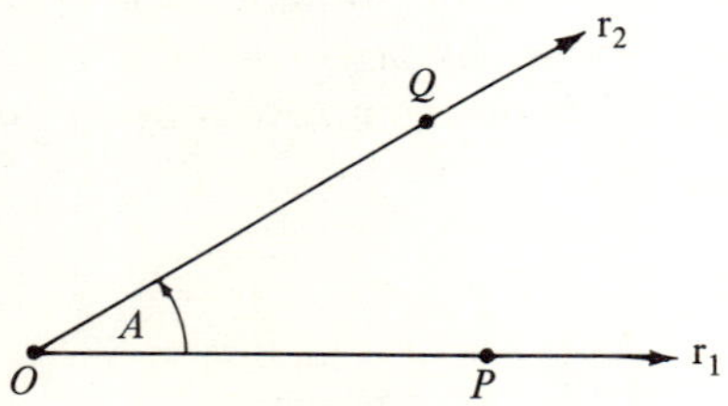

∠POQ or ∠POP

FIG. 10.15

Exercise 10.2

1. Using the whole number line, draw both open and closed line segments for the end points

   (a) $^-3, 0$ (c) $^-1, 4$
   (b) $2, 5$ (d) $^-4, 3$

   For these pairs of points, find all the points between them.

2. Do the above for the rational line, using the end points

   (a) $\frac{1}{2}, 3$ (c) $\frac{5}{2}, 6$

   (b) $\frac{^-4}{3}, \frac{1}{2}$ (d) $0, \frac{1}{2}$

For the above pairs of points, find some (at least three) points between them.

3. Draw the following angles:

(a) $-90°$	(d) $45°$
(b) $50°$	(e) $-25°$
(c) $90°$	(f) $-180°$

4. Find an angle complementary to each of the following angles:

(a) $43°$	(c) $27°$
(b) $89°$	(d) $52°$

5. Find an angle supplementary to each of the following angles:

(a) $43°$	(d) $49°$
(b) $110°$	(e) $162°$
(c) $123°$	(f) $18°$

6. If two supplementary angles are congruent, find their measurement.

7. If the measurement of an angle is three times its supplementary angle, find the measurement of the angle.

10.3 PARALLEL AND PERPENDICULAR LINES; SIMPLE CLOSED CURVES

We are faced now with the difficulty of trying to define a *plane*. Rather than attempt to develop a rigorous definition of a plane, we will instead appeal to the student's intuition.

Some physical representations of plane figures are desk tops, pages of books, blackboards, etc. In other words, anything *flat* is a model of a plane figure.

Now, a plane should be visualized as something flat, extending infinitely far in every direction.

In this section, we assume that all sets of points lie in a plane.

In the previous section we talked about lines, subsets of lines (such as rays, etc.) and the measurements of a line segment. As a matter of fact, we talked only about lines one at a time, without discussing any properties common to lines as a whole. We shall now begin a study of collections of lines.

We shall begin our discussion by introducing the concept of *parallel* lines.

Definition 10.3.1

If all the points of a set are contained in a line, this set of points is termed *collinear*.

For example, a set of two distinct points is collinear since there is a line passing through them, namely the line determined by them. Moreover, this line is unique; that is, there is only one line determined by the two points.

Suppose we consider two *distinct* lines. If these two lines had two or more points in common, the lines would, in reality, be identical since these two points determine exactly one line. Thus, two distinct lines have either one point or no points in common.

Definition 10.3.2

If two distinct lines m and n have *no* points in common, they are called *parallel*. We denote this by writing

$$m \parallel n. \tag{10.3.1}$$

If two distinct lines m and n have one point in common, they are called *intersecting lines*.

We will model our discussion of parallel lines after the work of Euclid. The reason for this is that Euclid's geometry is the one most often studied at the elementary and high school levels. In our informal approach to geometry, the following statement is taken as an axiom.

Playfair's Axiom

If a line and a point, not on the line, are given, there is exactly one line through the given point and parallel to the given line. (This is not the actual axiom as stated by Euclid, but is equivalent to it.)

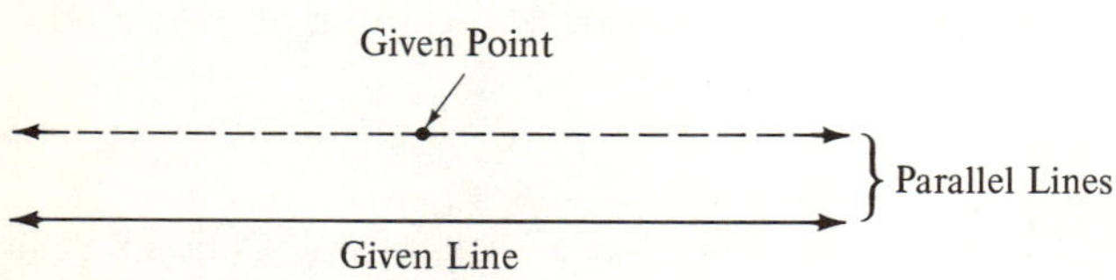

FIG. 10.16

Definition 10.3.3

Let m and n be two distinct lines. A line l is said to be a *transversal of m and n* if the intersection of l with m and n consists of exactly two points.

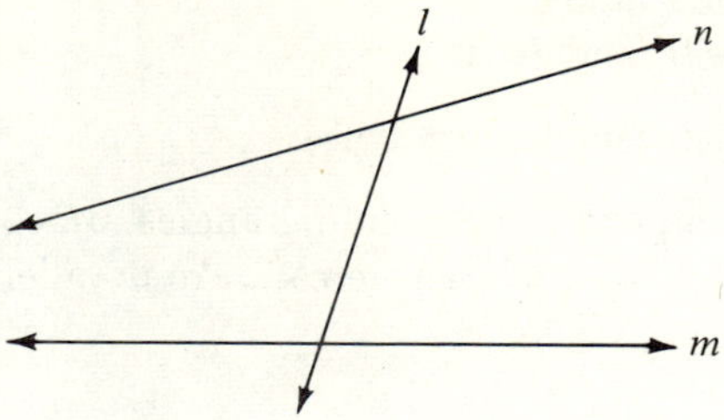

FIG. 10.17

Suppose *m* and *n* are two distinct lines and suppose *l* is a transversal of *m* and *n*.

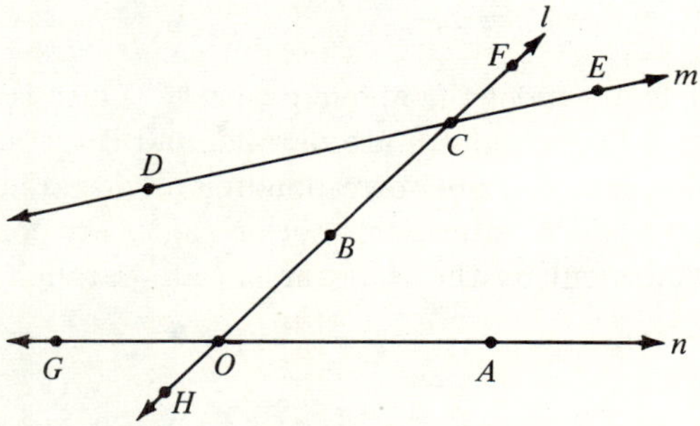

FIG. 10.18

The angles *AOB* and *ECF* are called *corresponding* angles. Other pairs of corresponding angles are

(1) angles *DCF* and *GOB*
(2) angles *GOH* and *DCB*
(3) angles *HOA* and *BCE*

The angles *BCE*, *BOA*, *GOB*, and *BCD* are called *interior angles.*

The angles *HOG*, *HOA*, *DCF*, and *ECF* are *exterior angles.*

The angles *DCB* and *BOA* are *alternate interior angles.* The pair of angles *GOB* and *BCE* are also *alternate interior angles.*

The angles *GOH* and *FCE* are *alternate exterior angles.* The angles *HOA* and *DCF* are also *alternate exterior angles.*

The angles *GOH* and *BOA* are *vertical angles.* Other pairs of vertical angles are

(1) angles *HOA* and *GOB*

(2) angles DCB and ECF
(3) angles DCF and BCE

The following statements are true.

(I) If any pair of corresponding angles are congruent, the lines m and n cut by the transversal l are parallel.

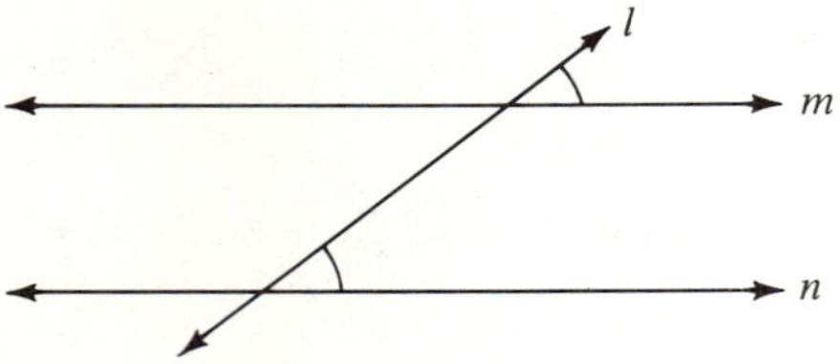

FIG. 10.19

The converse of the above property is also true.

(II) If two lines m and n are parallel and if l is a transversal of m and n, any pairs of corresponding angles are congruent.

(III) If any pair of alternate interior angles are congruent, the lines m and n cut by the transversal l are parallel.

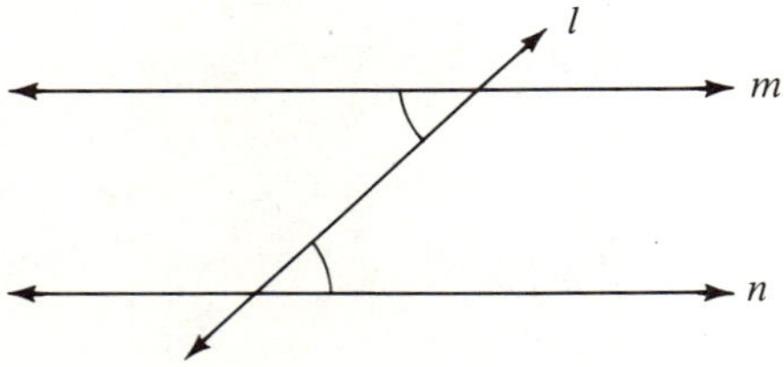

FIG. 10.20

The converse of property (III) is also true.

(IV) If two lines m and n are parallel and if l is a transversal of m and n, any pair of alternate interior angles are congruent.

(V) Any two vertical angles are congruent.

Definition 10.3.4

Two intersecting lines m and n are said to be *perpendicular*, denoted by

$$m \perp n \tag{10.3.2}$$

if and only if the angles formed by m and n are all right angles.

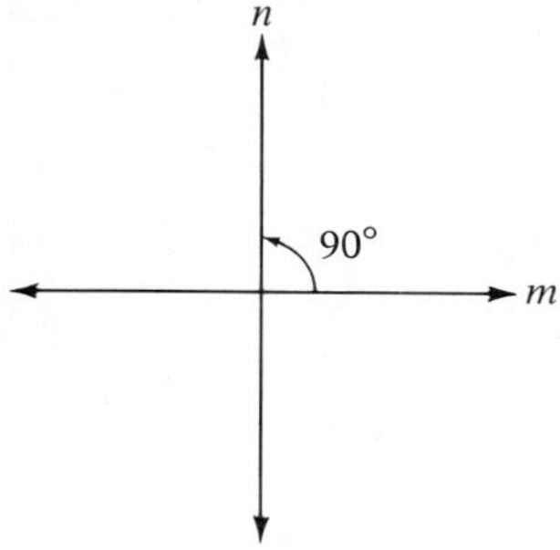

FIG. 10.21

There is one and only one line passing through a given point and perpendicular to a given line.

Exercise 10.3.1

1. Using Fig. 10.22, if we assume the two lines m and n are parallel, find

   (a) $m(ABE)$
   (b) $m(BCG)$
   (c) $m(GCD)$
   (d) $m(DCH)$
   (e) $m(HCB)$
   (f) $m(CBF)$

2. If in Fig. 10.22, $m(DCH) = 40°$, will m and n be parallel? Why?

3. If in Fig. 10.22, $m(GCD) = 50°$, will m and n be parallel? Why?

4. If in Fig. 10.22, $m(EBC) = 50°$, will m and n be parallel? Why?

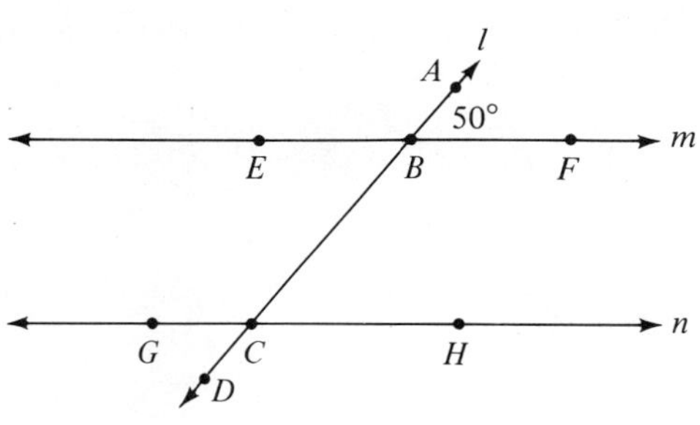

FIG. 10.22

5. If in Fig. 10.22, $m(FBC) = 130°$, will m and n be parallel? Why?

6. In Fig. 10.23, find the remaining three angles.

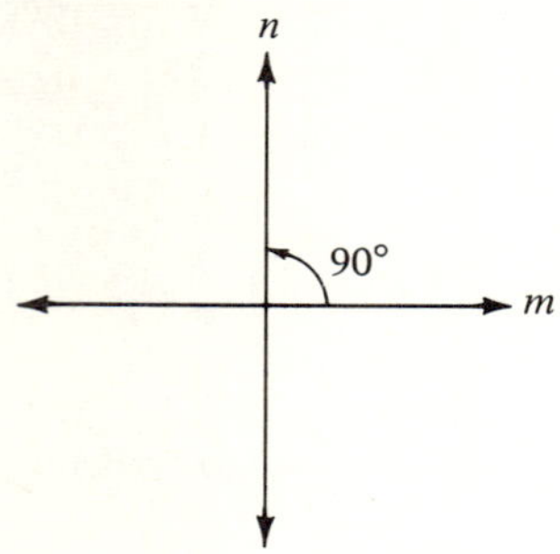

FIG. 10.23

The precise definition of a *simple closed curve* requires some knowledge of the branch of mathematics known as *topology*. However, for our purposes, the following intuitive description of a curve will have more appeal.

We may think of a *curve* as a nonempty set of points that can be traced by beginning at some point A of the curve and continuing until each point of the set has been touched, without lifting the tracing device.

Examples of some figures that are curves and some that are not are given in Fig. 10.24.

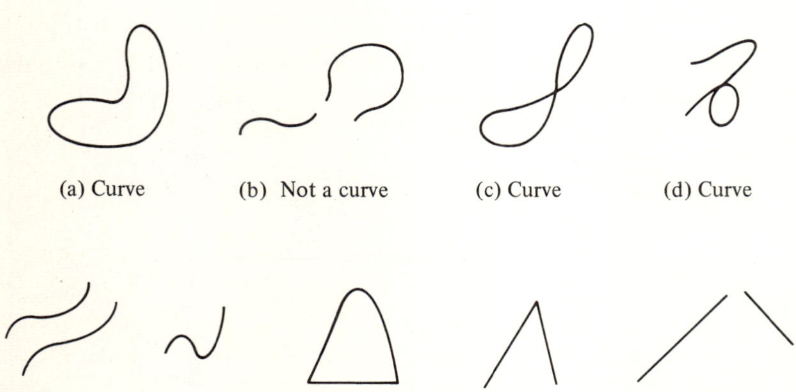

(a) Curve (b) Not a curve (c) Curve (d) Curve

(e) Not a curve (f) Curve (g) Curve (h) Curve (i) Not a curve

FIG. 10.24

A *simple* curve is a curve which has the property that it does not intersect itself.

A *closed* curve is a curve which can be traced by beginning at any point A of the curve and after all points have been traversed is back at A again.

A *simple closed* curve is a curve which is both simple and closed.

For the curves given in Fig. 10.24, curve (a) is both simple and closed; curve (c) is closed, but not simple; curve (d) is neither closed nor simple; curve (f) is simple but not closed; curve (g) is both simple and closed; and curve (h) is simple, but not closed.

The following fact is due to C. Jordan.

An important property of a simple closed curve is that it divides or partitions the plane into three nonempty, disjoint sets of points called (I) the interior, (II) the exterior, and (III) the curve itself.

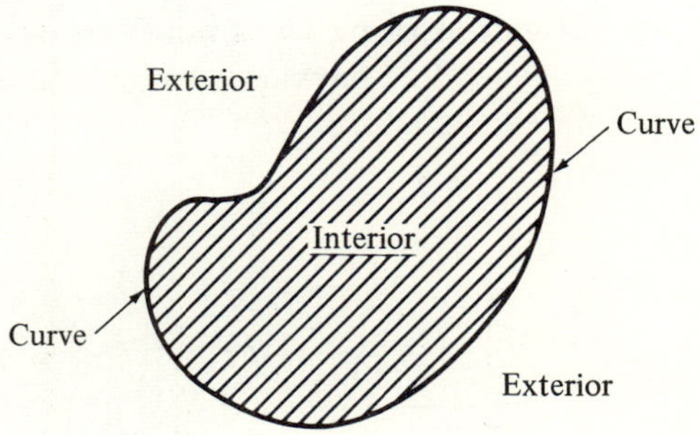

FIG. 10.25

A set A of points is said to be *convex* provided that the line segment joining *any* two points of the set lies entirely in the set.

In Fig. 10.24, the interior of curve (g) is convex. See Fig. 10.26 for further illustrations.

For example, a line segment is always convex. The exterior of a simple closed curve is never convex.

⋆As an application of trying to trace curves on paper without lifting the tracing device and without retracing any portion of the curve, we will discuss the following famous problem. The problem is to attempt to walk across all seven bridges that connect two islands with the shore without crossing any bridge twice. This situation occurred in the town of Königsberg, Germany, and the answer to the problem was given by Euler in 1736.

Euler gave a general solution which depends on the following definitions. A *node* is a point on a closed curve from which lines radiate. A *branch*

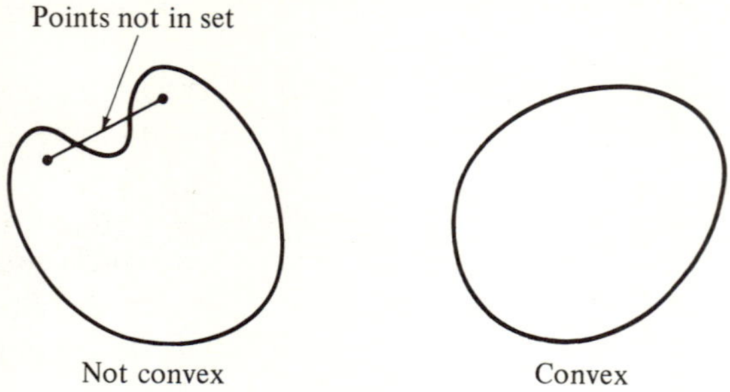

FIG. 10.26

is a line of a closed curve connecting two consecutive nodes. The number of branches emanating from a node is the *order of the node*, so that a node is either even or odd. A *path* consists of a number of branches that can be traced consecutively without crossing any branch twice. A closed curve is *unicursal* or *multicursal*, depending on whether it can be traced in one path or more than one path. See Fig. 10.28.

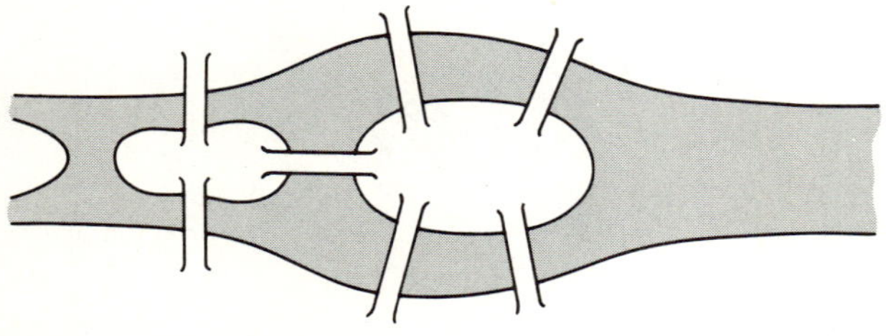

FIG. 10.27

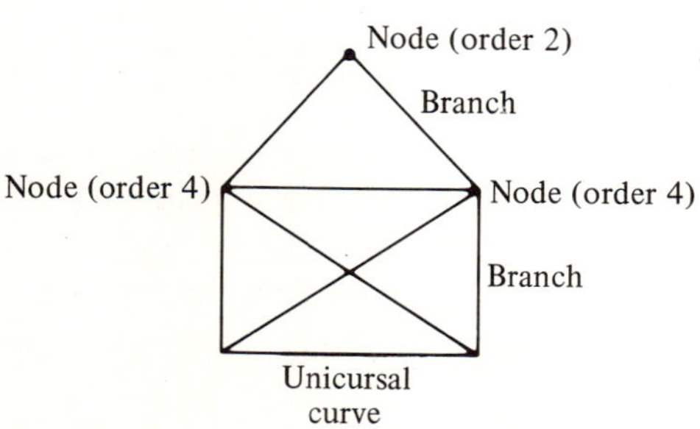

FIG. 10.28

Euler's solution is summarized as follows.

(a) For any closed curve, the number of odd nodes is even.

(b) A closed curve without any odd nodes can be traversed unicursally.

(c) A closed curve with exactly two odd nodes can be traversed unicursally by starting at one of the odd nodes and ending at the other.

(d) A closed curve with more than two odd nodes is multicursal.

Exercise 10.3.2

1. Which of the figures in Fig. 10.29 are curves? Which of the curves are simple? Which are closed? Which are simple and closed? Which simple closed curves have a convex interior?

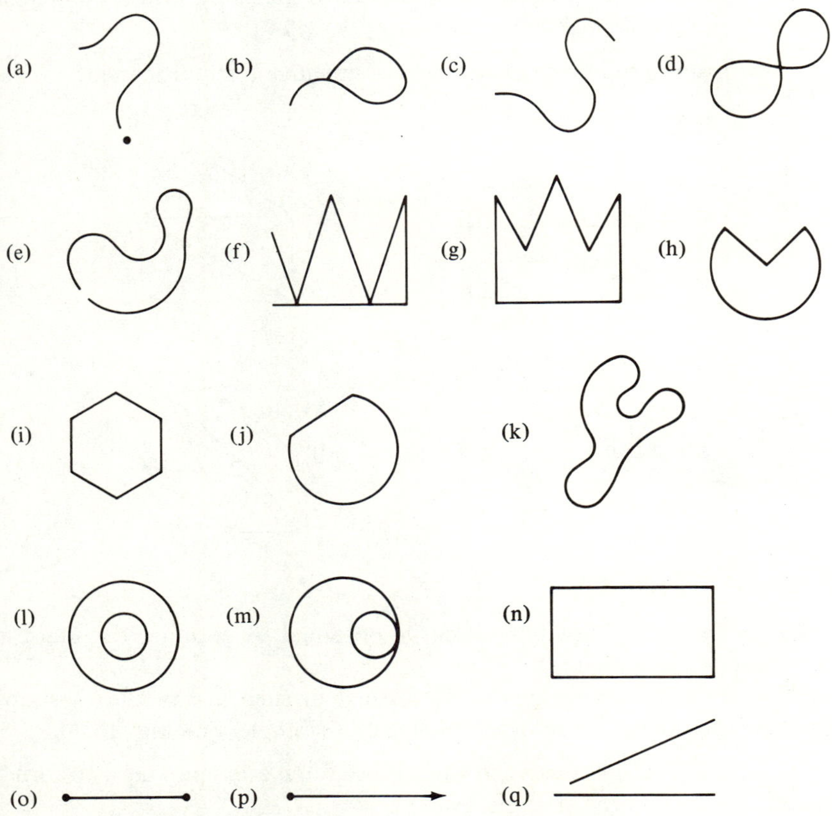

FIG. 10.29

2. Which of the following sets of points are convex?

   (a) the plane
   (b) a line
   (c) two points
   (d) a single point
   (e) the points on the boundary of a simple closed curve

3. If A and B are two convex sets,

   (a) is $A \cap B$ convex?
   (b) is $A \cup B$ convex?
   (c) is $\bar{A}$ convex?

   If the answer is no, demonstrate this by giving an example.

★4. Is the closed curve dealing with the Königsberg bridges unicursal or multicursal?

★5. Add an eighth bridge to make the Königsberg bridges unicursal. In fact, an eighth bridge was built.

★6. Are the following closed curves unicursal or multicursal?

(a)

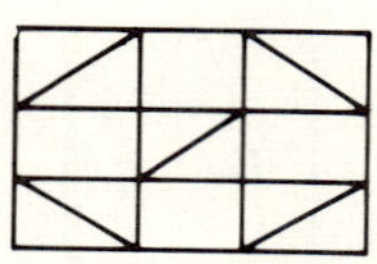

(b)

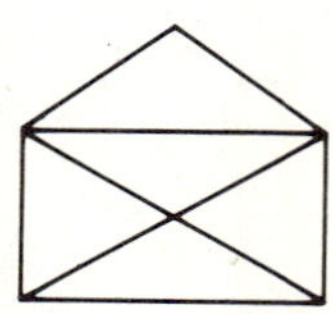

(c)

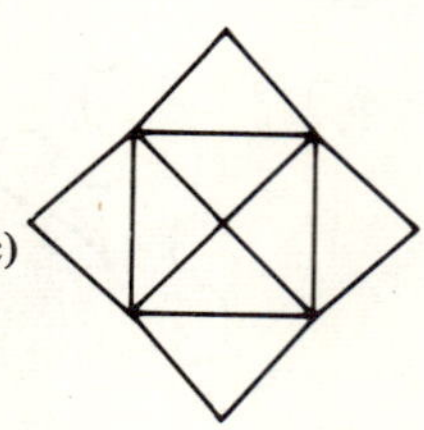

(d)

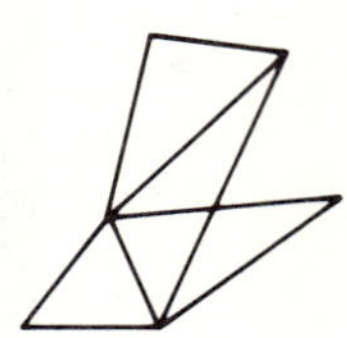

Definition 10.3.5

A *triangle* is a simple closed curve obtained by forming the union of three closed line segments.

Note that the union has to be formed in such a way that a simple closed curve is obtained in order to obtain a triangle. See Fig. 10.30.

If $\overline{AB}$, $\overline{BC}$, and $\overline{CA}$ are the three closed line segments used to form a triangle, we denote the triangle by

$$\triangle ABC \qquad (10.3.3)$$

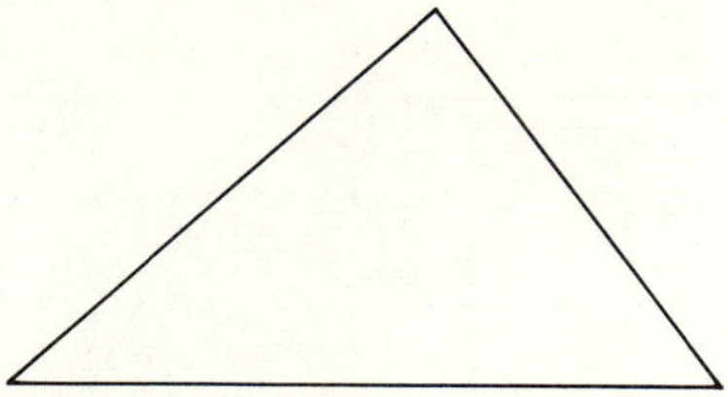

FIG. 10.30

Here the segments $\overline{AB}$, $\overline{BC}$, and $\overline{CA}$ are called the *sides* of the triangle $\triangle ABC$, and the points A, B, C are called its *vertices*.

For the triangle $\triangle ABC$ of Fig. 10.31, the *angles* of the triangle are CAB, ABC, and BCA. It is important to note that for the angle CAB we mean the angle formed by the rays $\overrightarrow{AB}$ and $\overrightarrow{AC}$.

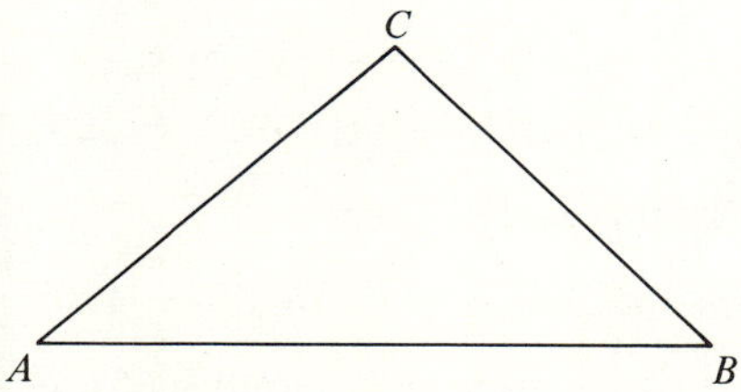

FIG. 10.31

Various types of triangles are defined below.

DEFINITION 10.3.6

An *isosceles triangle* is a triangle that has two of its sides congruent.

An *equilateral triangle* is a triangle which has all three of its sides congruent.

Some properties of isosceles triangles and equilateral triangles are as follows.

(I) *A triangle is isosceles if and only if it has two congruent angles.*

(II) *For an isosceles triangle, the two congruent sides are opposite the two congruent angles.*

(III) *A triangle is equilateral if and only if it has three congruent angles.*

An important property of triangles is the following:

The sum of the measures of the angles of a triangle is 180°.

This property can be demonstrated by using Playfair's axiom.

For, consider any triangle $\triangle ABC$. By Playfair's axiom, there is a line through the point A and parallel to the line determined by the points

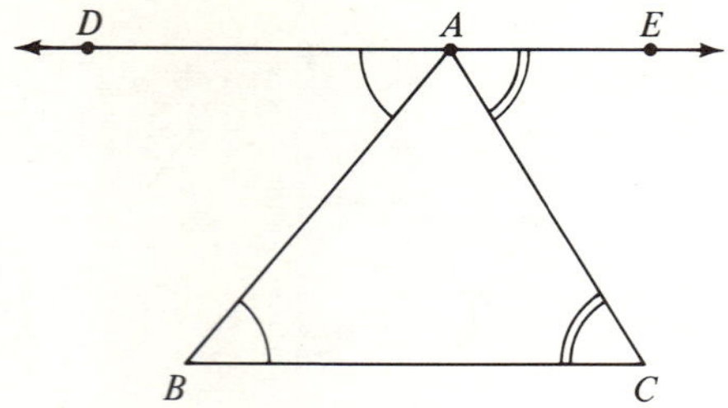

FIG. 10.32

B and C. See Fig. 10.32. By using the properties of transversals of parallel lines, we see that

$$m(DAB) = m(ABC); \qquad m(EAC) = m(BCA) \tag{10.3.4}$$

Now, we know that

$$m(DAB) + m(BAC) + m(EAC) = 180° \tag{10.3.5}$$

since they form a straight angle. Thus,

$$m(ABC) + m(BAC) + m(BCA) = 180° \tag{10.3.6}$$

That is, the sum of the measures of the angles of any triangle is 180°.

For the triangle $\triangle ABC$ in Fig. 10.33, the angle DCA is called an *exterior angle of* $\triangle ABC$.

We know that

$$m(DCA) + m(BCA) = 180° \tag{10.3.7}$$

since these angles form a straight angle. By the previous property, we know that

$$m(CAB) + m(ABC) + m(BAC) = 180° \tag{10.3.8}$$

Thus,

$$m(DCA) = m(CAB) + m(ABC) \tag{10.3.9}$$

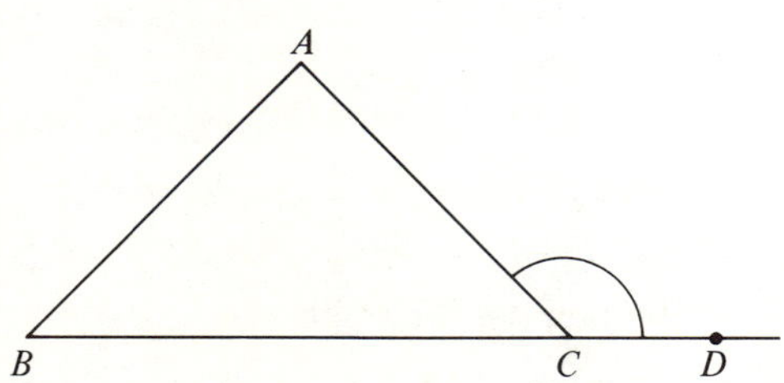

FIG. 10.33

This leads us to formulate the following property:

The measure of the exterior angle of a triangle equals the sum of the measures of the two nonadjacent angles.

A *right* triangle is a triangle in which one angle is a right angle. An *obtuse* triangle is a triangle in which the measure of one angle is greater than 90°. An *acute* triangle is a triangle in which the measure of all three angles is acute.

Now let us discuss two ways of classifying triangles—*congruence* and *similarity*.

DEFINITION 10.3.7

Two triangles $\triangle ABC$ and $\triangle A_1B_1C_1$ are said to be *congruent*, denoted by

$$\triangle ABC \cong \triangle A_1B_1C_1 \qquad (10.3.10)$$

provided the three sides of one triangle are congruent to the three sides of the other.

Again, we can say that two triangles are congruent if one of them can be placed on top of the other so that they match exactly. That is, the two triangles have the same size and shape.

To show that two triangles are congruent, we must find a way of matching their sides in such a way that they are congruent. If, after trying *all* possibilities, we cannot establish such a correspondence, the triangles are not congruent.

For example, in Fig. 10.34, there are two triangles $\triangle ABC$ and $\triangle A_1B_1C_1$.

Suppose

$$\overline{AC} \cong \overline{A_1C_1}, \qquad \overline{BC} \cong \overline{B_1C_1}, \qquad \overline{AB} \cong \overline{A_1B_1} \qquad (10.3.11)$$

Then

$$\triangle ABC \cong \triangle A_1B_1C_1 \qquad (10.3.12)$$

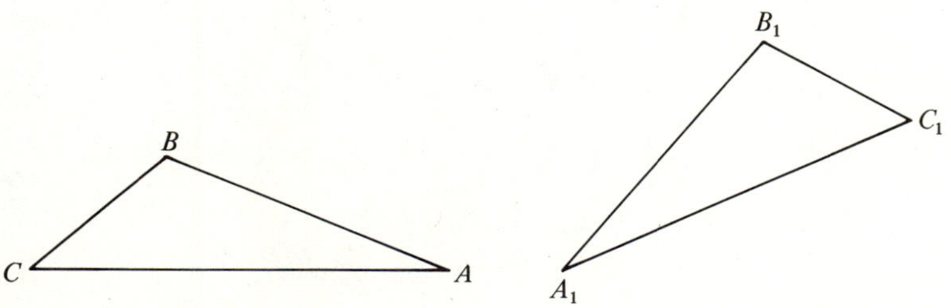

FIG. 10.34

In Fig. 10.35 there are two triangles $\triangle ABC$ and $\triangle A_1B_1C_1$. Suppose

$$\overline{AC} \cong \overline{A_1C_1}, \qquad \overline{AB} \cong \overline{A_1B_1} \tag{10.3.13}$$

Now, we still must have the other pair of sides congruent in order for the two triangles to be congruent.

We shall now state two facts concerning congruent triangles:

(I) *Two triangles are congruent if and only if two sides and the included angle of one triangle are congruent to two sides and the included angle of the other.* This is called the *side–angle–side* (*SAS*) *property.*

(II) *Two triangles are congruent if and only if one side and two angles of one triangle are congruent to a side and two angles of the other triangle.* This is called the *angle–side–angle* (*ASA*) *property.*

Of course, the definition of congruency of triangles is the *side–side–side* (*SSS*) *property.*

It should be clear that congruency is an equivalence relation between triangles. Hence the congruency relation partitions the set of triangles into equivalence classes; that is, any two triangles belong to the same equivalence class provided they are congruent. The student should verify this for himself.

DEFINITION 10.3.8

The triangles $\triangle ABC$ and $\triangle A_1B_1C_1$ are said to be *similar*, denoted by

$$\triangle ABC \sim \triangle A_1B_1C_1 \tag{10.3.14}$$

if and only if the angles of $\triangle ABC$ are congruent to the angles of $\triangle A_1B_1C_1$.

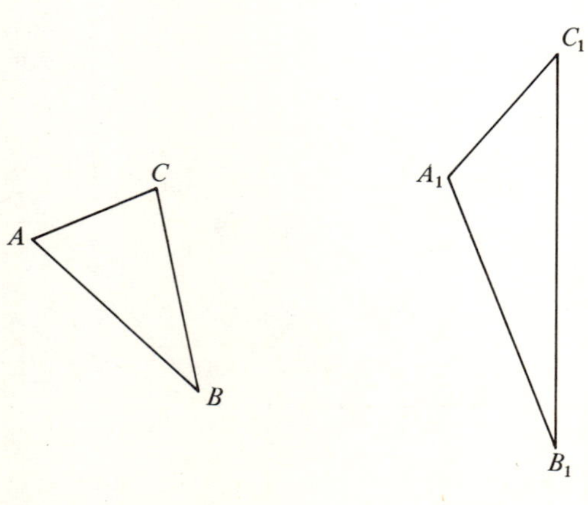

FIG. 10.35

This is called the *angle–angle–angle* (*AAA*) *property* for similarity of triangles.

Of course, the student should realize that because the sum of the measures of the angles of any triangle is always 180°, whenever two angles of a triangle are congruent to two angles of another triangle, then they must be similar.

DEFINITION 10.3.9

Let a, b, c and e, f, g be six real numbers. We shall say that a, b, c are *proportional* to e, f, g, respectively, if there can be found a nonzero real number k so that

$$a = ke, \qquad b = kf, \qquad c = kg \tag{10.3.15}$$

Here k is called the *factor of proportionality.*

For example, 2, 4, 6 are proportional to 1, 2, 3—the factor of proportionality being 2.

An important fact concerning similar triangles is the following:

Two triangles $\triangle ABC$ *and* $A_1B_1C_1$ *are similar if and only if the lengths of the three sides of one are proportional to the lengths of the three sides of the other.*

It should be apparent that whenever two triangles are congruent, they are also similar. However, the converse will not always be true. See Fig. 10.36.

Similarity between triangles is also an equivalence relation. What are the equivalence classes?

Exercise 10.3.3

1. Below are the measures of two angles of a triangle. Find the measure of the third angle.

(a) 60°, 110°	(d) 80°, 70°
(b) 40°, 60°	(e) 50°, 65°
(c) 90°, 30°	(f) 60°, 60°

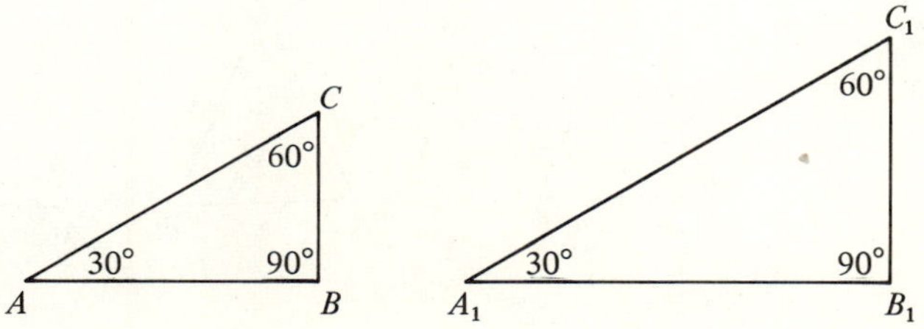

FIG. 10.36

2. Find the measure of the angles of an equilateral triangle.

3. Below is the measure of one angle of an isosceles triangle. Find the measure of the other two angles. (There may be more than one set of answers.)

(a)	30°	(d)	80°
(b)	120°	(e)	45°
(c)	90°	(f)	60°

4. For the triangles given in Fig. 10.37, determine which are congruent, which are similar, which are neither, and in which cases

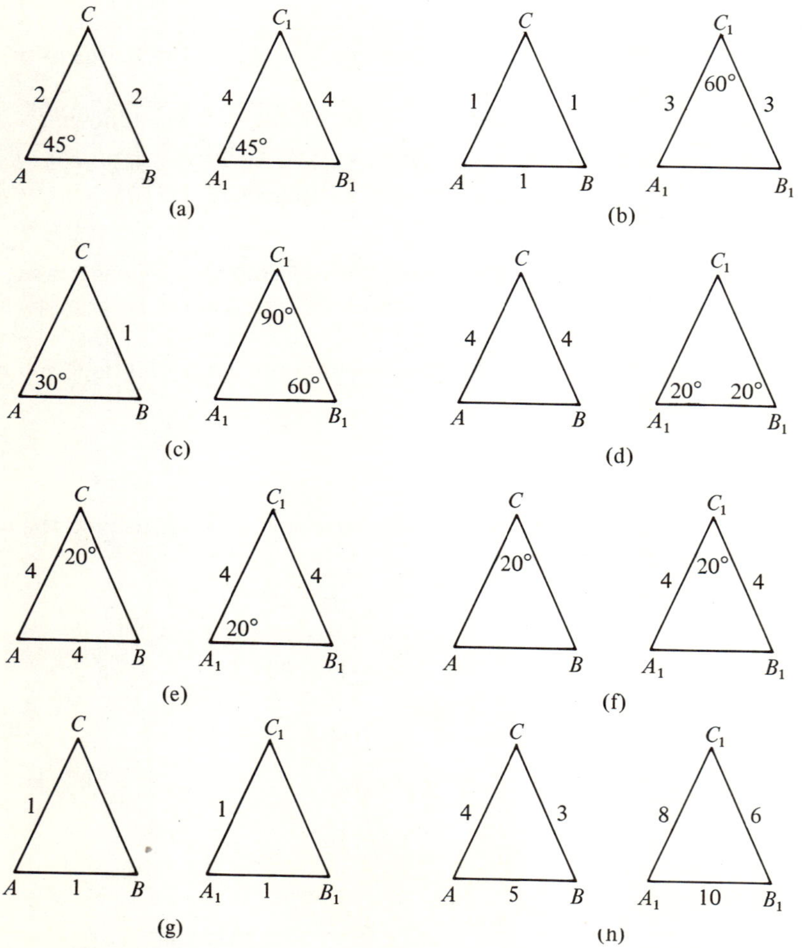

FIG. 10.37

not enough information is given. (Figures may not be drawn to scale.)

Next, we shall consider a very general type of simple closed curve.

DEFINITION 10.3.10

A *polygon* is a simple closed curve which is the union of a finite number of closed line segments.

The line segments are called the *sides* of the polygon and the points of intersection of consecutive line segments are called *vertices* of the polygon. The definition of a polygon does not permit more than two sides of the polygon to emanate from the same vertex.

For example, a triangle is a polygon with *three* sides.

Some special kinds of polygons are given in Fig. 10.38.

DEFINITION 10.3.11

A *regular polygon* is a polygon whose sides are all congruent.

Let us now turn to a study of polygons with four sides, namely, *quadrilaterals.*

DEFINITION 10.3.12

A *trapezoid* is a quadrilateral, two of whose sides are parallel, that is, the sides lie on parallel lines. See Fig. 10.39.

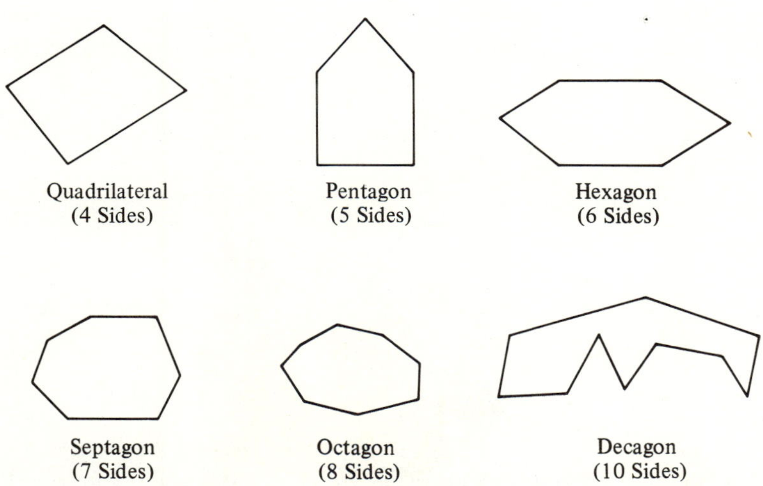

FIG. 10.38

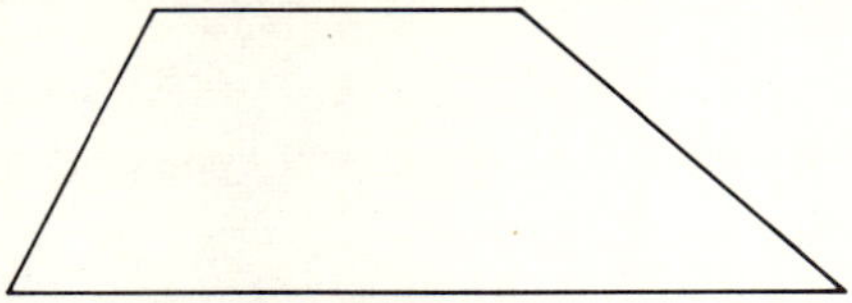

FIG. 10.39

Definition 10.3.13

A *parallelogram* is a quadrilateral in which each pair of sides is parallel. See Fig. 10.40.

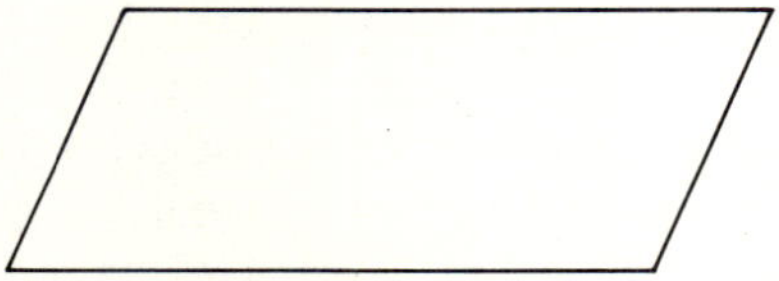

FIG. 10.40

An important property of parallelograms is that the parallel sides are congruent.

The line segments (not a side) joining two vertices of a parallelogram are called *diagonals* of the parallelogram. *The diagonals of a parallelogram bisect each other*. The student should verify this by using congruent triangles.

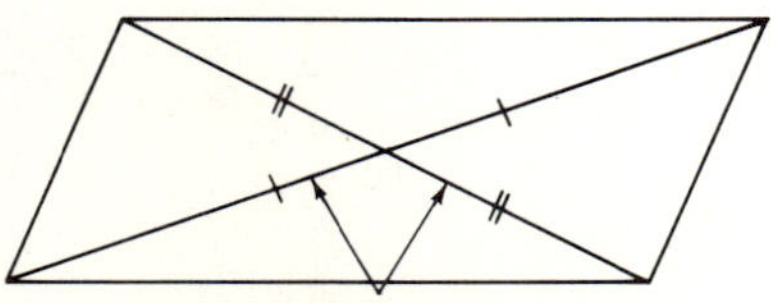

FIG. 10.41

Definition 10.3.14

A *rhombus* is a parallelogram in which all sides are congruent. See Fig. 10.42.

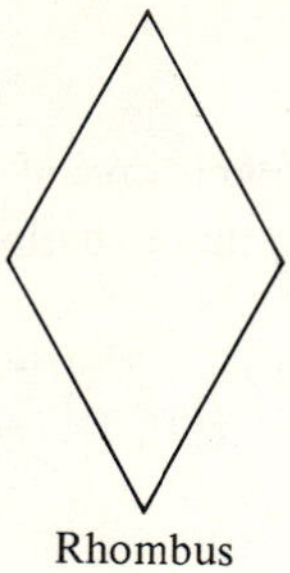

FIG. 10.42

DEFINITION 10.3.15

A *rectangle* is a parallelogram in which two sides meet at a right angle. See Fig. 10.43.

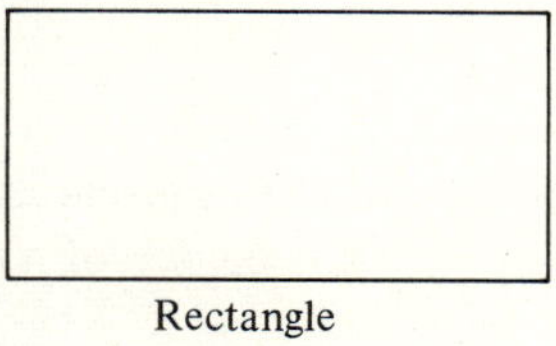

FIG. 10.43

DEFINITION 10.3.16

A *square* is a rectangle in which all sides are congruent. See Fig. 10.44.

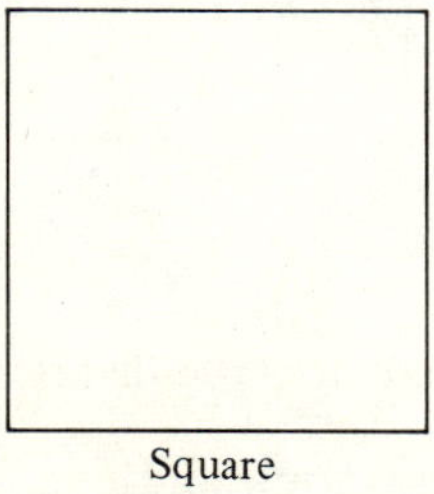

FIG. 10.44

Definition 10.3.17

A *circle* is a set of points each one of which is a fixed distance r from a fixed point P. The fixed distance is called the *radius* and the fixed point is called the *center*.

If L is a line containing P, then there are exactly two points, Q and R, *on* the circle so that $d(P, Q) = d(P, R) = r$.

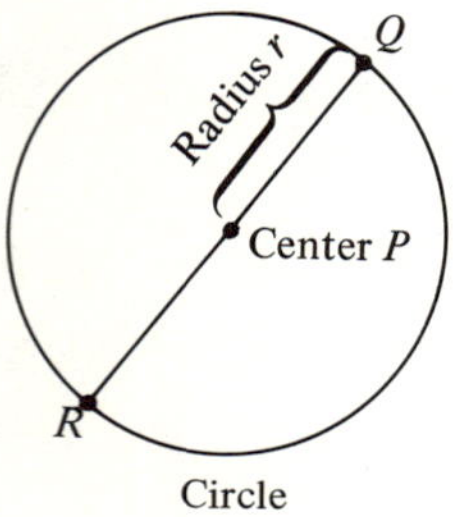

FIG. 10.45

A line segment joining two points on a circle is called a *chord*. A *diameter* is a chord which passes through the center of a circle.

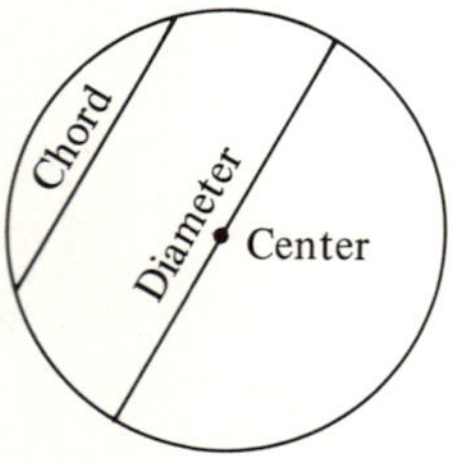

FIG. 10.46

Exercise 10.3.4

Are the following statements true or false? Give reasons.

1. Every rectangle is a trapezoid.
2. Every rhombus is a square.

3. Every square is a rectangle.

4. Every square is a parallelogram.

5. Every rectangle is a square.

6. Every square is a rhombus.

7. The diagonals of a rhombus bisect each other.

8. The diagonals of a square meet at right angles.

9. If the diagonals of a quadrilateral meet at a right angle, the quadrilateral is a rhombus.

From parallelogram, rectangle, trapezoid, rhombus, square, name those for which the following statements are true.

10. The diagonals bisect each other.

11. The diagonals are congruent.

12. The diagonals meet at a right angle.

13. Two sides are parallel.

14. Two sides are congruents.

15. The diagonals are congruent and meet at a right angle.

10.4 MEASUREMENT

The idea of measurement has always been an integral part of elementary and high school curricula. The student is asked to find the length of a line segment or the area of a closed curve, or the volume of a cylinder, etc. Still other kinds of measurement the student is asked about are the measurement of angles, the speed of a car, the time of an event, etc. In the subsequent sections, we will confine our study of measurement to the length, area, and volume of geometric figures.

Recall that in Chapter 3 we first discussed the notion of a *scale* or *unit*. Let us agree that an acceptable line segment has been chosen whose length will be one unit. For example, some units are the ruler (one foot), the yardstick (one yard), etc.

In the next section, when a number is given for a length, it is assumed that we mean that number of units or that number of unit line segments.

10.5 MEASUREMENT OF LENGTH

Given any arbitrary curve, such as the one in Fig. 10.47, when asked to find the "length" of this curve, the natural thing to do is to superimpose a string on top of the curve and then to stretch the string out and measure the length of the resulting line segment.

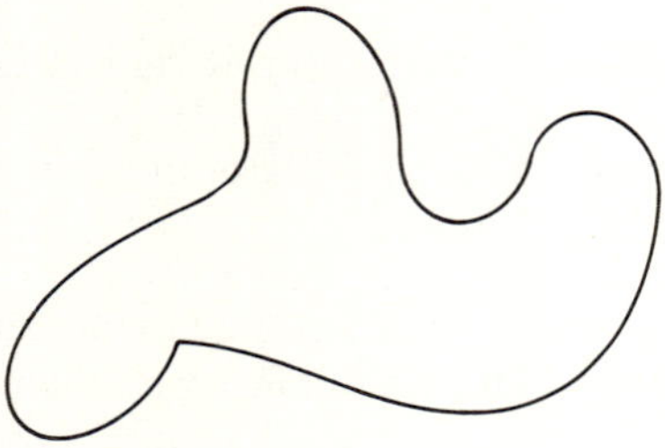

FIG. 10.47

This will hardly enable us to define the length of a curve. In fact, a definition of the length of a curve is beyond the scope of this text. However, we can define length for special kinds of simple closed curves, namely, polygons and circles.

DEFINITION 10.5.1

The *length* or *perimeter of a triangle* $\triangle ABC$ is the sum of the lengths of the sides of the triangle.

The above definition of the perimeter of a triangle is intuitively easy to see, for all we have done is to define the perimeter of a triangle by looking at the line segment formed when the triangle is straightened out.

In a right triangle the side opposite the right angle is called the *hypotenuse* of the triangle. An interesting property of right triangles is the Theorem of Pythagoras.

Theorem of Pythagoras

The sum of the squares of the lengths of two sides of a triangle equals the square of the length of the third side if and only if the triangle is a right triangle. In this case, the third side is the hypotenuse.

If a and b are the sides and c is the hypotenuse of a right triangle, then

$$a^2 + b^2 = c^2 \tag{10.5.1}$$

See Fig. 10.48.

Let us extend this definition of perimeter to polygon.

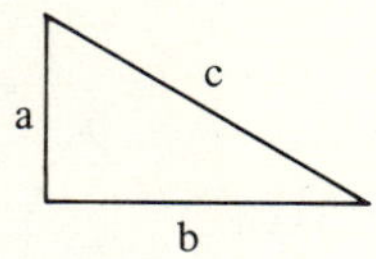

FIG. 10.48

DEFINITION 10.5.2

The *length or perimeter of a polygon* is the sum of the lengths of its sides.

For example, for a pentagon, suppose we know the sides are of lengths 2, 3, 2, 4, 2. The perimeter is $2 + 3 + 2 + 4 + 2 = 13$ units.

For a circle, the perimeter is termed the *circumference.*

If we let C denote the circumference of a circle, let us see if we can obtain a way of measuring C.

It seems reasonable to suppose that if you want to measure C approximately, you can do it by inscribing a regular polygon with a large number of sides and then measuring the perimeter of the polygon. In Fig. 10.49, we inscribed the same circle first by a six-sided polygon (a) and then by a 12-sided polygon (b). Thus we see that the perimeter p of the polygon becomes a better approximation of the circumference C when we increase the number of sides of the polygon. In other words, once we decide how close to C we want p to be, we should be able to get p this close to C merely by taking the number of sides large enough. In this way, p will approach C when n gets very large. Thus, *by the circumference of the circle we will mean the perimeter of an inscribed regular polygon with n sides for n sufficiently large.*

We now want to define the number π. Consider the three circles in Fig. 10.50, each one with a different diameter, and conduct the following experiment.

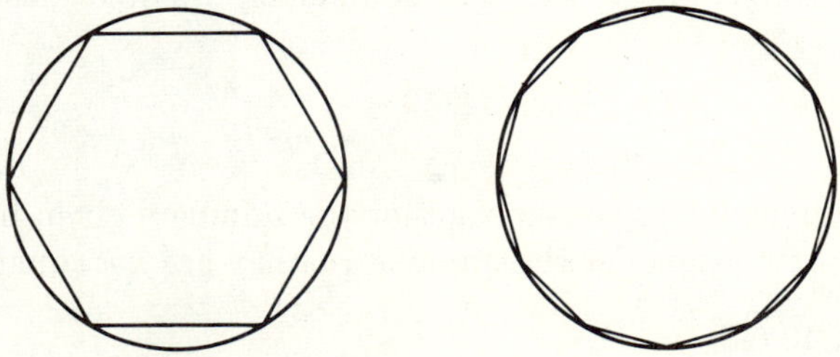

(a) (b)

FIG. 10.49

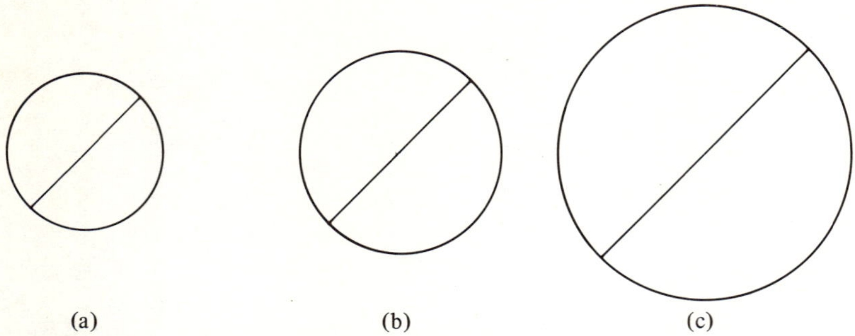

FIG. 10.50

Measure the diameter and the circumference of each circle and then take the ratio of the circumference to the respective diameters. You will "find" the following statement to be true.

The ratios of the circumference to the diameter for any two circles are equal.

Thus if d is the length of the diameter of a circle and if C is its circumference, then $\frac{C}{d}$ is some fixed real number. It has become traditional to represent this number by the symbol π (the Greek letter pi). Thus, we can write

$$\frac{C}{d} = \pi \quad \text{or} \quad C = \pi d \tag{10.5.2}$$

If r is the radius of the circle, then since $d = 2r$, we have

$$C = 2\pi r \tag{10.5.3}$$

It turns out that this real number π is not a rational number, that is, it is an irrational number. However, we can approximate its value by rational numbers. Some of the more common approximations used for π are

$$3.14,\ 3.1416,\ \frac{22}{7},\ 3.14159 \tag{10.5.4}$$

It is very important to remember that the numbers given in (10.5.4) are merely approximations for the number π—they are *not* equal to π.

DEFINITION 10.5.3

A *central angle of a circle* is an angle whose vertex is the center of the circle.

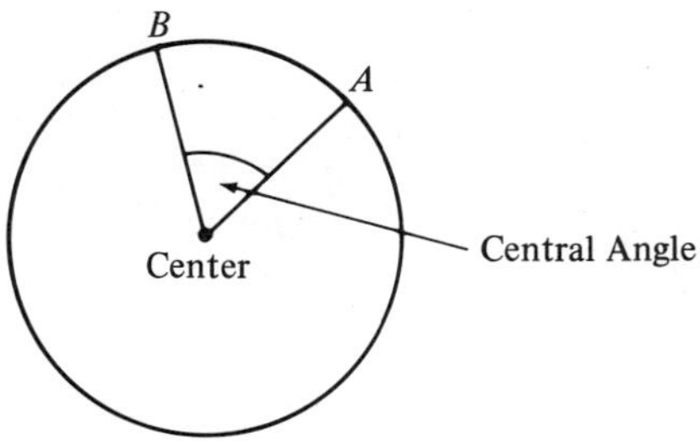

FIG. 10.51

The rays of a central angle intersect the circle in two points—say, A and B. These two points partition the circle into two parts, called *arcs*. The longer arc is called the *major arc* and the shorter arc is the *minor arc*. We denote either arc of a circle by AB.

We will say that two arcs AB and CD of the *same* circle are *congruent* if and only if they are of the same length.

A property of congruent arcs is the following:

Two minor arcs of the same circle are congruent if and only if their central angles are congruent.

To find the length of the minor arc of a circle, we must know the measure of its central angle. If θ is the measure of the central angle of an arc AB, to find the length K of the arc AB, form the ratio of θ to 360 and equate that quantity with the ratio of K to the circumference C. That is,

$$\frac{\theta}{360} = \frac{K}{C} \quad \text{or} \quad \frac{\theta}{360} = \frac{K}{2\pi r} \tag{10.5.5}$$

where r is the radius of the circle.

EXAMPLE 10.5.1

Find the length of the minor arc of a circle of radius 1 whose central angle has a measure of 90°.

Here,

$$\frac{90}{360} = \frac{K}{2\pi \cdot 1} \tag{10.5.6}$$

or

$$K = \frac{2\pi \cdot 90}{360} = \frac{\pi}{2} \tag{10.5.7}$$

Of course, we could have obtained this answer by noting that a central angle of 90° yields an arc equal to $\frac{1}{4}$ the circumference.

Exercise 10.5

Use 3.14 as an approximation for π.

1. Find the perimeters of the polygons of Fig. 10.52.

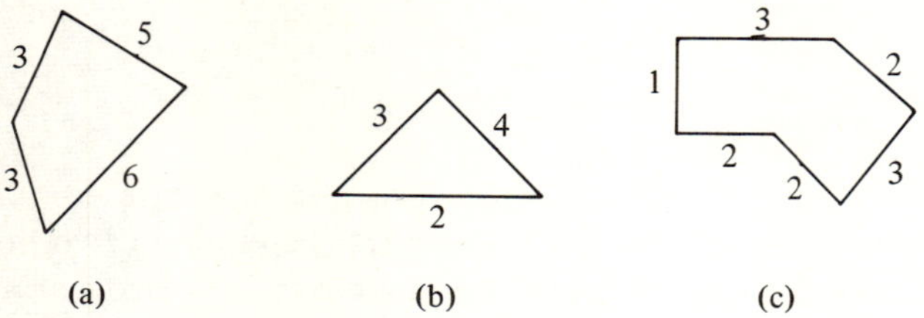

FIG. 10.52

2. Find the perimeter (or circumference) of

   (a) a parallelogram with sides of length 2 and 4
   (b) a square with side of length 6
   (c) a rhombus with side of length 6
   (d) a circle of radius 3
   (e) a circle with diameter 6
   (f) an equilateral triangle with a side of length 2

3. Find the length of the minor arc of a circle of radius 2 with the central angles of measure

   (a) 40° (c) 120° (e) 30°
   (b) 60° (d) 180°

4. The side of a square is of length 6. Find the circumference of its inscribed circle and of its circumscribed circle. What do you notice

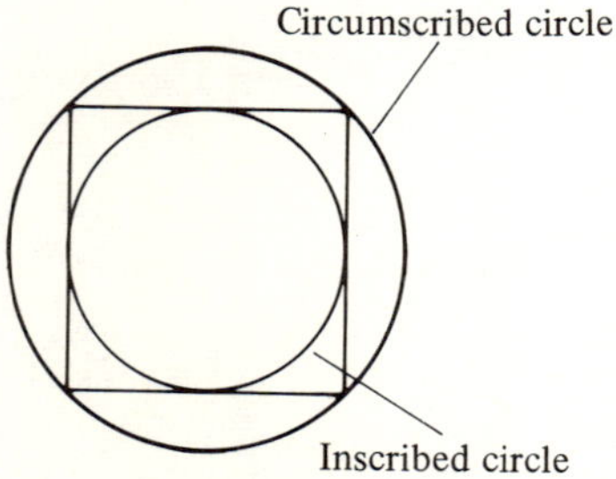

FIG. 10.53

about the perimeter of the square in relation to the circumference of the two circles? See Fig. 10.53.

5. Find the radius and diameter of a circle whose circumference is

(a) 2 (b) 3 (c) 6 (d) 8

10.6 MEASUREMENT OF AREA

In discussing simple closed curves it was mentioned that such a curve partitions the plane into three parts, the interior of the curve, the exterior, and the boundary. By a *region* we will mean the union of the interior and the boundary of a simple closed curve. It is our purpose in this section to discuss the measurement of a region, or express a relationship between the measure of the set of points lying on and in the region and the measure of the set of points lying on and in a square each side of length one. The number of such squares that exactly covers a region is termed the *area* of the region.

For the purposes of measuring area, an *area of one square unit* is defined as the area of a square whose sides are of length one unit.

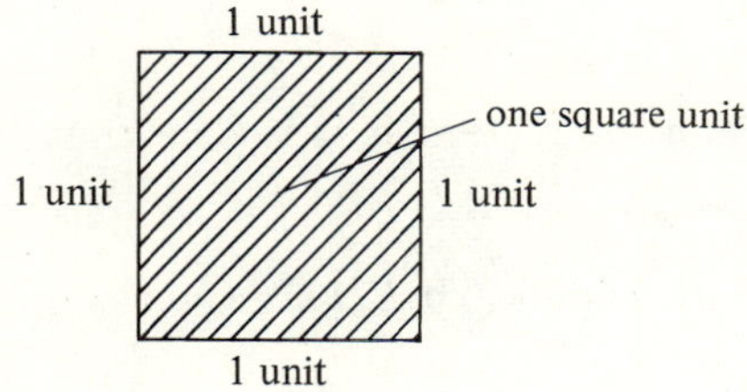

FIG. 10.54

To measure the area of a set of points within a simple closed curve, we must find out how many square units are necessary to cover this set.

Let us begin by considering a rectangle whose sides are of lengths a and b, respectively. The area A of this rectangle is

$$A = a \cdot b \text{ square units} \tag{10.6.1}$$

The lengths a and b of the sides of the rectangle may be rational or irrational numbers.

Before devising a formula for finding the area of a triangle, we need the following definition.

DEFINITION 10.6.1

For a triangle, a line segment from a vertex meeting the opposite side (or the extension of the opposite side) in a right angle is called an *altitude.* The opposite side is termed a *base.*

Clearly, every triangle has three altitudes—corresponding to each vertex. Also, to each altitude, there is a *base.*

Now we can state the following fact.

The area of a right triangle is one-half the product of the lengths of its two perpendicular sides.

This property can be demonstrated in the following way:

Consider the triangle $\triangle BCD$ with sides a and b in Fig. 10.55. Let its area be A. Form the rectangle $BEDC$. Let A_1 denote the area of $\triangle BED$.

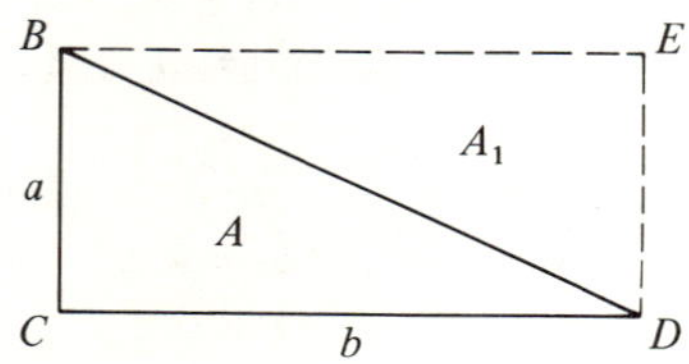

FIG. 10.55

Since $a \cdot b$ is the area of the rectangle, it is clear that

$$A + A_1 = a \cdot b \tag{10.6.2}$$

But the triangles $\triangle BCD$ and $\triangle BED$ are congruent. Hence,

$$A = A_1. \tag{10.6.3}$$

Thus,

$$A = \tfrac{1}{2}a \cdot b \tag{10.6.4}$$

Now we are in a position to find the area of any triangle.

The area of a triangle is one-half the product of the lengths of any base and the length of its corresponding altitude.

Thus, if A is the area, and b and h are the lengths of the corresponding base and altitude respectively, then

$$A = \tfrac{1}{2}b \cdot h \tag{10.6.5}$$

We can demonstrate this by considering three possibilities:

(a) the altitude is a side of the triangle
(b) the altitude meets the base at a point on the triangle
(c) the altitude meets the base at a point outside the triangle

We have already discussed (a). For (b), consider Fig. 10.56.

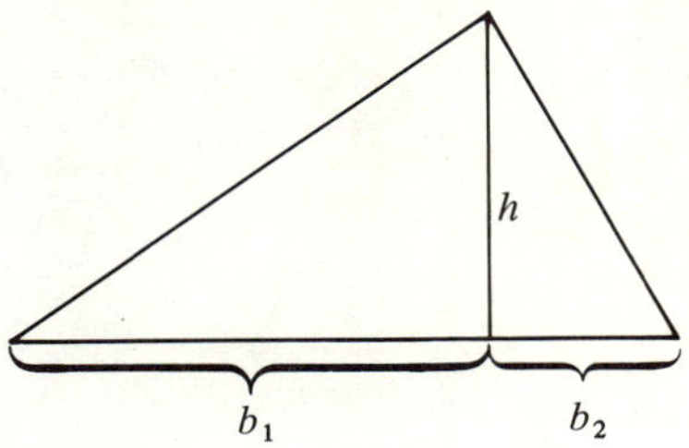

FIG. 10.56

Now,

$$A = \tfrac{1}{2}b_1 \cdot h + \tfrac{1}{2}b_2 \cdot h = \tfrac{1}{2}h(b_1 + b_2) = \tfrac{1}{2}b \cdot h \tag{10.6.6}$$

For (c), refer to Fig. 10.57.

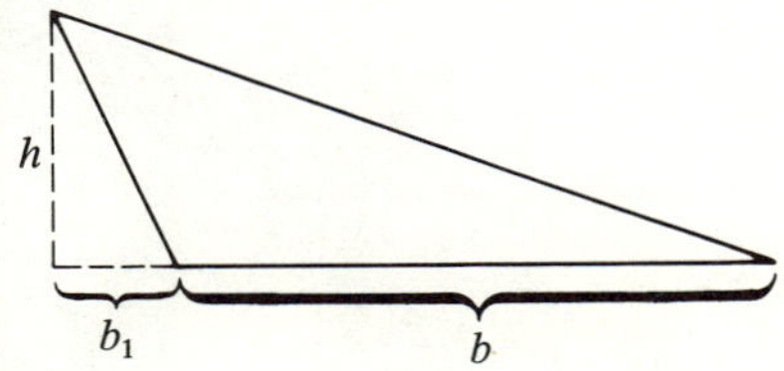

FIG. 10.57

Here,

$$A = \tfrac{1}{2}(b + b_1)h - \tfrac{1}{2}b_1h = \tfrac{1}{2}bh. \tag{10.6.7}$$

The area of a trapezoid is one-half the product of the length of its altitude and the sum of the lengths of its bases.

Consider Fig. 10.58.

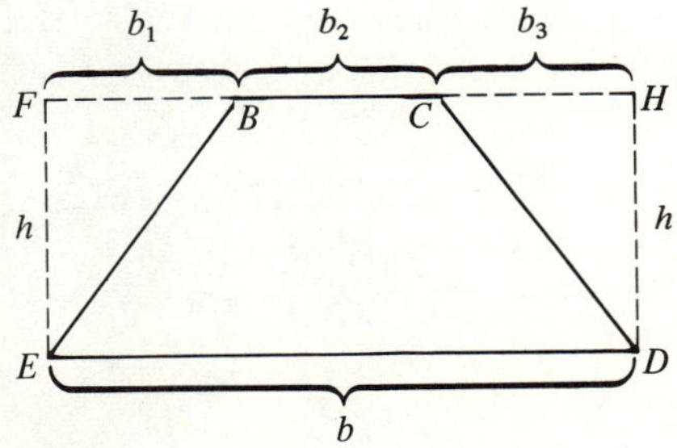

FIG. 10.58

We would like to compute the area $BCDE$. Let A denote this area. Extend the segment BC so as to make it congruent with ED. Then

$$\begin{aligned} A &= b \cdot h - \tfrac{1}{2}h(b - b_2) \\ &= b \cdot h - \tfrac{1}{2}hb + \tfrac{1}{2}hb_2 \\ &= \tfrac{1}{2}bh + \tfrac{1}{2}hb_2 \\ &= \tfrac{1}{2}h(b + b_2) \qquad (10.6.8) \end{aligned}$$

The area of a parallelogram is the product of any base and the corresponding altitude.

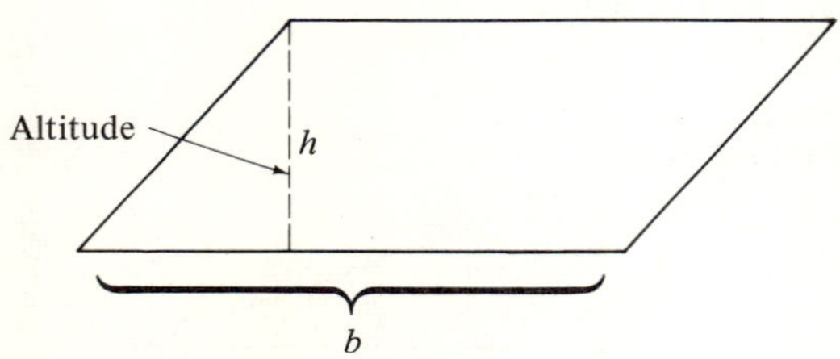

FIG. 10.59

Thus if A is the required area and h and b are the altitude and base, respectively,

$$A = h \cdot b \qquad (10.6.9)$$

The verification of this fact is left to the reader.

Exercise 10.6.1

1. Complete the table of measurement for a triangle in which b, h, and A stand for base, height, and area, respectively.

b	h	A
3	5	
$\frac{1}{2}$	$\frac{7}{8}$	
2		4
	5	17
	3	22

2. Complete the table of measurement for a trapezoid in which b_1 and b_2 are the bases and h is the altitude.

h	b_1	b_2	A
15	9	3.4	
7.8	3.8	$\sqrt{2}$	
	15	2	17
	7	3	22

3. Complete the table of measurement for a parallelogram.

b	h	A
0.7	6	
$\sqrt{5}$	3	
4		16
	9	$\sqrt{17}$

4. Verify 10.6.9.

5. An alternate technique for proving that the area of a triangle is one-half the product of a base by its altitude is outlined below:

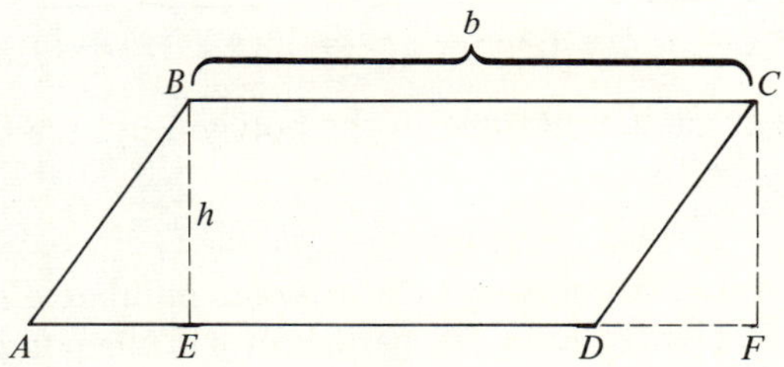

Let $ABCD$ denote a parallelogram and let BE be an altitude of the parallelogram. Since $\triangle ABE \cong \triangle DCF$ (why?), it follows that

$$\text{area of parallelogram} = b \cdot h$$

Now, consider

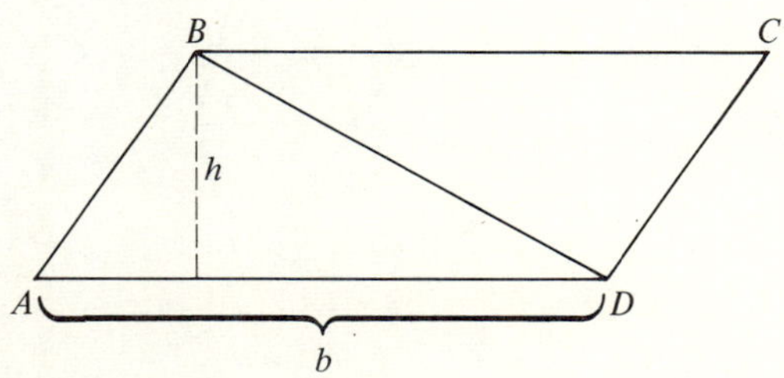

Here, $\triangle ABD \cong \triangle BCD$ (why?). Thus

$$\text{area of } \triangle ABD = \text{area of } \triangle BCD$$

Thus,

$$\text{area of } \triangle ABD = \tfrac{1}{2} \text{ area of parallelogram } = \tfrac{1}{2} b \cdot h.$$

We would like now to discuss a formula to find the area of a circular region.

By a *circular region* we mean the circle and its interior. Consider a circle with radius r and circumference C. In trying to find the area of a circular region we will try to use a technique similar to the one used to find the circumference of a circle, with the exception that now we will circumscribe the circular region with a regular polygon in such a way that each side touches the circle at exactly one point. In doing so, the radius r of the circle is perpendicular to the sides of the regular polygon. See Fig. 10.60. Let A be the desired area, A_n the area of the polygon, s a side of the regular polygon, and p the perimeter of the polygon. Then, by using (10.6.5)

$$A_n = \overbrace{\tfrac{1}{2}rs + \tfrac{1}{2}rs + \cdots + \tfrac{1}{2}rs}^{n \text{ times}} = \tfrac{1}{2}r\overbrace{(s + \cdots + s)}^{n \text{ times}} = \tfrac{1}{2}rp \qquad (10.6.10)$$

Since the circle is contained in the polygon, we see that

$$A < A_n \qquad (10.6.11)$$

However, if we use a polygon with a larger number of sides, that is, by letting n get sufficiently large, the perimeter p of the polygon approaches C, the circumference of the circle, and the area A_n *approaches* the desired area A. That is,

$$A_n \to A \text{ as } n \text{ gets sufficiently large} \qquad (10.6.12)$$

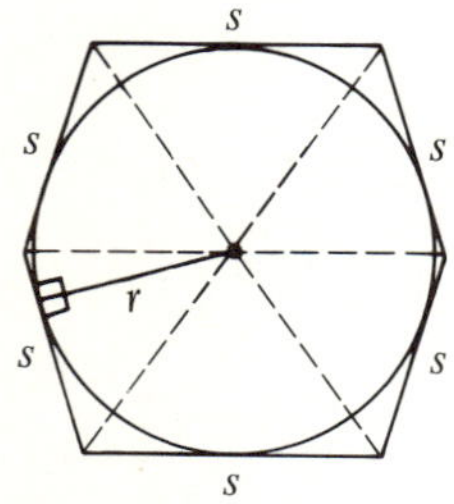

FIG. 10.60

Now

$$A_n = \tfrac{1}{2}rp \tag{10.6.13}$$

As n gets sufficiently large, A_n approaches A and p approaches C, so (10.6.13) can be written as

$$A = \tfrac{1}{2}r \cdot 2\pi r \tag{10.6.14}$$

$$A = \pi r^2. \tag{10.6.15}$$

The area of any circular region is πr^2.

A *sector* is the portion of the circle bounded by two radii and an arc of a circle. The area of a sector in a given circle varies directly as the measure of its central angle. If the measure of the central angle is θ and the radius is r, the area A of the sector is

$$A = \frac{\theta}{360}\pi r^2 \tag{10.6.16}$$

EXAMPLE 10.6.1

Find the area of a sector in which the measure of the central angle is 60° and the radius of the circle is 2. By (10.6.16)

$$A = \frac{60}{360}\pi 2^2 = \frac{2\pi}{3} \tag{10.6.17}$$

Exercise 10.6.2

1. What is the radius of a circle whose area is 4π, 9π, 17π?
2. Compute the area indicated by Fig. 10.61.

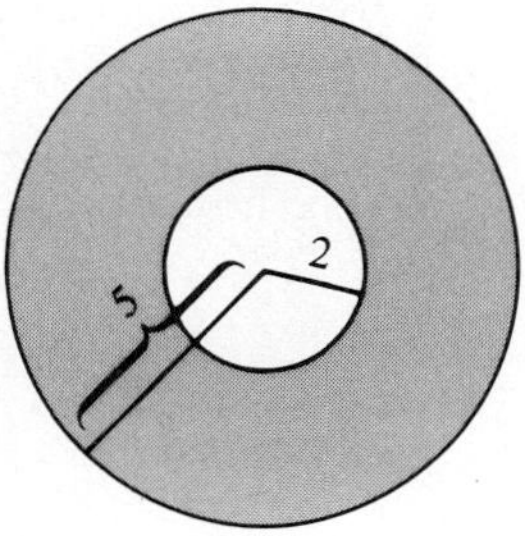

FIG. 10.61

3. Find the area of a circle if its diameter is 5.

10.7 MEASUREMENT OF VOLUME OF SOLIDS

Once again we are faced with introducing a concept that is rather difficult to define. The concept of a *surface* is a sophisticated one, and there may be no satisfactory way to define a *surface* in elementary mathematics. We will appeal to intuition and give some examples of surfaces from the physical world such as a donut, ice cream cone, loaf of bread, cylinder, and so on. The reader will notice that the physical objects have an interior and a surface. Thus, for example, the surface in Fig. 10.62 consists of six planes fused together.

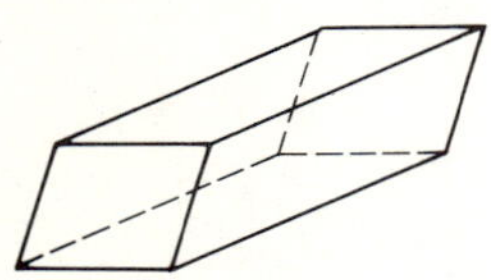

FIG. 10.62

Examples of geometrical figures that are surfaces are shown in Fig. 10.63.

By a *prism* we mean a surface with congruent polygons in two parallel planes as two of its faces, called *bases,* and with the remaining faces formed by parallel segments determined by connecting corresponding points of the polygons. The parallel segments are called *elements* and the faces formed are called *lateral faces* of the *prisms.*

A *triangular prism* is a prism whose bases are triangles. A *parallelepiped* is a prism whose bases are parallelograms. A *hexagonal prism* is one whose bases are hexagons. An *octagonal prism* is one whose bases are octagons. A *cube* is a parallelepiped whose faces are square.

Figure 10.64 depicts each of the above surfaces.

We would like to learn how to compute the volumes of various prisms.

The volume of a rectangular parallelepiped is the product of the length of the altitude and the area of the base. Thus, if V is the desired volume, A is the base area, and h is the length of the altitude, then

$$V = Ah, \qquad A = l \cdot w \tag{10.7.1}$$

EXAMPLE 10.7.1

Find the volume bounded by a rectangular parallelepiped which measures $2\frac{1}{2}$ by 5 by 8 feet.

$$V = 8 \cdot 2\tfrac{1}{2} \cdot 5 = 100 \tag{10.7.2}$$

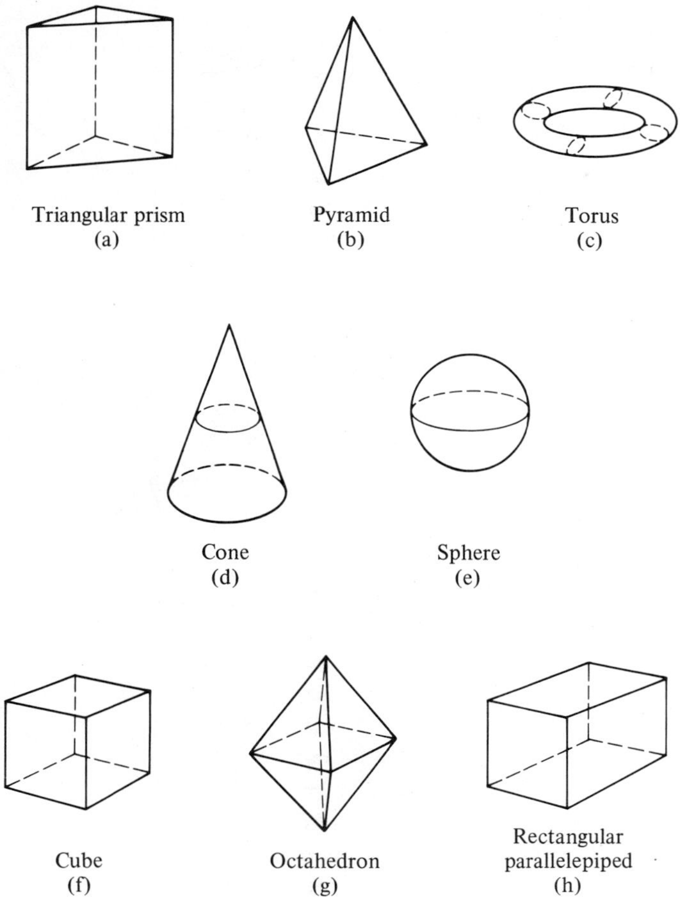

FIG. 10.63

In general the following statement can be made with regard to the volume of any prism.

The volume of any prism is the product of the length of its altitude and the area of its base.

A *pyramid* is a surface formed with a polygon as a base and with its remaining faces formed by joining the vertices of the polygonal base to a point not in the plane of the base. The point is called the *vertex* and the faces formed are called the *lateral faces*. The *altitude* of a pyramid is the perpendicular from the vertex to the plane containing the base.

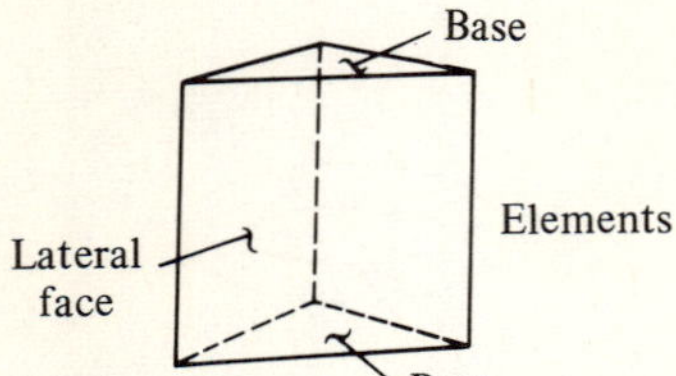

(a) Triangular prism

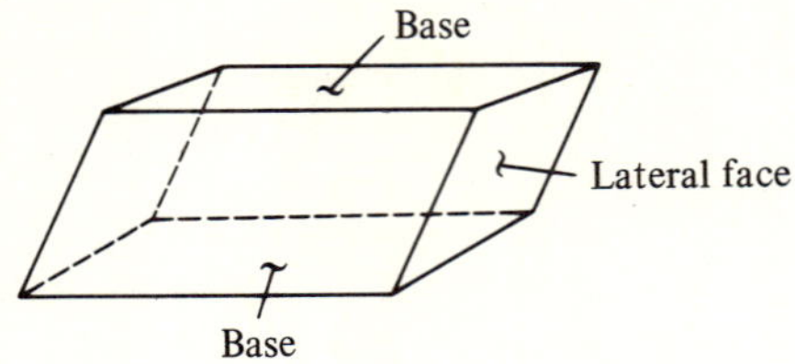

(b) Parallelpiped

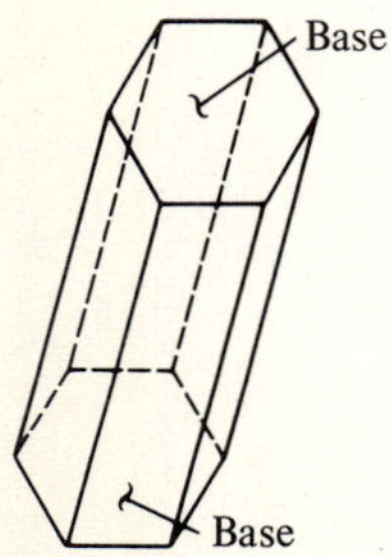

(c) Hexagonal prism

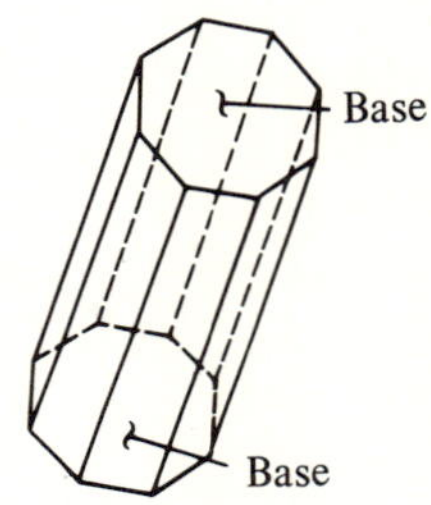

(d) Octagonal prism

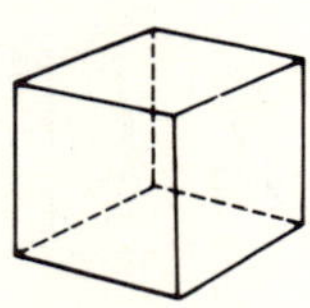

(e) Cube

FIG. 10.64

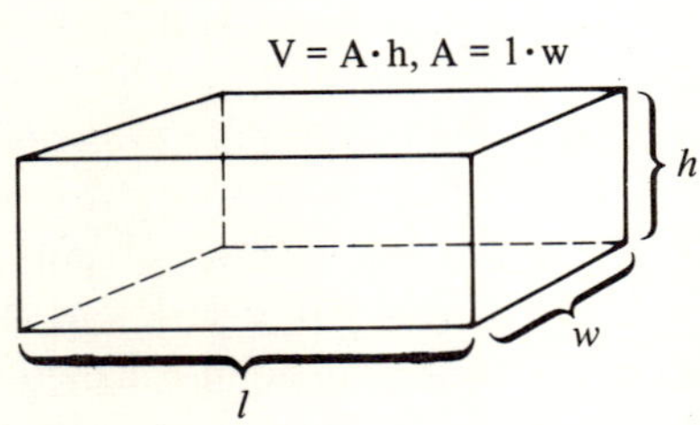

FIG. 10.65

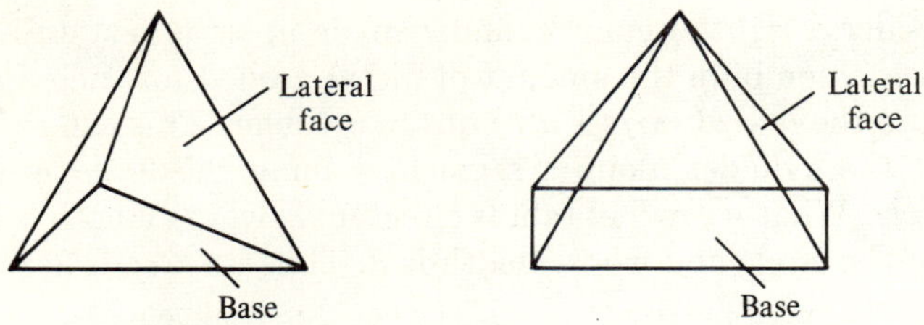

(a) Triangular pyramid (b) Rectangular pyramid

FIG. 10.66

A *triangular pyramid* is one whose base is a triangle. A *rectangular pyramid* is one whose base is a rectangle.

See Fig. 10.66.

The *volume of a pyramid* is one-third of the product of the length of its altitude and the area of its base.

So far we have discussed geometrical figures characterized by having bases as polygons. However, we would like to discuss geometrical figures whose bases are not polygons.

Two examples of such figures are the right circular cylinder and the cone, illustrated in Fig. 10.67.

The formula for the volume V of a cone is one-third the product of the length of the altitude by the area of the base. That is, if V is the volume, h the length of the altitude, and r the radius of the base, then

$$V = \tfrac{1}{3}h\pi r^2 \qquad (10.7.3)$$

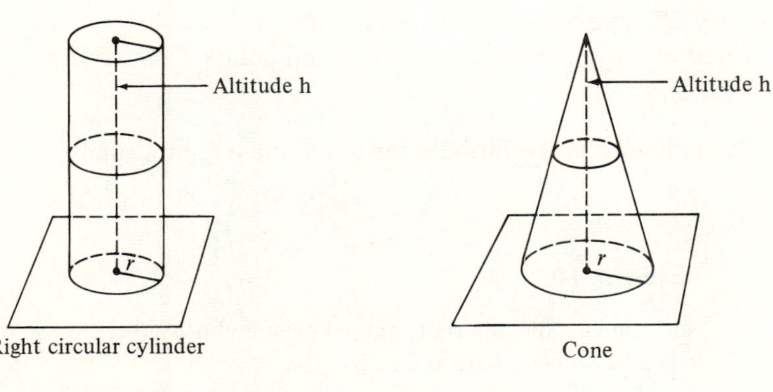

Right circular cylinder Cone

FIG. 10.67

Consider a right circular cylinder made up of a very thin metal. We would like to compute the amount of metal needed for such a cylinder, or to compute the *lateral surface area* of the cylinder. This can easily be done if we cut the cylinder along a segment joining the lower circle and the upper circle. What we will obtain is a rectangle whose length is the circumference of the circle and whose width is h. Thus its area is $2\pi rh$.

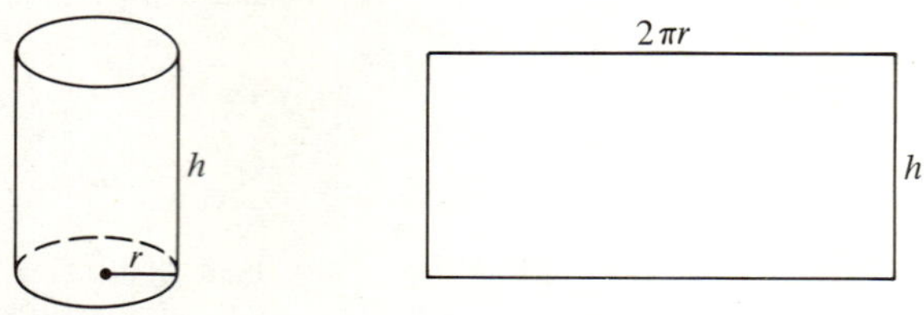

FIG. 10.68

The lateral surface area of a right cylinder is $2\pi rh$.

In the event the base and the bottom area are to be included in the lateral area, the formula is

$$2\pi rh + 2\pi r^2 \qquad (10.7.4)$$

The final geometrical figure that we would like to discuss is the *sphere*. A ball and a globe are nice examples of such a figure. What characterizes this figure is that there exists a point inside the sphere called the *center* such that all segments or radii from this point to the surface of the sphere are congruent. Thus, a sphere with radius r and center C consists of the set of all points P for which

$$d(C, P) = r \qquad (10.7.5)$$

where $d(C, P)$ stands for the length of $\overline{CP}$.

The interior of a sphere is the set of all points P satisfying

$$d(C, P) < r. \qquad (10.7.6)$$

The following is the formula for the volume V of a sphere of radius r.

$$V = \tfrac{4}{3}\pi r^3 \qquad (10.7.7)$$

Exercise 10.7

1. Find the volume of a rectangular prism whose sides of base measure 3, 4, and whose altitude is $7\frac{3}{4}$.

2. The altitude of a rectangular parallelepiped is 5, and the base measures are 6 and 8. Find the volume.

3. Find the volume of a square pyramid whose altitude is 14 and whose base has a side of 10. Also find the area of its lateral surface.

4. Find the volume of a right circular cone if the radius of the base is 1 and the altitude is 1.

5. The base of a cylinder is a circle of diameter 6. The altitude of the cylinder is 7. What is the volume of the cylinder?

6. Find the interior volume of Fig. 10.69.

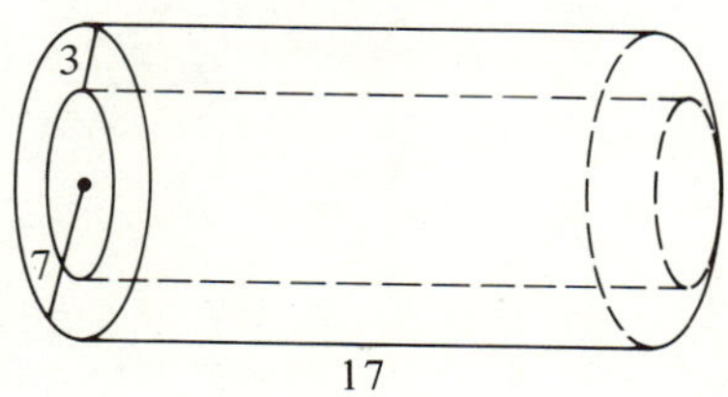

FIG. 10.69

7. Compute the volume and the lateral surface area of Fig. 10.70.

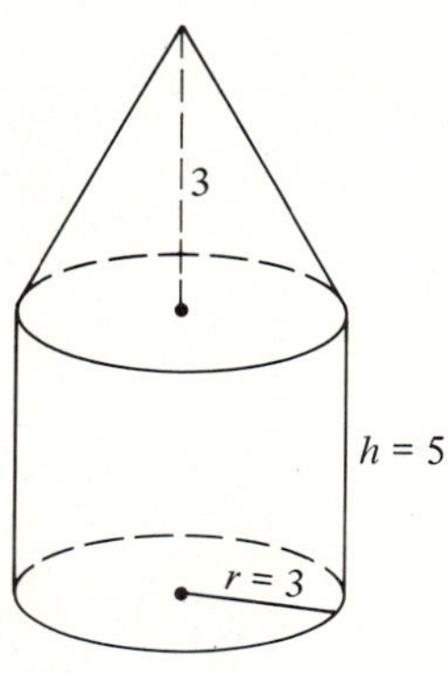

FIG. 10.70

8. Find the volume of a sphere whose radius is 3.

9. The volume of one sphere is one-half the volume of another sphere. What is the ratio of their radii?

11 Fundamental Concepts of Probability and Statistics

11.1 INTRODUCTION

So far, the reader has been introduced to mathematical systems which are basically deterministic. It is our purpose now to introduce the reader to a theory which is nondeterministic in nature.

We often hear statements of the following nature: "His chances of winning are fifty-fifty," or "He is likely to pass the mathematics course," or "Seventy-five percent of the time he is late." In each of the above statements we refer to a situation in which we are not certain of the outcome, but we express some degree of confidence (based on some data or other information) that our prediction will be verified. Thus we are constantly forced to make decisions based on *probable* information and incomplete evidence. The theory of probability provides a mathematical framework for such assertions. The purpose of this chapter is to familiarize the student with the fundamental ideas and to show how the theory may be derived from a few simple properties.

In order to illustrate the basic ideas involved in probability theory, let us consider the following experiments.

Let a green die and a red die be rolled. We would like to know, for instance, what are the "chances" of getting 7 as the sum of the faces? When two dice are thrown, the set of all possibilities consists of all the different ways that the dice may come to rest. This is referred to as set of *all logical possibilities* (this corresponds to the universal set in set theory). If we let

g and r denote, respectively, the number that comes up on the green die and the red die, then an *outcome* is represented by an ordered pair (g, r) where both g and r can assume all values of a set S whose members are 1, 2, 3, 4, 5, 6. Thus, we see that (g, r) is an element of $S \times S$, and since the cardinal number of S is 6, the cardinal number of $S \times S$ is 36. We represent the members of $S \times S$ graphically as shown in Fig. 11.1.

In asking what the chances are that the sum of the faces is seven, we are in essence asking first that all those members (g, r) of $S \times S$ with the property that $g + r = 7$ be found, and they are $(1, 6)$, $(2, 5)$, $(3, 4)$, $(4, 3)$, $(5, 2)$, and $(6, 1)$ or 6 members out of the total of 36. Due to long experience with "fair" dice, we know that each of the possible 36 outcomes occurs equally. Thus it seems "reasonable" to say that chances of any one of them occurring is 1/36. And hence it is "reasonable" to say that the chances that the sum of the faces is seven are $6 \cdot (1/36) = 6/36 = 1/6$.

Suppose that a card is drawn from a deck of 52 cards. What chance is there that the card is a king? To answer this question, we first make the following observations. It is known that none of the 52 cards is "more likely" to be drawn than any other. Thus, the drawing of a card from the

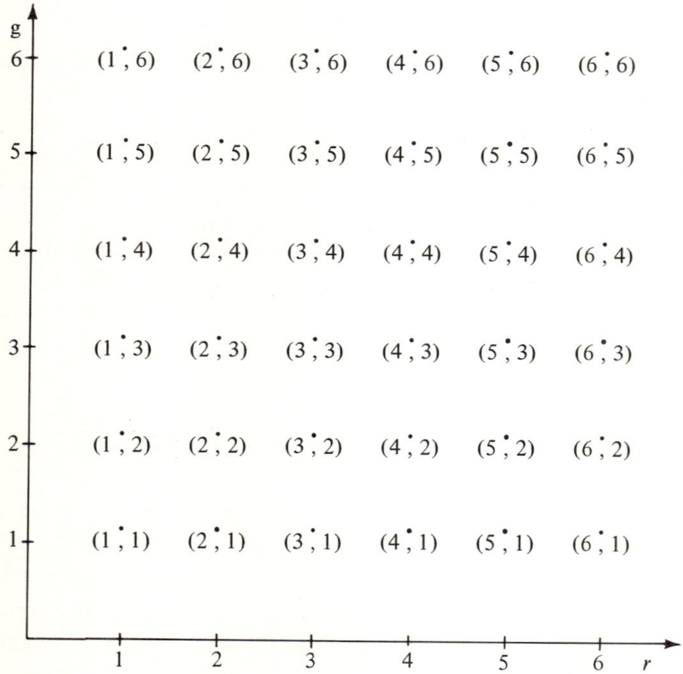

FIG. 11.1

deck is a *trial* that can result in one of 52 "equally likely" outcomes. If a king is drawn, the trial is considered a *success*. If any other card is drawn, the trial is a *failure*. Hence there are 4 possible successful outcomes out of 52 "equally likely" outcomes. It seems "reasonable" to expect that out of 52 such drawings, a king would be drawn *approximately* 4 times. We say the chances of drawing a king are 4 *in* 52 and express this relationship as a ratio $4/52 = 1/13$. The *ratio* is the probability that a card drawn from a full deck is a king.

11.2 PROBABILITY AND SET THEORY

Such reasoning as demonstrated in the two previous examples leads intuitively to the following statement which is taken as a definition of probability.

DEFINITION 11.2.1

If an experiment has n equally likely outcomes among which m are considered successes, then the *probability* of success is m/n.

This statement sometimes offers difficulties as a definition because it is circular. We are using the term *equally likely* to define probability without defining what *equally likely* means. To say that an experiment has n equally likely outcomes is just saying that all outcomes have the same probability. Thus we are using *probability* to define *equally likely* and *equally likely* to define *probability*. However, when the set of all logical possibilities is *finite*, the above definition is a proper one.

In the above examples we already touched on some fundamental concepts. Let us now introduce them precisely as definitions.

DEFINITION 11.2.2

A *sample space* S, associated with a real or conceptual experiment, is the set of all logical possibilities.

Thus, in the above examples, our sample spaces are the set of all ordered pairs (g, r), where $1 \leqq g \leqq 6$ and $1 \leqq r \leqq 6$ in the first example, and the set of all 52 cards in the second example.

In this book, we confine our attention to those cases where the sample space is finite, that is, to those situations where it is possible to have only a finite number of outcomes.

The reader should note that in our definition we say *a* sample space, rather than *the* sample space, since an experiment can be described in many different ways. In general, it is a safe guide to include as much detail as possible in the description of the outcomes of the experiment to enable us to answer all pertinent questions concerning the result of the experiment.

Definition 11.2.3

Each element of a sample space S of an experiment is called an *outcome*. A set of outcomes may all be successes. We define an *event* to be a subset of the sample space.

In Definition 11.2.3 we defined an event as a subset of the sample space S. Because of this definition, the language and notation of set theory can be expected to find an extensive use in probability theory.

example 11.2.1

In a two-dice experiment, involving a green and red die, let A be the event that the green die registers less than or equal to 3 and the red die is 5 and let B be the event that the green die is 2 and the red die is 5 or 6. Describe event A *or* event B.

Here,

$$A = \{(g, r) \mid g \leqq 3, r = 5\}, \qquad B = \{(g, r) \mid g = 2, r = 5, 6\} \quad (11.2.1)$$

Clearly, A has three members and B has two members. The event $A \cup B$ is merely the union of A with B. Thus,

$$A \cup B = \{(1, 5), (2, 5), (3, 5), (2, 6)\} \quad (11.2.2)$$

That is, $A \cup B$ has four members. Of course, $A \cap B$ has one member. The number of members of $A \cup B$ is the sum of the members in A with those in B less the members of $A \cap B$.

The probability of obtaining event A is 3/36 and of obtaining event B is 2/36. The probability of obtaining A or B is 4/36.

example 11.2.2

In a two dice experiment, we present the following situations in terms of events:

(i) the sum of the faces is 3
(ii) the sum of the faces is 7
(iii) the sum of the faces is 7 or 3
(iv) the sum of the faces is 7 and 3

solution: (i) The sum of the faces is 3, if and only if the outcome is a member of the event $A = \{(1, 2), (2, 1)\}$.

(ii) The sum of the faces is 7 if and only if the outcome is a member of the event $B = \{(1, 6), (2, 5), (3, 4), (4, 3), (5, 2), (6, 1)\}$.

(iii) The sum of the faces is 7 or 3 if and only if the outcome is a member of $A \cup B$.

(iv) The sum of the faces is 3 and 7 if and only if the outcome is a member of $A \cap B$. Since $A \cap B = \varnothing$, we say the event is *impossible*.

Exercise 11.2

1. If a die is thrown, what is the probability that the face shows 3? 4?
2. From a class of 40 students with 25 girls, one student is chosen by lot. What is the probability that a boy is chosen?
3. One card is drawn at random from a well-shuffled bridge deck of 52 cards. What is the probability of drawing a heart? An ace? A black card?
4. A letter is chosen at random from the word *equations*. What is the probability that the letter is a vowel? A consonant? The letter *m*?
5. Given a box containing 3 red balls, 4 black balls, 8 white balls, and 2 blue balls, what is the probability that if 1 ball is drawn from the box it is blue? It is red? It is white or it is black?
6. A rat is put in a maze with 10 possible and equally likely goal boxes for him to run to. What is the probability of the rat running to a particular box?

11.3 PROPERTIES AND ELEMENTARY THEOREMS OF PROBABILITY

Let P be a function whose domain is the sample space S and the range is the set of all real numbers between 0 and 1 inclusive. P is then a set function; it associates each subset of S with a real number.

Thus in Example 11.2.1, we write

$$P(A \cup B) = \frac{4}{36} = \frac{1}{9} \qquad (11.3.1)$$

In Example 11.2.2, we can write

$$P(A \cup B) = \frac{8}{36} = \frac{2}{9} \qquad (11.3.2)$$

We are now able to use all the concepts above to formulate the definition of a probability of an event of a sample space.

DEFINITION 11.3.1

Let A be an event of a sample space S. The *probability of the event* A, written as $P(A)$, is a value assigned to A having the following properties:

Property I: *Positiveness* $0 \leqq P(A) \leqq 1$ for every subset A of S.
Property II: *Certainty* $P(S) = 1$.
Property III: *Union* $P(A \cup B) = P(A) + P(B)$ if $A \cap B = \varnothing$.

A function P is called a *probability function* if it obeys Property I, assuming values between 0 and 1. Property II is called the *certainty property* because it says, in effect, that the probability of an event that is bound to occur is 1. The entire sample space is a certain event because it contains all possible outcomes of the experiment. The third property, concerning the probability of the union of two disjoint (mutually exclusive) events permits us to focus attention on separate events when we are assigning probabilities.

Using the above properties, we will be able to define the probability of an event. In assigning such probabilities, we either make assumptions that appear to fit the physical requirements, or we base our knowledge on some past data.

Let A be a subset of S, such that the cardinal number of A is m. Let the cardinal number of S be p. Since A is a subset of S, we can write A as

$$A = \{a_1\} \cup \{a_2\} \cup \cdots \cup \{a_m\} \tag{11.3.3}$$

where $a_1, \cdots, a_m$ are members of A.

Assume that the elements of S are mutually exclusive and "equally likely" (that is, for some reason we consider good, we decide to assign the *same* probability $1/p$ to each of the elements $a_1, \cdots, a_p$ of S). Thus,

$$P(S) = p(1/p) = 1 \tag{11.3.4}$$

By property III and associativity of the operation union we can write,

$$P(A) = \underbrace{1/p + 1/p + \cdots + 1/p}_{m \text{ of these}} = m/p \tag{11.3.5}$$

But

$$m = \text{cardinal number of } A$$

$$p = \text{the cardinal number of } S \tag{11.3.6}$$

Hence,

$$P(A) = \frac{\text{the cardinal number of } A}{\text{the cardinal number of } S} \tag{11.3.7}$$

It is easy to verify that the definition above satisfies all three properties—for example, if $A = S$, then

$$P(S) = \frac{\text{the cardinal number of } S}{\text{the cardinal number of } S} = \frac{p}{p} = 1 \tag{11.3.8}$$

EXAMPLE 11.3.1

In the dice experiments of Example 11.2.2, the cardinal number of S is 36 and to each element we assign the probability 1/36. The answer in terms of probability to Examples 11.2.2(i) and 11.2.2(ii), respectively, are

$$\begin{aligned} P(A) &= P[\{(1,2)\} \cup \{(2,1)\}] = P[\{(1,2)\}] + P[\{(2,1)\}] \\ &= 1/36 + 1/36 = 2/36 = 1/18 \end{aligned} \tag{11.3.9}$$

and similarly

$$P(B) = 6/36 = 1/6 \tag{11.3.10}$$

Theorem 11.3.1

Let $\varnothing$ be the empty set. Then $P(\varnothing) = 0$.

PROOF: Write

$$S = S \cup \varnothing \tag{11.3.11}$$

Then

$$\begin{aligned} P(S) &= P(S \cup \varnothing) \\ &= P(S) + P(\varnothing) \end{aligned} \tag{11.3.12}$$

since $S \cap \varnothing = \varnothing$. But $P(S) = 1$. Hence,

$$1 = 1 + P(\varnothing) \quad \text{or} \quad P(\varnothing) = 0 \tag{11.3.13}$$

Theorem 11.3.2

Let A be a subset of S. Then

$$P(\bar{A}) = 1 - P(A) \tag{11.3.14}$$

PROOF:

$$S = A \cup \bar{A}$$

$$P(S) = P(A) + P(\bar{A}) \tag{11.3.15}$$

since $A \cap \bar{A} = \varnothing$. Or

$$1 = P(A) + P(\bar{A})$$

$$P(\bar{A}) = 1 - P(A) \tag{11.3.16}$$

EXAMPLE 11.3.2

In a two dice experiment, find

(i) the probability that the sum of the faces is less than or equal to 7
(ii) the probability that the sum of the faces is greater than 7.

The cardinal number of the set A defined for (i) is 21. The cardinal number of the set S is 36. Thus,

$$P(A) = 21/36 = 7/12$$

$$P(\bar{A}) = 1 - 7/12 = 5/12 \tag{11.3.17}$$

Theorem 11.3.3

For any two subsets A and B of S:

$$P(A \cup B) = P(A) + P(B) - P(A \cap B) \tag{11.3.18}$$

Exercise 11.3

1. In a throw of two ordinary dice, what is the probability that the numbers on their upper faces add up to 3? to 4? to 7 or 11?

2. If a die is thrown, what is the probability that the upper face shows more than 3? Less than 3? An even number? An odd number?

3. When two coins are tossed, what is the probability that both show heads? That they show one head and one tail?

4. In a family of three children, what is the probability that all three are boys?

5. What is the probability that the number on one die is double the number on the other?

6. What is the probability that one die gives a 5 and the other die a number less than 5?

7. What is the probability that the green die gives a number less than 3 and the red die a number greater than 3?

8. Let r = red die and g = green die. Evaluate

   (a) $P(r + g = 6)$
   (b) $P(r + g = 8)$
   (c) $P(r + g < 5)$
   (d) $P(r + g \geqq 9)$
   (e) $P(r < g + 4)$

9. A die is rolled. Let E be the event *die shows* 4, and F be the event *die shows even number*. Are events E and F mutually exclusive?

10. What is the probability of not throwing a double in a two dice experiment?

11. Give algebraic descriptions of the following verbally described events.

    (a) not throwing a double
    (b) red die shows two less than green die

(c) green die shows number at least 2 greater than red die
(d) number on red die twice that on green die

12. Give verbal descriptions of the following algebraically described events.

(a) $r = 3g$	(d) $r + g = 8$
(b) $r - g = 1$	(e) $g = r^2$
(c) $r \neq g$	(f) $r < g$

13. A bag contains five times as many red marbles as black marbles (identical except for color). One marble is drawn at random. What is the probability that it is red?

14. What is the probability of throwing a 1 or a 2 or a 3 with a single fair die?

15. If the probability that A wins a game is 0.6, what is the probability that A loses? Are the two events A *wins* and A *loses* mutually exclusive? Are they complementary? What about a tie?

11.4 PROBABILITY OF INDEPENDENT EVENTS

Consider the experiment of three throws of an ordinary coin, where we assume the eight possibilities to be equally likely.

Let G = a head turns up on the first throw

Q = a tail turns up on the second throw

$G = \{(HHH), (HHT), (HTT), (HTH)\}$

$Q = \{(TTT), (HTH), (HTT), (TTH)\}$

Note that

$$P(G) = P(Q) = 4/8 = 1/2 \qquad (11.4.1)$$

Also

$$G \cap Q = \{(HTH), (HTT)\} \qquad (11.4.2)$$

Hence

$$P(G \cap Q) = 2/8 = 1/4 \qquad (11.4.3)$$

Note that

$$P(G \cap Q) = P(G) \cdot P(Q) = 1/2 \cdot 1/2 = 1/4 \qquad (11.4.4)$$

In general, for any two events A and B of a sample space S, if $P(A \cap B) = P(A) \cdot P(B)$, then we say that A and B are *independent*, otherwise we say that A and B are *dependent*. Thus, events are independent if they have no effect on each other.

11.5 MEASURES OF CENTRAL TENDENCY

The idea of *average* is a familiar concept to practically everyone. Quite often we hear people talk about average salaries, average height, average grades, average families, and so on. If we have a set of real numbers, we can try to represent this set of values by the one measure, say average, that is most representative of the members of the set. However, the idea of averages is so commonly used that it is not surprising that several kinds of averages have been introduced in mathematics.

Averages will be called *measures of central tendency*. The three measures in common use are the *arithmetic mean*, the *median*, and the *mode*.

The *arithmetic mean* or *mean* of a set of N numbers $X_1, X_2, \cdots, X_N$ is denoted by $\bar{X}$ and is defined as

$$\bar{X} = \frac{X_1 + X_2 + \cdots + X_N}{N} \tag{11.5.1}$$

EXAMPLE 11.5.1

The grades of a student on eight examinations were 70, 65, 69, 85, 94, 62, 79, 100. The mean of the grades is

$$\bar{X} = \frac{70 + 65 + 69 + 85 + 94 + 62 + 79 + 100}{8} = 78$$

$N = 8$, being the number of items under consideration.

Note: (1) That the sum of deviations of each number from the mean is zero. Thus, in Example 11.5.1, $\bar{X} = 78$ and $(70 - 78) + (65 - 78) + (69 - 78) + (85 - 78) + (94 - 78) + (62 - 78) + (79 - 78) + (100 - 78) = 0$.

(2) If $\bar{y}$ is any guessed or assumed mean (which may be any number) and if $d_j = X_j - \bar{y}$, i.e., d_j is the deviation of each member of the data X_j from the assumed mean, then the real mean is

$$\bar{X} = \bar{y} + \frac{d_1 + d_2 + \cdots + d_N}{N} \tag{11.5.2}$$

Consider again Example 11.5.1. We know that the real mean is 78.

However, let us assume a mean, say 52; then using formula (11.5.2), we get

$$\bar{X} = 52 + \frac{\begin{array}{c}(70 - 52) + (65 - 52) + (69 - 52) + (85 - 52) \\ + (94 - 52) + (62 - 52) + (79 - 52) + (100 - 52)\end{array}}{8}$$

$$= 52 + 26 = 78$$

which indeed is what we found before.

The purpose of introducing (11.5.2) is that if the numbers to be added in finding the mean are large, (11.5.2) can simplify the computation.

DEFINITION 11.5.1

The *median* of a set of numbers arranged in order of magnitude is the middle value or the arithmetic mean of the two middle values. Thus the median is the average *position*.

DEFINITION 11.5.2

The *mode* of a set of numbers is that value which occurs with the greatest frequency, that is, the most common value. The mode need not exist, and if it does exist, it may not be unique.

EXAMPLE 11.5.2

The data 2, 2, 3, 3, 4, 5, 7, 7, 7, 11, 15 have a mode of 7 and median 5.

EXAMPLE 11.5.3

The data 2, 3, 4, 5, 7, 11, 15 have no mode and median 5.

EXAMPLE 11.5.4

The data 2, 2, 2, 3, 3, 5, 7, 7, 7, 11, 15 have mode 2 and 7, and is called *bimodal*. The median is 5.

Exercise 11.5

For the following data calculate (1) the mean, (2) the median, and (3) the mode (if possible).

1. 50, 54, 49, 62
2. 24, 30, 27, 26
3. 9, 3, 7, 4, 5, 9, 5, 3
4. 364, 358, 362, 363, 362, 259

11.6 MEASURES OF DISPERSION

Consider the following sets of scores:

$$A = \{4, 6, 8, 10, 12, 14, 16\}$$

$$B = \{4, 7, 9, 10, 11, 13, 16\} \qquad (11.6.1)$$

Note that the means of A and B and the medians of A and B are 10, yet the scores in each set are different and those in B seem to be more closely clustered around 10 than those in A.

Thus we need, in addition to the measures of central tendency, a statistical measure to indicate the extent to which the varieties spread out. Such measures are called *measures of dispersion*.

The simplest measure of dispersion is the *range*. It is the difference between the largest and the lowest scores. For A and B the range is $16 - 4 = 12$. We can see that the range is a poor measure of dispersion since it depends on two measures while telling us nothing about the rest of the scores.

Also, if we try to use the deviation from the mean as a measure of dispersion, and add it up, the result is zero.

EXAMPLE 11.6.1

Score X	Deviation from the Mean $X - \bar{X}$
4	-6
6	-4
8	-2
10	0
12	2
14	4
16	6
	0

We need, therefore, a measure that will give us an idea of how much deviation is involved without having these deviations add up to zero. We can obtain this by squaring each deviation. Each squared deviation is a nonnegative number, and these, when added together and divided by the number of items, give an average squared deviation, which is called the *variance* of the set of scores. In order to apply this measure in practical situations (we cannot talk about "squared dollars" for instance) we will find the square root of the variance: this measure is called *standard deviation* of a set of scores. The standard deviation is represented by s and defined as follows:

$$s = \sqrt{\frac{\sum_i (X_i - \bar{X})^2}{N}} \tag{11.6.2}$$

where X_i is the ith score in the list of scores, and N is the number of items in the list and $\sum_i$ means to add up.

For the data of Ex. 11.6.1, the standard deviation is

$$s = \sqrt{\frac{36 + 16 + 4 + 0 + 4 + 16 + 36}{7}}$$

$$= \sqrt{\frac{112}{7}} = \sqrt{16} = 4 \qquad (11.6.3)$$

EXAMPLE 11.6.2

Tables 11.1 and 11.2 show the scores on two tests. Let us find the standard deviation of each and interpret the result.

TABLE 11.1

Score X	Deviation from the Mean $X - \bar{X}$	Deviation Squared $(X - \bar{X})^2$
100	30	900
90	20	400
90	20	400
85	15	225
80	10	100
75	5	25
75	5	25
75	5	25
70	0	0
70	0	0
65	−5	25
65	−5	25
60	−10	100
40	−30	900
40	−30	900
40	−30	900
Mean = 70		$\sum_i(X_i - \bar{X}) = 4950$
$N = 16$		

The standard deviation is

$$s = \sqrt{\frac{\sum_i(X_i - \bar{X})^2}{N}} = \sqrt{\frac{4950}{16}}$$

$$= \sqrt{\frac{4950}{4}}$$

$$= \frac{70.3}{4} = 17.6 \qquad (11.6.4)$$

TABLE 11.2

Scores X	Deviation from the Mean $X - \bar{X}$	Deviation Squared $(X - \bar{X})^2$
80	10	100
80	10	100
80	10	100
80	10	100
75	5	25
75	5	25
75	5	25
75	5	25
70	0	0
70	0	0
65	−5	25
65	−5	25
60	−10	100
60	−10	100
55	−15	225
55	−15	225
Mean = 70 $N = 16$		$\sum_i (X_i - \bar{X})^2 = 1200$

The standard deviation is

$$s = \sqrt{\frac{\sum_i (X_i - \bar{X})^2}{N}} = \sqrt{\frac{1200}{16}}$$

$$= \sqrt{\frac{300}{4}} = \sqrt{\frac{300}{2}}$$

$$= \frac{17.3}{2} = 8.7 \tag{11.6.5}$$

These two results show that, while the samples have the same mean, 70, and the same sample size, 16, the scores in the first test deviate further from the mean than do the scores in the second test.

In general, *a relatively small standard deviation indicates that the measures tend to cluster close to the mean, and a relatively high one shows that the measures are widely scattered from the mean.*

Exercise 11.6

1. Compute the standard deviation for the following sample:

   4, 6, 9, 9, 10, 14, 25

2. Find the mean, mode, range, median, and standard deviation of the following sample of a test in biology:

   99, 91, 83, 77, 65, 60, 40, 35, 20

3. A group of 25 applicants for admission to Indiana University made the following scores on the quantitative part of an aptitude test:

591	570	425	472	555
409	415	479	517	570
606	614	542	607	441
502	506	603	488	460
550	551	420	590	482

   Find the mean, range, and standard deviation of these scores.

11.7 NORMAL DISTRIBUTION

Consider the following IQ scores obtained from entering freshmen:

78	91	96	101	105	111
82	92	97	102	106	113
84	92	98	102	107	114
86	94	100	103	107	116
87	95	100	104	108	119
89	96	100	105	110	123

We will now group these scores into classes with intervals of 5.

Class Interval	Frequency of Occurrence of Scores in the Class Interval
76–80	1
81–85	2
86–90	3
91–95	5
96–100	7
101–105	7
106–110	5
111–115	3
116–120	2
121–125	1

Now plot these scores on a graph with the frequency shown on the vertical axis and the class intervals on the horizontal axis. When the points are connected with a smooth curve, we find that we have a symmetrical, bell-shaped curve. This is called the *normal distribution,* or normal curve.

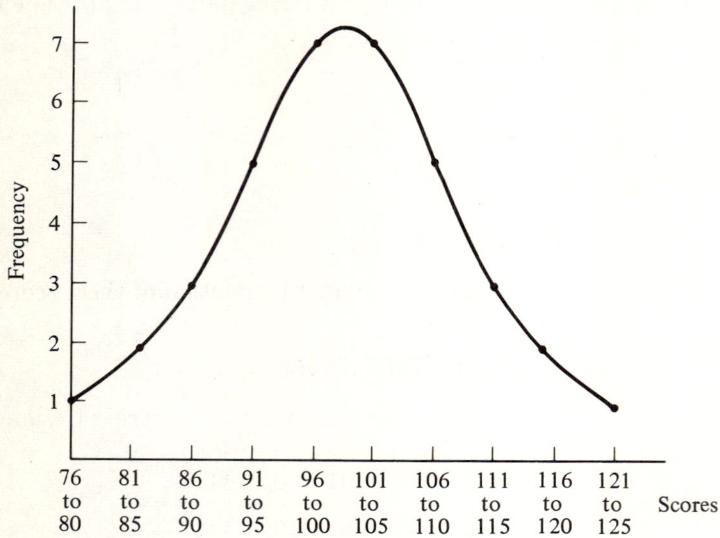

FIG. 11.2

Intelligence test scores are just one of many examples of normal distributions. Others include times in foot races, height and weight of humans of the same sex, and ages of students in schools.

The significance of the normal distribution lies in the fact that it can be used to estimate the probability of *one* score in a sample being within a certain range when the mean and standard deviation are known. Consider the following graph of a normally distributed characteristic.

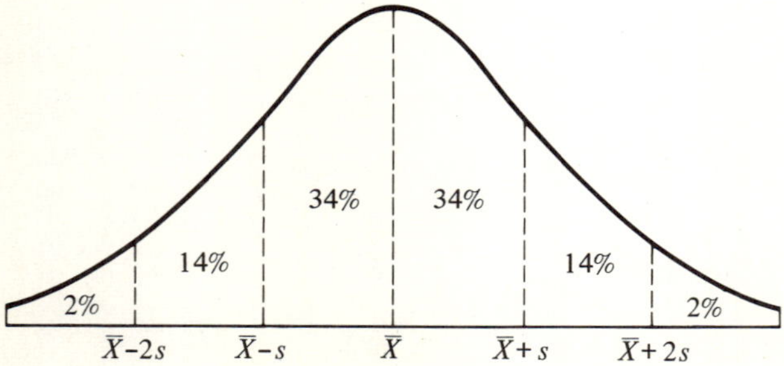

FIG. 11.3

In this distribution, $\bar{X}$ is the mean and s is the standard deviation. Then if we guess that any one score will be within one standard deviation of the mean, we will be right 68% of the time; the probability that the score will be one standard deviation less than the mean is 34%, the probability that the score will be within one standard deviation greater than the mean is 34%: 34% + 34% = 68%. This is the same as saying

$$P[\{(\bar{X} - s), (\bar{X} + s)\}] = P[\{\bar{X} - s\}] + P[\{\bar{X} + s\}]$$
$$= 0.34 + 0.34 = 0.68 = 68\% \qquad (11.7.1)$$

EXAMPLE 11.7.1

On standard IQ tests, the mean is 100 and the standard deviation is 16. The probability that any score will be between $84 = \bar{X} - s$ and $100 = \bar{X}$ is 0.34 or 34%. The probability that any one score will be above 100 is 0.50 or 50% (0.34 + 0.14 + 0.02).

DEFINITION 11.7.1

The *coefficient of rank correlation* r is a measure of the reliability of one score in predicting another score, and is defined as

$$r = 1 - \frac{6 \sum d^2}{n(n^2 - 1)} \qquad (11.7.2)$$

where n denotes the number of elements in the data and d is the deviation from the mean.

As the dean of admissions at a university, you wish to determine if IQ scores are a reliable predictor of achievement in college. Following is a random sample of students with IQ scores and grade point averages at the end of the freshman year.

Student	IQ Test Score	GPA
A	100	2.1
B	105	2.4
C	95	2.0
D	97	2.7
E	110	2.5
F	120	2.2
G	115	3.0
H	130	3.4
I	125	3.8
J	116	2.6

To find the coefficient of rank correlation, we first rank the students according to IQ and GPA.

Student	Rank in IQ	Rank in GPA	Deviation d	d^2
A	8	9	-1	1
B	7	7	0	0
C	10	10	0	0
D	9	4	$+5$	25
E	6	6	0	0
F	3	8	-5	25
G	5	3	$+2$	4
H	1	2	-1	1
I	2	1	$+1$	1
J	4	5	-1	1
$n = 10$				$\sum d^2 = 58$

The coefficient of rank correlation is

$$r = 1 - \frac{6 \sum d^2}{n(n^2 - 1)} = 1 - \frac{6(58)}{10(100 - 1)}$$

$$= 1 - \frac{348}{990}$$

$$= 1 - 0.35$$

$$= 0.65 \qquad (11.7.3)$$

Notice that the students who scored high on the IQ tests tended to have high GPA's. This is borne out by the 0.65 coefficient of rank correlation. The highest possible coefficient of correlation is 1.0 and is *perfect* rank correlation. The lowest possible coefficient of correlation is -1.0 and is called *perfect negative* rank correlation. Negative correlation indicates that high scores of one characteristic predict low scores of another. The closer the coefficient of correlation gets to zero, the less reliable the characteristic is as a predictor. Correlations greater than 0.4 and less than -0.4 are generally considered significant.

Exercise 11.7

1. If the scores on a standardized achievement test had mean 75 and standard deviation 5, what percentage of the people who took the test should have achieved scores above 80? above 85? between 60 and 75? You would expect two-thirds to achieve scores in what range?

2. On a distribution of normal scores the mean is 100 and the standard deviation is 20. What percentage of scores would be expected to be

above 120? 140? 80? 60? What percentage of scores would be expected between 60 and 100? 80 and 120? 100 and 140?

3. In a normal distribution with mean 75, if $1\frac{1}{2}$ standard deviations above the mean is equal to 90, find the *range* of scores of the following:

   (a) 1 standard deviation to $1\frac{1}{2}$ standard deviations above the mean
   (b) $\frac{1}{2}$ standard deviation below the mean to 1 standard deviation above the mean
   (c) less than $\frac{1}{2}$ standard deviation above the mean
   (d) greater than $\frac{3}{4}$ standard deviation above the mean

4. Find the rank correlation between the following students on the two tests indicated below. Is it significant?

Student	Test I	Test II
A	75	78
B	89	85
C	27	35
D	52	69
E	81	69
F	23	32
G	98	75
H	100	100
I	86	85
J	94	87

5. Find the rank correlation between music and physical education grades of the following elementary school students. How do the grades correlate?

Student	Music	Gym
A	C	A+
B	B+	C
C	A	A−
D	B−	C−
E	B	A
F	C−	B+
G	C+	B
H	A−	D
I	A+	C+
J	D	B−

Answers to Selected Problems

CHAPTER 1

Exercise 1.2

1. yes
2. yes
3. no
5. yes
7. no
10. yes
12. John is not tall.
13. Jack is neither smart nor ugly.
17. Some soldiers are not blood-thirsty.
21. (a) Jack studies mathematics or Jack studies physics.
    (e) Jack does not study mathematics or Jack does not study physics.
    (g) Jack studies either mathematics or physics, but not both.

Exercise 1.3.1

1.

p	q	$p \vee \sim q$
T	T	T
T	F	T
F	T	F
F	F	T

9.

p	q	$(p \vee \sim q) \wedge p$
T	T	T
T	F	T
F	T	F
F	F	F

6.

p	q	$(p \vee \sim q) \wedge \sim p$
T	T	F
T	F	F
F	T	F
F	F	T

12.

p	q	$(p \wedge \sim q) \vee (q \wedge \sim p)$
T	T	F
T	F	T
F	T	T
F	F	F

Exercise 1.3.2

1. (b)

p	q	$p \vee q$	$q \vee p$	$p \wedge q$	$q \wedge p$
T	T	T	T	T	T
T	F	T	T	F	F
F	T	T	T	F	F
F	F	F	F	F	F

2. (c)

p	$\sim p$	$\sim(\sim p)$
T	F	T
F	T	F

3. (b)

p	q	$(p \wedge q) \vee p$
T	T	T
T	F	T
F	T	F
F	F	F

Exercise 1.4

2. *Converse*: If the grass is wet, then it rains.
   Contrapositive: If the grass is not wet, then it is not raining.
5. *Converse*: If it is not raining, then it is not cloudy.
   Contrapositive: If it is raining, then it is cloudy.

Exercise 1.5

1\. (e)

p	q	$\sim p \rightarrow q$
T	T	T
T	F	T
F	T	T
F	F	F

(f)

p	q	$p \vee q \rightarrow p$
T	T	T
T	F	T
F	T	F
F	F	T

3\. (a) $p \rightarrow q$
(c) $\sim q \wedge \sim p$
(e) $q \rightarrow p$

CHAPTER 2

Exercise 2.1

1. $A = \{6, 7, 8, 9\}$ or $A = \{x \mid x \text{ is a digit from 6 to 9 inclusive.}\}$
4. $\in$
5. neither
6. $\notin$
7. $\in$
8. $\notin$

Exercise 2.2

1. $\supset; \supseteq$
3. $\sim$
5. $\subset; \subseteq$
6. none of these
7. $\supset; \supseteq$
9. $=; \sim; \leq; \geq$

Exercise 2.3

1. (a) A (c) U
2. (a) $\{0, 1, 2, 3, 5, 7, 8\}$ (c) $\{0, 1, 7\}$ (f) $\{0, 1, 5, 7\}$
   (h) $\{0, 1, 2, 3, 7, 8\}$
3. (b) $\{2, 3\}$ (d) $\{2, 3, 5\}$ (f) $\{4\}$ (h) $\{1, 2, 4\}$
4. (a) $A = B$ (b) $A = B$ (e) $B \subseteq A$
5. (b) $X = \bar{A}$ (d) $X = U$ (e) $X \subseteq \bar{A}$
9. (d) $\{3\}$ (g) $\{3\}$
10. (a) $\{1, 2, 3, 4, 5, 7\}$ (e) $\{1, 3, 5, 7\}$ (i) $\{5, 7\}$

Exercise 2.4

1. $A \cup (B \cup A) = A \cup (A \cup B) = (A \cup A) \cup B = A \cup B$
6. $A \cap (\bar{\bar{A}} \cup B) = A \cap (A \cup B) = A$
8. $\overline{(\bar{A} \cup \bar{B})} = \bar{\bar{A}} \cap \bar{\bar{B}} = A \cap B$
10. $\overline{(A - B)} = \overline{A \cap \bar{B}} = \bar{A} \cup \bar{\bar{B}} = \bar{A} \cup B$

Exercise 2.5

1. (f)

(A∪B)∩C

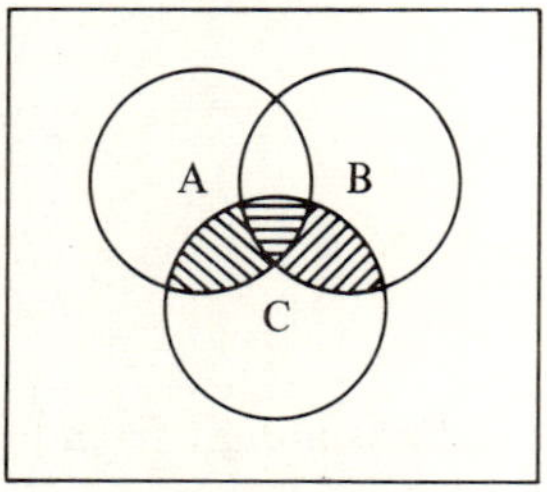

2. (a) 43
   (b) 9
   (c) 54
   (d) 31

Exercise 2.6

1. (a) 10 (c) 3 (d) $\aleph_0$
2. (c) 0 (d) 7 (f) 3
5. (a) 4 (d) 5 (e) 2
6. $c(A \cap B) = 0$

Exercise 2.7

1. (a) $\{(d, 1)\ (d, 2)\ (8, 1)\ (8, 2)\ (5, 1)\ (5, 2)\}$
   (b) $\{(1, c),\ (a, c)\ (x, c)\}$
   (c) $\{(x, x)\}$
   (d) $\varnothing$
2. (b) $\{(1, 4)\ (1, 5)\ (2, 4)\ (2, 5)\ (3, 4)\ (3, 5)\ (4, 4)\ (4, 5)\}$
   (c) $\{(4, 1)\ (5, 1)\ (4, 2)\ (5, 2)\ (4, 3)\ (5, 3)\ (4, 4)\ (5, 4)\}$
   (d) $\{(4, 4)\ (4, 5)\ (5, 4)\ (5, 5)\}$
3. (a) $\varnothing$
   (d) $\{(1, 1)\ (1, 4)\ (2, 1)\ (2, 4)\ (3, 1)\ (3, 4)\}$
   (f) $\varnothing$
4. $A = B$
5. (a) 6 (e) 0

CHAPTER 3

Exercise 3.2

5. (a) 36 (c) 19 (f) 33
6. (b) 14 (d) 54 (e) 15

Exercise 3.3

1. (a) 5 (c) 18 (e) 36
   (g) 18 (i) 4 (j) 17
2. (a) $2 \cdot (6 + 4) \cdot 6$ (d) $2 + (4 \cdot 7 + 8)$
3. (a) $x = 4$ (b) No such whole number exists.
   (d) All whole numbers.

Exercise 3.4

1. (a) $\{3, 4, 5\}$ (b) $\{3, 4, 5, 6, 7\}$ (d) $\{7, 8\}$
   (e) $\varnothing$ (g) $\{0\}$
2. (a) $3 < 5$ or $3 \leq 5$ (b) $6 > 2$ or $6 \geq 2$
   (d) $4 > 3$ or $4 \geq 3$ (f) $7 < 8$ or $7 \leq 8$
   (g) $0 < 3$ or $0 \leq 3$ (h) $0 < 1$ or $0 \leq 1$

Exercise 3.5

1. (a) $\{2, 4\}$ (c) $\{1, 2, 4\}$ (d) $\varnothing$
2. (a) Not comparable (d) $4 \mid 8$

CHAPTER 4

Exercise 4.2

1. (a) $^{-}7$ (d) $^{-}1$
2. $x = {}^{-}5$

Exercise 4.3

1.	(a) $^{-}3$	(d) $^{-}6$	(e) $^{-}8$	(f) $^{-}8$
2.	(a) $^{-}4$	(d) $^{-}1$	(e) $^{-}5$	(f) $^{-}17$
3.	(a) $x = 8$	(c) $x = 0$	(e) $x = 2$	(f) $x = {}^{-}5$
8.	(a) $x = 3$	(b) $x = {}^{-}2$	(d) $x = {}^{-}3$	(f) $x = 2$
	$x = {}^{-}3$	$x = {}^{-}7$	$x = {}^{-}2$	$x = {}^{-}4$
			$x = 4$	$x = 3$

Exercise 4.4

1. (a) $^{-}11$ (d) 14 (e) 11
2. (a) 15 (c) $^{-}60$ (e) 8 (f) 10
3. (a) $x = {}^{-}1$ (c) $x = {}^{-}2$ (d) $^{-}2$
4. (a) $x = 8$ (c) $x = {}^{-}11$ (d) No solution
9. Cost of hat $9. Cost of coat $36.

CHAPTER 5

Exercise 5.1

1. (a) $>, \geqq$ (c) $<, \leqq$ (d) $<, \leqq$
   (f) $>, \geqq$ (g) $=, \leqq, \geqq$ (i) $>, \geqq$

Exercise 5.3

1. (a) $7 > 3$ (b) $^{-}4 < {}^{-}3$ (c) $^{-}1 \leqq x \leqq 1$
   (d) $x > 0$ (f) $2 < 5$ (i) $x \leqq 0$
3. (a) $x \leqq 10$ or $\{10, 9, 8, \cdots\}$
   (c) $x \geqq {}^{-}1$ or $\{^{-}1, 0, 1, 2, \cdots\}$
   (d) $x \leqq {}^{-}4$ or $\{^{-}4, {}^{-}5, \cdots\}$
4. (a) $0 < x \leqq 3$ or $\{1, 2, 3\}$
   (b) $^{-}3 \leqq x \leqq 8$ or $\{^{-}3, {}^{-}2, {}^{-}1, 0, 1, 2, \cdots, 7, 8\}$
5. (a) $\{2, 3, 4, 5\}$ (c) $\{3, 4, 5, 6\}$
   (d) $\{8, 9, 10\}$ (e) $\{^{-}4, {}^{-}3, {}^{-}2, \cdots, 0, 1, 2, \cdots, 13\}$

6. (b) $x > 4$ or $\{5, 6, 7, \cdots\}$
(c) $x < 4$ or $\{1, 2, 3\}$
(d) $x > 5$ or $\{6, 7, 8, \cdots\}$
(f) $x \geqq 6$ or $\{6, 7, 8, \cdots\}$
(g) $x > 0$

Exercise 5.4

1. (a) $x = 2$ or $x = {}^-2$ (b) $x = 0$
(c) $x = 5$ or $x = {}^-5$ (d) no solution
2. (c) ${}^-2 < x < 2$ or $\{{}^-1, 0, 1\}$
(d) $x > 4$ or $x < {}^-4$
3. (a) ${}^-4 \leqq x \leqq 10$ or $\{{}^-4, {}^-3, {}^-2, {}^-1, 0, 1, 2, \cdots, 8, 9, 10\}$
(c) $x > {}^-1$ or $x < {}^-5$
(d) no solution
4. (a) ${}^-6 \leqq x \leqq {}^-2$ or $\{{}^-6, {}^-5, {}^-4, {}^-3, {}^-2\}$
(b) no solution (c) all integers
(e) $x > {}^-4$ or $x < {}^-18$
5. (a) x is any integer (c) ${}^-6 \leqq x \leqq 10$ (d) all integers
6. (a) no solution
(c) ${}^-6 \leqq x \leqq 6$ or $\{{}^-6, {}^-5, {}^-4, {}^-3, {}^-2, {}^-1, 0, 1, 2, 3, 4, 5, 6\}$
7. (a) $x = 3$ (c) $x = {}^-7$ (d) $x = 11$
8. (a) ${}^-5 \leqq x \leqq 9$

Exercise 5.5.1

1. 2, 3, 5, 7, 11, 13, 17, 19, 23, 29, 31, 37, 41, 43, 47, 51, 53, 59, 61, 67
6. 8: ${}^-8, {}^-1, {}^-2, {}^-4, 1, 2, 4, 8$
12: ${}^-12, {}^-1, {}^-2, {}^-4, {}^-3, {}^-6, 1, 2, 3, 4, 6, 12$
13: ${}^-1, {}^-13, 1, 13$
7. (a) 29, 31
(b) 59, 61
8. There is none.

Exercise 5.5.2

2. (a) $2 \cdot 2 \cdot 2 \cdot 3$ (c) $7 \cdot 5$ (e) $5 \cdot 13$
(h) $2 \cdot 2 \cdot 2 \cdot 19$ (j) $2 \cdot 2 \cdot 7$
3. (b) $2 \cdot 2 \cdot 43$ (d) $2 \cdot 2 \cdot 2 \cdot 2 \cdot 2 \cdot 5 \cdot 13$ (e) $2 \cdot 23 \cdot 127$
(i) $2 \cdot 2 \cdot 2 \cdot 2 \cdot 3 \cdot 3 \cdot 3$ (j) $2 \cdot 2 \cdot 2 \cdot 3 \cdot 3 \cdot 3 \cdot 3$
4. (a) $q = 1, r = 16$ (b) $q = 18, r = 1$
(e) $q = 21, r = 9$ (f) $q = {}^-3, r = 12$
(h) $q = 4, r = 22$ (j) $q = 22, r = 0$
(k) $q = 0, r = 0$

5. (a) $\gcd\{14, 35\} = 7$
$\operatorname{lcm}\{14, 35\} = 70$
(b) $\gcd\{11, 15\} = 1$
$\operatorname{lcm}\{11, 15\} = 165$
(h) $\gcd\{1480, 648\} = 8$
$\operatorname{lcm}\{1480, 648\} = 119,880$
(i) $\gcd\{2080, 1220\} = 20$
$\operatorname{lcm}\{2080, 1220\} = 126,880$

10. (a) $a \cdot b$ (b) a (e) $\operatorname{lcm}\{a, b\} = 40$

Exercise 5.6

1. (a) odd (b) even
(c) even (g) odd

6. $4 = 2 + 2$ $16 = 13 + 3$ $26 = 23 + 3$
$6 = 3 + 3$ $18 = 11 + 7$ $28 = 23 + 5$
$8 = 5 + 3$ $20 = 13 + 7$ $30 = 23 + 7$
$10 = 5 + 5$ $22 = 19 + 3$ $32 = 19 + 13$
$12 = 7 + 5$ $24 = 19 + 5$ $34 = 17 + 17$
$14 = 7 + 7$

8. 6 and 28 are perfect

CHAPTER 6

Exercise 6.2

1. (a) $\frac{6}{10}, \frac{12}{20}$ (c) $\frac{1}{35}, \frac{4}{140}$

(i) $\frac{2}{3}, \frac{4}{6}$ (k) $\frac{^-2}{14}, \frac{^-3}{21}$

2. (a) $\frac{29}{15}$ (c) $\frac{65}{33}$ (d) $\frac{31}{14}$

(e) $\frac{^-8}{3}$ (g) $\frac{4}{65}$ (h) $\frac{^-25}{39}$

(k) $\frac{1}{60}$ (l) $\frac{21}{135}$

3. (a) $\frac{^-6}{5}$ (c) $\frac{13}{6}$ (e) $\frac{75}{32}$

(g) $\frac{^-27}{16}$ (i) $\frac{80}{27}$

6. 3 gallons
7. 300 miles
10. (a), (b), (d)

Exercise 6.3

3. (a) additive inverse $\frac{^-7}{5}$ (c) additive inverse $\frac{4}{5}$

multiplicative inverse $\frac{5}{7}$ multiplicative inverse $\frac{^-5}{4}$

(g) additive inverse 0
no multiplicative inverse
(h) additive inverse $^-1$
multiplicative inverse 1

4. (a) $\frac{32}{35}$ (b) $\frac{124}{105}$

(c) $\frac{5}{12}$ (d) $\frac{146}{105}$

5. (a) 0 (b) $\frac{22}{7}$ (c) 0 (d) $\frac{^-1}{8}$

Exercise 6.4

1. (a) $x = \frac{9}{8}$ (b) $x = \frac{23}{9}$

(c) $x = \frac{4}{3}$ (d) $x = 0$

2. (a) $x = 1$ (b) $x = \frac{3}{2}$

(c) $x = 0$ (d) $x = 0$

5. (a) $x = \frac{3}{2}$; $x = \frac{^-3}{2}$ (c) $x = \frac{^-3}{4}$; $x = \frac{11}{7}$ (e) $x = \frac{3}{5}$; $x = \frac{7}{5}$

6. (a) $x = \frac{^-1}{2}$ (c) $x = \frac{^-13}{12}$ (d) $x = \frac{9}{44}$

7. (b) $\frac{^-15}{8}$ (c) $\frac{1}{7}$ (d) $\frac{623}{132}$

8. (a) $x = \frac{5}{3}$ (b) $x = {^-16}$

Exercise 6.5

1. (a) $\frac{1}{2}$ (d) $\frac{28}{39}$ (e) $\frac{57}{51}$

   (f) $\frac{^-2}{3}$ (h) 0 (i) 1

2. (a) $\frac{2}{49}$ (b) $\frac{5}{2}$ (c) $\frac{5}{6}$

   (d) $\frac{3}{5}$ (f) $\frac{4}{3}$

3. (a) $x = \frac{14}{9}$ (c) $x = \frac{2}{3}$

4. (a) $\frac{15}{14}$ (d) 8 (f) $\frac{11}{3}$

5. (a) 1 (b) $\frac{4}{9}$ (d) $\frac{3}{4}$

   (e) 0 (h) 8

8. 12 years old
9. $1.20

11. $\frac{5}{8}, \frac{8}{9}$

Exercise 6.6.1

1. (a) negative
   (h) positive
   (j) negative

Exercise 6.6.2

1. (a) $x = \frac{5}{2}; x = {}^{-}\frac{5}{2}$

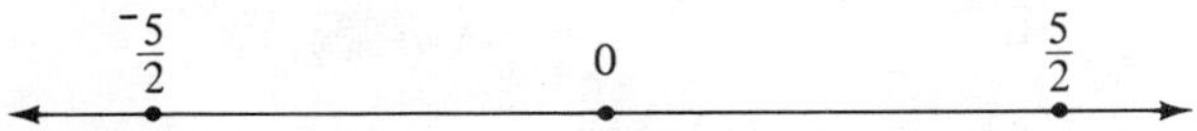

(c) $x = 0$

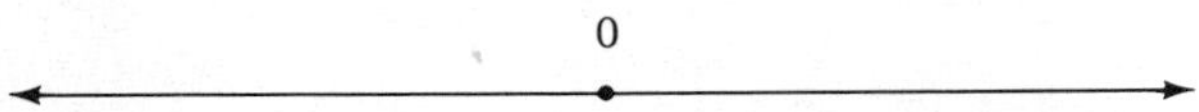

(d) no solution

2. (a) $\left\{x \middle| \frac{1}{2} \leqq x \leqq \frac{5}{2}\right\}$

(b) $\left\{x \middle| x = \frac{1}{2}; x = {}^{-}1\right\}$

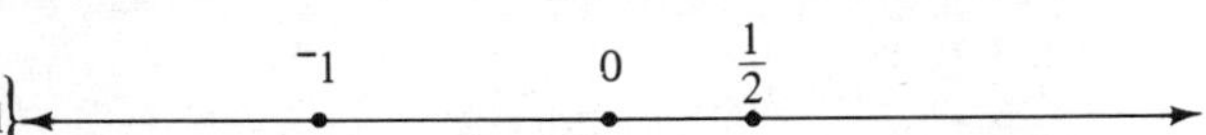

3. (a) $x \leqq {}^{-}12$

(c) $x \leqq \frac{33}{46}$

(f) $x \leqq \frac{8}{3}$

(g) $x \leqq \frac{7}{8}$

4. (a) $x \leqq \frac{2}{3}$

(b) $x \leqq 8$

(c) $x \geqq \frac{11}{6}$

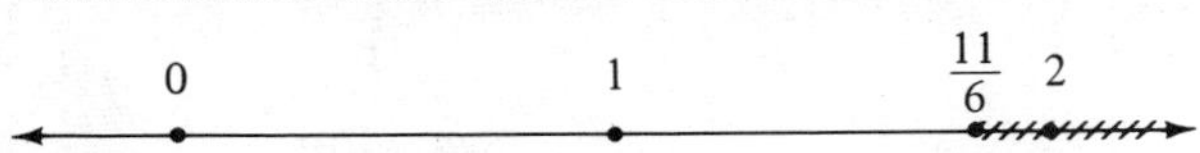

5. $x \geqq {}^{-}\frac{25}{7}$

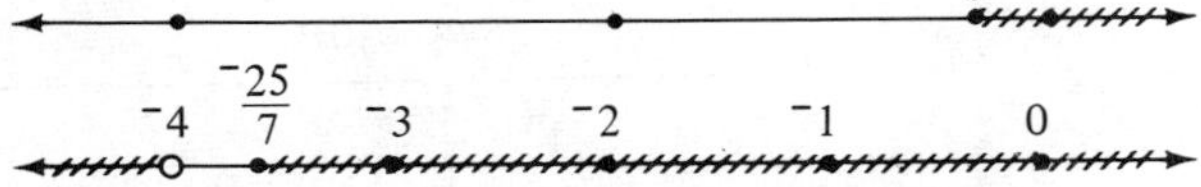

8. (a) $x \leqq \frac{13}{2}$

(b) $x \geqq \frac{1}{2}$

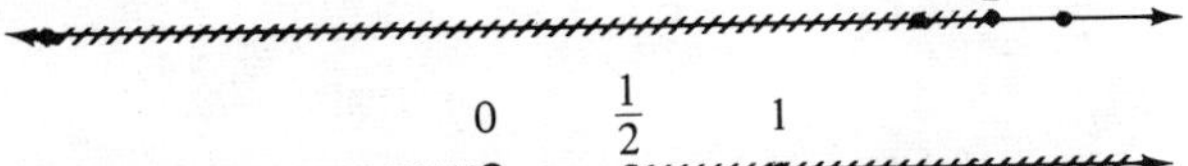

(d) $x < {}^{-}\frac{8}{3}$

10. (a) $\frac{-8}{3}$ (c) $\frac{8}{3}$ (f) $\frac{25}{18}$

(g) $\frac{5}{24}$ (h) $\frac{-3}{35}$

13. $\frac{2}{3} < \frac{17}{24} < \frac{3}{4}$

if $\frac{a}{b} < \frac{c}{d}$ then $\frac{a}{b} < \frac{ad + bc}{2bd} < \frac{c}{d}$

CHAPTER 7

Exercise 7.1

1. (a) $\frac{13}{6}$ (b) $\frac{1}{2}$ (d) $\frac{5}{6}$ (e) $\frac{37}{10}$

2. (a) no

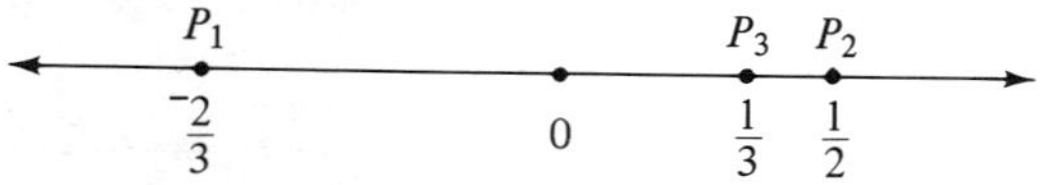

(b) yes

(e) no

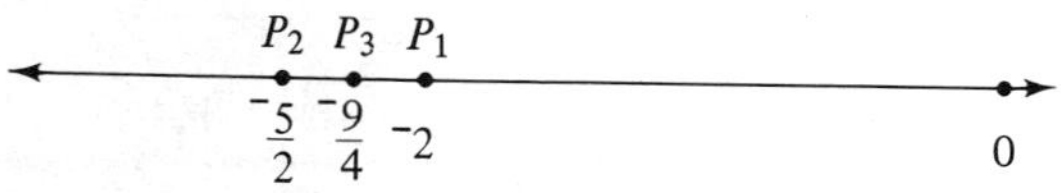

3. (a) $\frac{5}{8}$ (c) $\frac{1}{8}$ (e) 0

4. (a) no

(b) no

Exercise 7.2

1. (a) no absolute maximum or minimum

$B = \{x \mid x \geqq 3\}$ lub $[x] = 3$

$A = \{x \mid x \leqq \frac{2}{3}\}$ glb $[x] = \frac{2}{3}$

(c) no absolute maximum or minimum
$B = \{x \mid x \geqq 0\}$ lub $[x] = 0$
$A = \{x \mid x \leqq {}^-1\}$ glb $[x] = {}^-1$

(e) no absolute maximum or minimum
$B = \{x \mid x \geqq \frac{5}{4}\}$ lub $[x] = \frac{5}{4}$
The set of lower bounds is empty. glb $[x] = -\infty$

(f) no absolute maximum or minimum
The set of upper bounds is empty. lub $[x] = +\infty$
$A = \{x \mid x \leqq 0\}$ glb $[x] = 0$

Exercise 7.3

1. $\frac{13}{8} < x < \frac{7}{4}$

Exercise 7.4

3. yes to both questions
4. (a) no absolute minimum lub $[x] = 3$
absolute maximum: 3 glb $[x]$ none
(b) no absolute maximum lub $[x]$ none
no absolute minimum glb $[x] = -\infty$
(d) glb $[x] = {}^-1$
lub $[x] = \frac{2}{3}$

Exercise 7.5

1. (a) yes $\frac{6}{10} = .6$
(c) no
(d) no
(g) yes $\frac{25}{10} = 2.5$
(i) yes $\frac{8}{100} = .08$
2. (a) sum 8.11
product 16.419
(c) sum 4.2323
product .009726
(e) sum 91.8011
product .10098
(g) sum 100.3
product 215.82
(h) sum 2.3131
product .007161

3. (a) 4.5 (b) $.22\overline{2}\cdots$
(d) .125 (f) $.83\overline{3}\cdots$
(h) $1.66\overline{6}\cdots$ (i) $.857142\overline{857142}\cdots$
(l) $.888\overline{8}\cdots$

4. (a) 1.73205 (c) 3.63510 (d) 150.17024
(e) 7.48999 (g) 20.52364 (h) .05667

5. (a) $\frac{1}{4}$ (c) $\frac{7}{10}$ (d) $\frac{1}{2}$
(e) $\frac{509}{999}$ (g) 6 (h) $\frac{4}{9}$
(i) $\frac{143}{450}$

Exercise 7.6

1. (a) .32; 32% (b) .16; 16%
(d) .36; 36% (e) .75; 75%
(f) .24; 24% (g) .65; 65%
(h) .27; 27% (i) .07; 7%

2. (a) $.22; \frac{11}{50}$ (c) $.12; \frac{3}{25}$ (d) $1.25; \frac{5}{4}$
(f) $4.22; \frac{211}{50}$ (h) $.85; \frac{17}{20}$

3. (a) 13.5 (b) 640 (c) 12.5%
(d) 2.5% (e) 507 (f) 14.64

4. after 1 year: \$26.00
after 2 years: \$27.04

CHAPTER 8

Exercise 8.4

1. (a) $5^3 = 125$ (b) $3^2 \cdot 2 = 18$ (c) 1
(d) 72 (e) 1 (f) $\frac{1}{3} = 3^{-1}$

2. (a) 1.73×10^6 (b) 2.82×10^8
(c) 7.8×10^{-5} (d) 4.98×10^{-4}

3. (a) 350, 000 (d) .0000081

4. (a) 810 (b) 4

(d) $\frac{32}{375}$ (f) $\frac{1024}{9}$

5. (c) 2.3605×10^4 (d) 8.25×10^{-8}
6. (a) $.000043$ (d) $.00000009892$

Exercise 8.5

1. (a) $6023_{(\text{ten})} = 6 \cdot 10^3 + 0 \cdot 10^2 + 2 \cdot 10^1 + 3 \cdot 10^0$
   (b) $5421_{(\text{ten})} = 5 \cdot 10^3 + 4 \cdot 10^2 + 2 \cdot 10^1 + 1 \cdot 10^0$
2. (a) $321_{(\text{five})} = 3 \cdot 5^2 + 2 \cdot 5^1 + 1 \cdot 5^0$
   (b) $3123_{(\text{five})} = 3 \cdot 5^3 + 1 \cdot 5^2 + 2 \cdot 5^1 + 3 \cdot 5^0$
   (d) $4321_{(\text{five})} = 4 \cdot 5^3 + 3 \cdot 5^2 + 2 \cdot 5^1 + 1 \cdot 5^0$
   (f) $34_{(\text{five})} = 3 \cdot 5^1 + 4 \cdot 5^0$
3. (a) $t_{(\text{twelve})} = t \cdot 12^0$
   (c) $501_{(\text{twelve})} = 5 \cdot 12^2 + 0 \cdot 12^1 + 1 \cdot 12^0$
   (e) $1e_{(\text{twelve})} = 1 \cdot 12^1 + e \cdot 12^0$
   (f) $t05_{(\text{twelve})} = t \cdot 12^2 + 0 \cdot 12^1 + 5 \cdot 12^0$
4. (a) $203_{(\text{four})} = 2 \cdot 4^2 + 0 \cdot 4^1 + 3 \cdot 4^0$
   (c) $568_{(\text{nine})} = 5 \cdot 9^2 + 6 \cdot 9^1 + 8 \cdot 9^0$
   (e) $101_{(\text{two})} = 1 \cdot 2^2 + 0 \cdot 2^1 + 1 \cdot 2^0$
   (f) $426_{(\text{seven})} = 4 \cdot 7^2 + 2 \cdot 7^1 + 6 \cdot 7^0$
5. (a) $320_{(\text{five})}$ (b) $600_{(\text{seven})}$
   (c) $805t_{(\text{twelve})}$ (f) $6105_{(\text{eight})}$

Exercise 8.6

1. (a) $250_{(\text{ten})}$ (c) $139_{(\text{ten})}$
2. (a) $1573_{(\text{ten})}$ (c) $86_{(\text{ten})}$
3. (a) $43_{(\text{ten})}$ (c) $11_{(\text{ten})}$
4. (a) $937_{(\text{ten})}$ (c) $25_{(\text{ten})}$
5. (a) $46_{(\text{ten})}$ (c) $926_{(\text{ten})}$

Exercise 8.7

1. (a) $4103_{(\text{five})}$ (b) $1020_{(\text{eight})}$ (c) $380_{(\text{twelve})}$
3. (a) $2523_{(\text{twelve})}$ (b) $10153_{(\text{eight})}$ (c) $15153_{(\text{seven})}$
5. (a) $10011101_{(\text{two})}$ (b) $421_{(\text{six})}$ (c) $1112_{(\text{five})}$

Exercise 8.8.1

1. (a) $11404_{(\text{five})}$ (b) $10231_{(\text{five})}$
   (d) $1300_{(\text{five})}$ (f) $2000_{(\text{five})}$

2. (a) $2144_{(\text{five})}$ (c) $20433_{(\text{five})}$
   (e) $233032_{(\text{five})}$ (f) $21234_{(\text{five})}$
3. (a) $4023_{(\text{five})}$ (c) $2104_{(\text{five})}$
   (e) $1143_{(\text{five})}$ (f) $2141_{(\text{five})}$

Exercise 8.8.2

1. (a) $231_{(\text{five})}$ (c) $2314_{(\text{five})}$
   (e) $2244_{(\text{five})}$ (f) $14_{(\text{five})}$
2. (b) $303_{(\text{five})}$ (c) $3044_{(\text{five})}$ (e) $21_{(\text{five})}$

Exercise 8.9.1

1. $11111_{(\text{two})}$
3. $1011011_{(\text{two})}$
5. $1001_{(\text{two})}$
7. $1101_{(\text{two})}$

Exercise 8.9.2

1. $703_{(\text{twelve})}$
3. $e20_{(\text{twelve})}$
5. $183_{(\text{twelve})}$
7. $1820_{(\text{twelve})}$
9. $274_{(\text{twelve})}$

Exercise 8.9.3

1. (a) $45879_{(\text{twelve})}$ (c) $345t_{(\text{twelve})}$ (d) $49_{(\text{twelve})}$
2. (a) $e3628_{(\text{twelve})}$ (c) $2e057t5_{(\text{twelve})}$
4. (a) $774_{(\text{eight})}$ (b) $106_{(\text{eight})}$ (c) $537_{(\text{eight})}$
6. (a) $15052_{(\text{eight})}$ (b) $14020_{(\text{eight})}$ (c) $2020_{(\text{eight})}$

CHAPTER 9

Exercise 9.2

1. $\{(1, 2), (1, 3), (3, 2), (3, 3), (5, 2), (5, 3)\}$
3. $\{(3, 0), (3, 1)\}$
5. $\{(^-1, ^-3), (^-1, ^-4), (^-2, ^-3), (^-2, ^-4)\}$

Exercise 9.3

1. a, b, e
2. (a) domain $= \{1, 3, 4\}$ range $= \{1, 2\}$
   (c) domain $= \{1, 3\}$ range $= \{1, 2, 3\}$

3. (a) $R^{-1} = \{(1, 1), (2, 3), (1, 4)\}$
   (c) $R^{-1} = \{(2, 1), (3, 3), (1, 3)\}$
5. (a) $\{(1, 2), (2, 3)\}$
   (b) $\{(1, 2), (1, 3), (2, 3)\}$
6. a, d, e

Exercise 9.4

1. (a) T
   (b) (i) $R^{-1} = \{(1, 4), (2, 4), (3, 4), (5, 4)\}$
   (ii) $S^{-1} = \{(1, 1), (2, 1), (2, 2), (4, 1)\}$
   (iii) $T^{-1} = \{(2, 2), (4, 3), (3, 4), (1, 1)\}$
   (c) R^{-1}, T^{-1}
3. f is one — one

Exercise 9.5.1

1. (a) none (b) all three (c) transitive
   (e) symmetric (g) symmetric
3. R is reflexive, symmetric, and transitive
5. R is reflexive and transitive
6. (a) yes
   (b) yes
7. (a) yes
9. (a) yes
   (b) yes

Exercise 9.6.1

1. b, d, e
4. yes

Exercise 9.6.2

1. (a)

$+$	$\bar{0}$	$\bar{1}$
$\bar{0}$	$\bar{0}$	$\bar{1}$
$\bar{1}$	$\bar{1}$	$\bar{0}$

$\cdot$	$\bar{0}$	$\bar{1}$
$\bar{0}$	$\bar{0}$	$\bar{0}$
$\bar{1}$	$\bar{0}$	$\bar{1}$

2. yes
3. yes
5. (a) 7 (c) 11 (e) 9
8. for modulo 8, no as $4 \cdot 2 = 0$
   for modulo 5, yes
11. m is composite

Exercise 9.6.3

1. (a) absolutely wrong
   (c) probably correct
   (e) probably correct

Exercise 9.7.1

1. a, c, d, g
2. (a) (7, 4), (4, 1), (8, 5)
   (c) (7, 8), (8, 9), (9, 10)

Exercise 9.7.2

1. (a) 4 (b) $^-2$ (c) 18 (d) $^-18$
2. (a) $\{(1, 4), (2, 5), (3, 6), \cdots\}$
   (c) $\{(1, 1), (2, 2), (3, 3), \cdots\}$
   (e) $\{(1, 7), (2, 8), (3, 9), \cdots\}$
   (g) $\{(1, 16), (2, 17), (3, 18), \cdots\}$

Exercise 9.8.1

1. a, b, e, f, g
2. (a) (10, 4), (20, 8), (40, 16)
   (c) $(2, 4), (^-1, ^-2), (^-2, ^-4)$
   (e) $(^-2, 2), (2, ^-2), (1, ^-1)$
   (g) $(2, ^-4), (^-1, 2), (^-2, 4)$

Exercise 9.8.2

1. (a) 5 (c) $\frac{1}{5}$ (e) $\frac{4}{5}$
2. (a) $(^-4, 1), (4, ^-1), (8, ^-2)$
   (c) $(9, 4), (18, 8), (^-9, ^-4)$
   (e) $(5, 1), (^-5, ^-1), (10, 2)$

Exercise 9.8.3

1. (a) sum: $\frac{17}{3}$ product: $\frac{10}{3}$

   (b) sum: $\frac{^-11}{40}$ product: $\frac{^-1}{20}$

   (c) sum: $\frac{14}{9}$ product: $\frac{8}{27}$

CHAPTER 10

Exercise 10.2

4. (a) 47° (b) 1° (c) 63°
5. (a) 137° (b) 70° (c) 57°
6. 90°
7. 135°

Exercise 10.3.1

1. (a) 130° (b) 130° (c) 50°
3. yes
4. not necessarily
6. 90°

Exercise 10.3.2

2. *a*, *b*, *c*, *d*
3. (a) yes (c) no

Exercise 10.3.3

1. (a) 10° (b) 80° (c) 60°
2. 60°
3. (a) 75°, 75° or 30°, 120° (b) 30°, 30° (c) 45°, 45°

Exercise 10.3.4

1. true
2. false
3. true
4. true
6. true
9. false
11. rectangle, square
13. all
15. square, rhombus

Exercise 10.5

1. (a) 17 (b) 9
2. (a) 12 (c) 24
   (e) 6π (f) 6
3. (a) $\frac{4\pi}{9}$ (c) $\frac{4\pi}{3}$ (e) $\frac{\pi}{3}$

4. inscribed circle: 6π
circumscribed circle: $6\pi\sqrt{2}$

5. (a) $r = \frac{1}{\pi}$ $\quad d = \frac{2}{\pi}$

(c) $r = \frac{3}{\pi}$ $\quad d = \frac{6}{\pi}$

Exercise 10.6.1

1.

b	h	A
3	5	$\frac{15}{2}$
$\frac{1}{2}$	$\frac{7}{8}$	$\frac{7}{32}$
2	4	4
$\frac{34}{5}$	5	17
$\frac{44}{3}$	3	22

2.

h	b_1	b_2	A
15	9	3.4	93
7.8	3.8	$\sqrt{2}$	$14.82 + 3.9\sqrt{2}$
2	15	2	17
4.4	7	3	22

3.

b	h	A
.7	6	4.2
$\sqrt{5}$	3	$3\sqrt{5}$
4	4	16
$\frac{\sqrt{17}}{9}$	9	$\sqrt{17}$

Exercise 10.6.2

1. 2, 3, $\sqrt{17}$
2. 21π
3. $\frac{25\pi}{4}$

Exercise 10.7

1. 93
3. $466\frac{2}{3}$
5. 21π
8. 36π

CHAPTER 11

Exercise 11.2

1. $\frac{1}{6}, \frac{1}{6}$

2. $\frac{15}{40} = \frac{3}{8}$

5. $\frac{2}{17}, \frac{3}{17}, \frac{12}{17}$

Exercise 11.3

1. $\frac{2}{36} = \frac{1}{18} \qquad \frac{3}{36} = \frac{1}{12} \qquad \frac{8}{36} = \frac{2}{9}$

3. $\frac{1}{4}, \frac{1}{2}$

5. $\frac{6}{36} = \frac{1}{6}$

10. $\frac{30}{36} = \frac{5}{6}$

13. $\frac{5}{6}$

Exercise 11.5

1. mean = 53.75
   median = 52
   no mode

Exercise 11.6

1. $\sqrt{\frac{299}{7}} = 6.5$

2. mean $= 63\frac{1}{3}$
   no mode
   range $= 79$
   median $= 65$

   standard deviation $= \sqrt{\frac{570}{9}} = 7.9$

INDEX